CONJUGATED POLYMERS FOR ORGANIC ELECTRONICS

Focusing on how conjugated polymers can be designed and made for use in efficient organic electronic devices, this book covers the tools for future development of more environmentally and economically friendly devices. Including examples of interdisciplinary science, it exemplifies how chemists and physicists work together to enable the design and synthesis of high-performance material in devices, allowing polymer-based electronic devices to become viable commercial products. It provides the main classes of conjugated polymers and their applications in organic electronic devices such as transistors, light-emitting diodes and solar cells, making this a comprehensive introduction. This complete guide includes the methods for making conjugated polymers, the properties and specific structures that make them suitable for use and how their synthesis can be optimised to improve device performance.

Written by experts in the field, this is the ideal guide for researchers and practitioners across materials science, physics, chemistry and electrical engineering.

ANDREW GRIMSDALE, Associate Professor at Nanyang Technological University, has worked on the synthesis of conjugated materials for organic electronic devices for more than 30 years at leading institutions including Cambridge, Mainz and Melbourne. Grimsdale has written the most widely cited and comprehensive reviews on luminescent polymers and is internationally recognised as a leading authority in the field.

PAUL DASTOOR, Professor at University of Newcastle, is internationally known for his work on organic electronic devices, developing the science, engineering and manufacturing of printed solar and biosensor technologies based on semiconducting polymers. Dastoor has a strong track record in research commercialisation and has spun out several companies, including one which was listed on the NASDAQ.

CAMBRIDGE MOLECULAR SCIENCE

As we move further into the twenty-first century, chemistry is positioning itself as the central science. Its subject matter, atoms and the bonds between them, is now central to so many of the life sciences on the one hand, as biological chemistry brings the subject to the atomic level, and to condensed matter and molecular physics on the other. Developments in quantum chemistry and in statistical mechanics have also created a fruitful overlap with mathematics and theoretical physics. Consequently, boundaries between chemistry and other traditional sciences are fading and the term *Molecular Science* now describes this vibrant area of research.

Molecular science has made giant strides in recent years. Bolstered both by instrumental and theoretical developments, it covers the temporal scale down to femtoseconds, a time scale sufficient to define atomic dynamics with precision, and the spatial scale down to a small fraction of an Angstrom. This has led to a very sophisticated level of understanding of the properties of small molecule systems, but there has also been a remarkable series of developments in more complex systems. These include: protein engineering; surfaces and interfaces; polymers; colloids; and biophysical chemistry. This series provides a vehicle for the publication of advanced textbooks and monographs introducing and reviewing these exciting developments.

CONJUGATED POLYMERS FOR ORGANIC ELECTRONICS

Design and Synthesis

ANDREW GRIMSDALE
Nanyang Technological University

PAUL DASTOOR
University of Newcastle

CAMBRIDGE
UNIVERSITY PRESS

CAMBRIDGE
UNIVERSITY PRESS

Shaftesbury Road, Cambridge CB2 8EA, United Kingdom

One Liberty Plaza, 20th Floor, New York, NY 10006, USA

477 Williamstown Road, Port Melbourne, VIC 3207, Australia

314–321, 3rd Floor, Plot 3, Splendor Forum, Jasola District Centre, New Delhi – 110025, India

103 Penang Road, #05-06/07, Visioncrest Commercial, Singapore 238467

Cambridge University Press is part of Cambridge University Press & Assessment, a department of the University of Cambridge.

We share the University's mission to contribute to society through the pursuit of education, learning and research at the highest international levels of excellence.

www.cambridge.org
Information on this title: www.cambridge.org/9781107008168

DOI: 10.1017/9781139035262

First published 2024

A catalogue record for this publication is available from the British Library

A Cataloging-in-Publication data record for this book is available from the Library of Congress

ISBN 978-1-107-00816-8 Hardback

Contents

Preface *page* ix
List of Abbreviations xi

1 Introduction 1
1.1 Introduction 1
1.2 Developing Flexible Displays: Organic Light-Emitting Diodes 6
1.3 Developing Solar Paint: Organic Photovoltaics 11
1.4 Developing Disposable Integrated Circuits: Organic Transistors 16

2 Polyacetylenes 20
2.1 Polyacetylene 20
2.2 Electrical Properties of Polyacetylene 23
2.3 Substituted Polyacetylenes 24

3 Poly(arylene vinylene)s 29
3.1 Quinodimethane Routes to PAVs 29
3.2 Other Routes to PAVs 38
3.3 Structure–Property Relationships in Optical Properties of PAVs 41
3.4 Applications of PAVs: FETs 45
3.5 Increasing Emission Efficiency of PAVs by Minimising Defects 46
3.6 Increasing Emission Efficiency by Attaching Charge Transporting Units 47
3.7 PAVs as Electron Donors and Acceptors in OPVs 49
3.8 Poly(thienylene vinylene)s as Low Bandgap Polymers for OFETs and OPVs 52

4 Poly(arylene ethynylene)s 54
4.1 Synthesis of PAEs 54

4.2 Optical and Electronic Properties of PAEs 56
4.3 PAE–PAV Mixed Polymers 58

5 Poly(phenylene)s 61
5.1 Poly(para-Phenylene) (PPP) 61
5.2 Soluble PPPs 63
5.3 Ladder-Type PPPs 65
5.4 Application of Poly(phenylene)s in Organic Electronic Devices 72

6 Polyfluorenes and Related Polymers 74
6.1 Polyfluorenes 74
6.2 Polycarbazoles and Other Heterofluorene-Based Polymers 89
6.3 Other Stepladder-Type Polyphenylenes 92

7 Polythiophenes 104
7.1 Polythiophene 104
7.2 Poly(3-alkylthiophene)s 104
7.3 Routes to Regioregular P3ATs 106
7.4 Effects of Regioregularity upon Electronic Properties of P3ATs 108
7.5 PEDOT as an Electrically Conducting Polymer and Its Application in Transistor-Based Sensors 110
7.6 Optical Properties of P3ATs and Their Applications in OLEDs and Sensors 111
7.7 Polythiophenes as Donor Materials for OPVs 113
7.8 Polythiophenes as Active Layers in Sensors 114
7.9 Fused Thiophene-Based Materials 115

8 Other Arylene-Based Polymers 117
8.1 Block Copolymers 118
8.2 Copolymers for LEDs 120
8.3 Copolymers for FETs 124
8.4 Copolymers for OPVs 133

9 Hyperbranched Polymers, Star Polymers and Dendrimers 149
9.1 Hyperbranched Emissive Polymers 149
9.2 Emissive Dendrimers 155
9.3 Emissive Star Polymers 159

10 Polymers with Molecular-like Chromophores 160
10.1 Polymers with Emissive Side Groups 160
10.2 Copolymers with Isolated Chromophores in the Main Chain 164

11 Polymers for Phosphorescent LEDs 173
11.1 Emissive Polymers as Hosts for Phosphors 175
11.2 Phosphorescent Polymers 177

12 Polymers for White-Emitting PLEDs 184
12.1 White EL from Blends 185
12.2 White EL from Single Polymers 188

13 Polymers for Other Luminescent Devices 196
13.1 Light-Emitting Devices with Non-standard Configurations 196
13.2 Light-Emitting Transistors 197
13.3 Light-Emitting Electrochemical Cells (LECs) 198
13.4 Electrogenerated Chemiluminescence Cells 200
13.5 Polymer Microcavities and Lasers 201
13.6 Integrated Polymer Devices and Other Devices 203

14 Conclusion and Outlook 204

References 208
Index 263

Preface

The development of organic electronics from a scientific curiosity to a rapidly growing commercial reality has been one of the most important advances in materials science in the last few decades. This advance began with the discovery in 1977 by Hideki Shirakawa, Alan Heeger and the late Alan MacDiarmid that polyacetylene could be made conductive by doping, for which they received the Nobel Prize in Chemistry in 2000. Until then the idea of electronics based upon organic materials was seen as a subject perhaps more suitable for science fiction than for serious scientific investigation. Since this seminal discovery, it has been shown that conjugated organic oligomers and polymers can be used as the active materials in a variety of electronic devices such as organic light-emitting diodes (OLEDs), organic thin-film transistors (OTFTs), organic photovoltaic devices (OPVs) and sensors. In 2000, displays based on OLEDs were already entering the commercial marketplace, while prototypes of the other classes of organic electronic devices had been fabricated in laboratories and were under development as potential commercial products. While organic electronic devices still have a long way to go until they can match the performance of the established inorganic semiconductor-based technologies, they are improving rapidly and their market share is increasing dramatically.

Much of the pioneering work on organic electronics and also the first commercial devices used small conjugated molecules or oligomers. However, these molecular materials tend to be crystalline and so require relatively expensive vacuum deposition methods in order to form high-quality films. By contrast, conjugated polymers are generally amorphous in the solid state and so can be processed by cheap solution-based methods such as spin-coating or even printing. As a result, polymer-based devices are expected to have a significant cost advantage over those using small molecules, provided their performance can be improved sufficiently.

While a number of extensive reviews, monographs and books on various aspects of the synthesis of conjugated polymers and their applications in devices have appeared, there is currently no reasonably comprehensive yet accessible overview of this subject.

Anyone interested in learning more about the field, whether student or researcher, has to attempt to assimilate information from a wide variety of primary, secondary and tertiary sources, much of which deals with only some aspects of it and is usually written primarily for people who are already experts in one discipline (usually chemistry or physics) involved in what is an intensely multidisciplinary subject. In this book we seek to provide the sort of overview that we ourselves would have wanted to have available when we were starting out in the area. When we started working in the area of organic electronics there were already hundreds of important publications on conjugated polymers synthesis and applications. The number is now in the thousands or even the tens of thousands. No one new to the field can possibly hope to be au fait with more than a fraction of them. But fortunately, whereas previously one proceeded by trial and error as one never knew for sure whether or not new structures or device designs were likely to prove better than old ones, now it is possible to pick a rational pathway towards one's goals. In this book we hope to provide both newcomers and old hands with guidelines on how to do just that.

One of the key messages we hope to deliver is that advances in this field depend upon active and effective collaboration between chemists, physicists and device engineers, with advances in one aspect leading to new developments in the others and with timely feedback from one set of researchers to the others being crucial to making progress.

To this end we shall start by giving an overview of the types of devices using conjugated polymers, describing how they work and indicating what properties it is important for a material to have if it is to produce satisfactory performance in them. From there we move on to consider each of the major types of conjugated polymer in turn. Here we will focus on how structure–property relationships have been developed and in particular on how understanding the roles of impurities or defects arising from side reactions during synthesis or from interactions of the materials with air or electrode materials has led to advances in synthetic methodology and/or device fabrication techniques producing devices with improved efficiency and/or lifetime. We will not be attempting to be in any way comprehensive in our coverage (for that the reader is referred to the reviews we just mentioned) but aim to present the most important and informative examples. Nor do we intend to focus primarily on what is the current state of the art in device performance as this is subject to continual improvement and whatever is the best material for a given device type at the time of writing will probably have been surpassed by the time this book is published. But the general principles by which one can begin to rationally design a high-performing material should remain the same, so we hope this work will remain relevant for some time with only minor updating required periodically. Finally, we will present our views on the future of this field, on the work that remains to be done to bring organic electronics to its full potential and on new devices and challenges (e.g., polymer lasers) that remain to be tackled.

Abbreviations

BDT	benzodithiophene
ECL	electrochemiluminescence
EL	electroluminescence
FET	field-effect transistor
GRIM	Grignard metathesis
HOMO	highest occupied molecular orbital
HT	head-to-tail
IR	infrared
ITO	indium tin oxide
LCD	liquid crystal display
LEC	light-emitting electrochemical cell
LED	light-emitting diode
LPPP	ladder-type poly(para-phenylene)
LUMO	lowest unoccupied molecular orbital
OFET	organic field-effect transistor
OLED	organic light-emitting diode
OLET	organic light-emitting transistor
OPV	organic photovoltaics
P3HT	poly(3-hexylthiophene)
PA	polyacetylene
PAE	poly(arylene ethynylene)
PAV	poly(arylene vinylene)
PCBM/$PC_{61}BM$	phenyl-C_{61}-butyric acid methyl ester
$PC_{71}BM$	phenyl-C_{71}-butyric acid methyl ester
PDAF	poly(9,9-dialkylfluorene)
PEDOT	poly(3,4-ethylenedioxythiophene)
PL	photoluminescence
PLED	polymer light-emitting diode

PPP	poly(para-phenylene)
PPV	poly(para-phenylene vinylene)
PSS	poly(styrene-4-sulphonate)
PTV	poly(2,5-thienylene vinylene)
PVK	poly(N-vinylcarbazole)
TADF	thermally activated delayed fluorescence
TFT	thin-film transistor
UV	ultraviolet

1
Introduction

1.1 Introduction

The development of plastic in the late 1800s revolutionised manufacturing technology and led to what is now commonly known as the Plastics Age [1]. These new materials were flexible, durable and formable in ways that natural materials were not. The first plastic material was developed by Alexander Parkes and was revealed publicly at the 1862 Great Exhibition held at South Kensington in London. Parkesine, as the new material was dubbed, was a cellulose-based material akin to celluloid that could be moulded when heated [2]. In 1899, Arthur Smith patented the use of phenol-formaldehyde resins for use as an ebonite substitute in electrical insulation and this invention led to the development of Bakelite by Leo Baekeland in 1907; probably the best known early plastic [3]. Further developments in the twentieth century led to the invention of Vinyl (polyvinylchloride) by Walter Semon in 1926, Teflon (polytetrafluoroethylene) by Roy Plunkett in 1938 and Nylon by Wallace Carothers in 1939. These materials, and other related plastics, share many similar properties: they are inexpensive, flexible and easy to process. Another property that appeared to be systemic to plastics up until the 1960s was that they were all electrical insulators. Indeed, one of the first domestic applications for Bakelite was as electrical insulation.

However, there are a group of organic, polymeric materials that exhibit electrical conductivity that date back to the pre-historical era. Known as *carbon blacks*, these materials are typically formed from the partial burning of organic substances (such as oils or wood) and have been an important commercial product for centuries, primarily due to their colour (used in pigments and dyes), high electrical conductivity (~50 Ω^{-1} cm^{-1}) and excellent thermal and mechanical properties [4].

Prior to the 1950s, these so-called carbonaceous polymers were thought to consist of three-dimensional cross-linked units with chemical structures that were complex and ill-defined [5]. In order to address these problems, several researchers started to work on developing synthetic routes with the goal of developing controlled conducting polymers, such as polyvinylenes and polyphenylenes [6]. However, the first demonstration of the synthesis of a substantially conductive polymer was achieved in Australia in 1963 by Donald Weiss and co-workers with the thermal polymerisation of tetraiodopyrrole to produce polypyrrole with resistivities of 11–200 Ω cm [7–9].

Conducting polymer research changed significantly in 1977 with the discovery by Alan Heeger, Hideki Shirakawa and the late Alan MacDiarmid that polyacetylene becomes a metallic conductor upon iodine doping, with conductivities as high as 500 S cm^{-1} [10]. This discovery ultimately created an entirely new field of organic electronics and has led to extremely rapid developments in the areas of organic light emitting diodes (OLEDs), organic photovoltaics (OPVs) and organic field effect transistors (OFETs), to name but a few. The rapid development of organic electronics as a research area has paved the way for the rapid development of commercial devices based on this new technology. The first applications have come from the commercialisation of OLED-based display technology, with initially small screens for consumer products ultimately resulting in the development of the OLED-based XEL-1 TV released by Sony in 2008. These new OLED-based displays offer enhanced contrast ratios, wider viewing angles and much thinner formats than comparable LCD or plasma displays (Figure 1.1).

The defining feature of organic electronics is that it allows us to combine the flexibility, formability and economy of plastics with the technological capability of electronics. We are now in the exciting position of being able to envisage a world where consumer electronics devices can be printed at such low cost as to be disposable, where every building has a flexible photovoltaic coating on its roof and where integrated bioelectronic sensors and transistors are routinely available [11].

1.1.1 Origins of Semiconductivity

Inorganic semiconductors, such as silicon, are crystalline solids with an electronic structure that can be described in terms of energy bands, where each energy band is actually an array of allowed energy levels (or states) that can be occupied by an electron (Figure 1.2). Just as audience members have to occupy seats in a theatre, electrons have to occupy states within the energy band. In the idealised case, the electronic structure consists of a higher energy conduction band and a lower energy valence band separated by an energy gap, the size of which depends upon the material. Indeed, one can stretch the theatre analogy a little further to think of the valence and conduction bands as corresponding to the stalls and upper circle seats, respectively. This concept of an energy gap, which is entirely a consequence

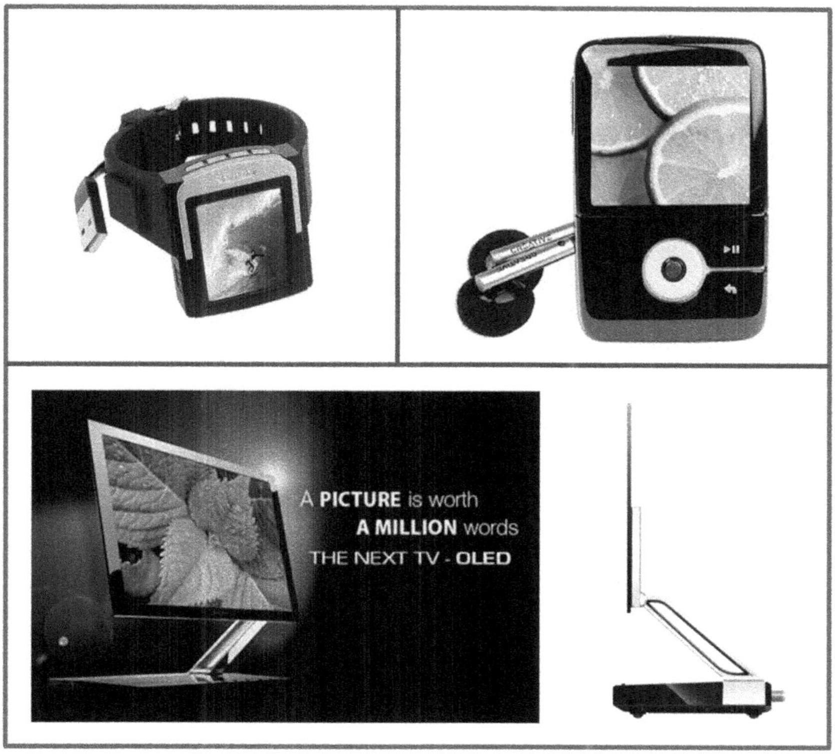

Figure 1.1 An OLED electronic display watch (top left), mp3 player (top right) and OLED television (bottom panel)

of quantum mechanics, is crucial to our understanding of conductivity in materials and its subsequent exploitation in a vast range of electronic devices. The high electrical conductivity of metals is a consequence of the fact that the highest occupied band is only half full and, hence, there are many empty states within the band that electrons can move to. In the case of insulators, however, the highest occupied band is full and, thus, for current to flow the electrons have to cross an energy gap of several electron volts, hence their conductivity is very low. Inorganic semiconductors tend to have band gaps typically in the energy range of 0.1–2.2 eV. In the case of silicon, for example, the band gap is 1.12 eV, whereas for gallium arsenide it is 1.4 eV. It is energetically feasible, therefore, that there is sufficient thermal energy at room temperature to excite an electron from the valence band into the conduction band of the material, thus producing two charge carriers: an electron in the conduction band and a so-called hole in the valence band. Although the hole that is produced is simply an empty electronic state, which can be occupied by other electrons in the valence band, it behaves to all intents and purposes as though it is an independent carrier of positive charge. However, to

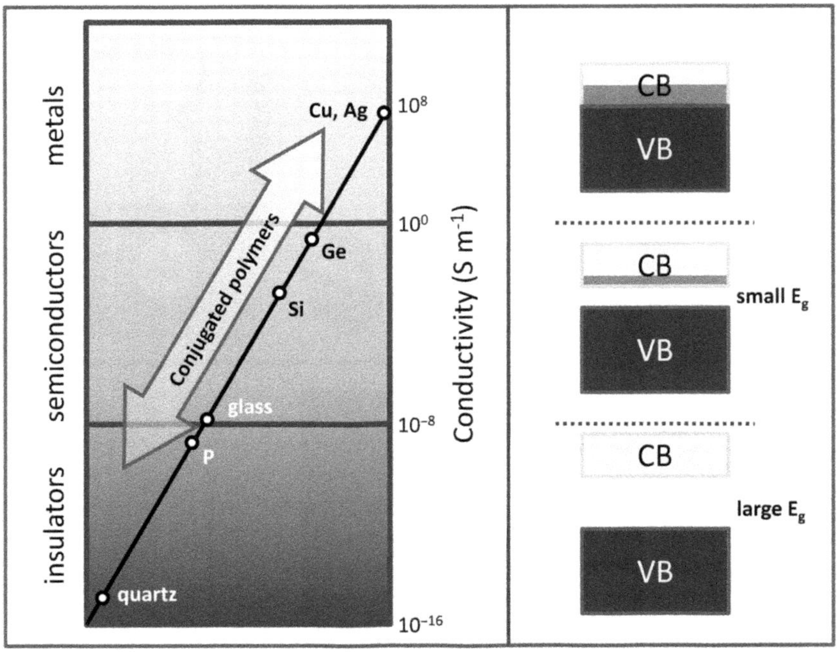

Figure 1.2 Left hand panel: Electrical conductivity of conjugated polymers encompasses the range of conductivities from insulators to semiconductors to metals. Right hand panel: The corresponding band structures for metals, semiconductors and insulators. In an insulator, the valence band (VB) is full and no electrons can cross the band gap (E_g) to occupy the conduction band (CB). For a metal there is no E_g and the CB is partially filled. In a semiconductor, E_g is small enough to permit thermally activated occupancy of CB.

make practical devices, inorganic semiconductors typically need to be doped in order to increase their conductivity by making them rich in either electrons (n-type) or holes (p-type).

Organic semiconducting materials are not limited to polymers but also encompass the small molecule systems, such as the arenes, pthalocyanines and the vast range of organic dye materials. The common feature of all of these organic semiconductors is that they possess chains of alternating single and double carbon bonds, producing what is known as a conjugated π–electron system that is delocalised over the entire molecular system. Effectively, the extra electrons from the double bonds are only loosely held and can move along the polymer chain (intra-chain transport) or between chains (inter-chain transport). More accurately, it is the characteristics of the π-bonds that actually lead to the electronic properties of conjugated polymers. Firstly, the π-bonds are delocalised over the entire molecule and, secondly, the quantum mechanical overlap of p_z orbitals actually produces *two* orbitals, a bonding (π orbital) and an anti-bonding (π^* orbital). The lower-energy

π orbital produces the valence band while the higher-energy π^* orbital forms the conduction band, with the difference in energy between the two levels producing the band gap that determines the optical properties of the material. Interestingly, most semiconducting polymers appear to have a band gap that lies in the range of 1.5–3 eV, making them ideally suited as optoelectronic devices working in the optical light range.

Rather than the usual electron wave picture that is used to describe conductivity in crystalline inorganic semiconductors, electron transport in the more amorphous conjugated polymers is better described by a hopping transport model. As such, carrier mobilities in conjugated polymers are not as high as those observed in the best crystalline inorganic materials and, thus, it is unlikely that these organic materials will be able to truly compete with inorganic semiconductors in terms of speed or ultimate device miniaturisation. Indeed, the initial mobility values in these organic materials were very low (~10^{-5} cm^2 V^{-1} s^{-1}) in comparison with silicon-based inorganic devices (0.1–1 cm^2 V^{-1} s^{-1}). However, with the development of new organic semiconductors these mobilities have improved, with, for example, OFETs made from pentacene exhibiting mobilities in excess of 1 cm^2 V^{-1} s^{-1}. Indeed, mobilities in excess of 10 cm^2 V^{-1} s^{-1} have been reported for organic molecular crystal materials, with DC mobilities as high as 43 cm^2 V^{-1} s^{-1} measured for rubrene [12].

The vacuum deposition of short chain oligomers on the other hand tends to lead to polycrystalline or single crystalline films. These materials tend to assume $\pi-\pi$ stacked structures parallel with the OFET substrate, thus allowing efficient charge transport in the plane of the film. However, defect scattering occurs at the grain boundaries in these materials, thus lowering the mobility of the polycrystalline films with respect to the single crystal materials. The polymers that show the most promise in this area contain thiophene units. Indeed, solution processed films of poly(3-hexylthiophene) (P3HT) have high field effect mobilities of between 10^{-2} and 1 cm^2 V^{-1} s^{-1} and, hence, are well within the range of amorphous silicon-based devices and, as we will discuss in Chapter 8, copolymers of thiophene and other units have been made with reported mobilities between 1 and 20 cm^2 V^{-1} s^{-1}. The mobilities of these more ordered thiophene-based polymers are much higher than those found in amorphous π-conjugated polymers such as polyacetylene and polythiophene and, thus, it appears that the degree of polymer crystallinity plays an important role in device performance.

Another important question is whether the performance of a given organic transistor material is dominated by extrinsic factors, such as oxidation, the presence of moisture or chemical impurities, or whether the intrinsic structural and energetic disorder of the organic semiconductor and/or specific structural defects are responsible for the measured device properties, particularly with regard to their stability [13]. For example, large hysteresis effects and threshold voltage shifts are often observed for polar un-cross-linked polymer dielectrics such as polyvinylphenol (PVP) [14], which are then markedly reduced in poly(4-vinylpyridine) (PVPy)-based organic thin film transistors [15].

Overall, therefore, even considering their lower transport properties, solution processed polymers still offer the greatest potential for the production of low-cost so-called soft electronics, since they can be easily processed as liquid, unlike the organic crystals and short chain oligomers, which are typically vapour deposited.

1.2 Developing Flexible Displays: Organic Light-Emitting Diodes

The initial application focus for organic electronic device development was in the area of organic light-emitting diodes (OLEDs) [16]. The small area of these devices, coupled with the high value of consumer products such as cameras, mobile phones and televisions, meant that research into electroluminescent displays (ELDs) based on OLEDs rapidly gained traction in the early 1990s.

Although ELDs based on an organic molecule (anthracene) were first demonstrated in the early 1960s [17], it wasn't until the late 1980s that Tang demonstrated the first real OLED based on 8-hydroxyquinoline aluminium (Alq) [18]. The first ELD based on solution deposited polymers was reported by the Cambridge group in 1990 using a polyphenylenevinylene derivative [19]. These two seminal papers provided the foundation for the modern development of the field of OLED research, both for systems based on small organic molecules (deposited primarily via vapour-phase deposition) and semiconducting polymers (fabricated from solution-based processing such as spin-coating and printing).

1.2.1 Basic Operation

In an OLED device, an applied voltage results in charge injection at the electrodes, with electrons injected at the low work function cathode and holes injected at the high work function anode (Figure 1.3). Charge transport occurs via a hopping mechanism and in an OLED device culminates in the recombination of differing charges to form a neutral exciton. As will be discussed in more detail in Section 1.3, the exciton is a bound electron-hole pair that can diffuse a short distance through the polymer system, primarily via a Förster resonant energy transfer (FRET) mechanism [20]. The recombination of the exciton results in light emission and thus the OLED design has to be tailored to ensure efficient charge injection, charge transport, exciton formation and subsequent light emission.

The mobility of electrons in organic materials is typically much lower than holes (especially in semiconducting polymers), primarily due to the presence of trapping sites (such as oxygen-containing functional groups or other electronegative

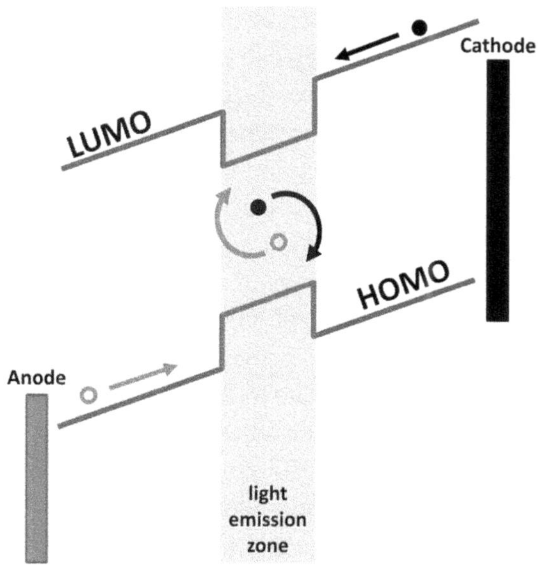

Figure 1.3 Basic operation of an OLED. Holes are injected from the high work function anode while electrons are injected from the low work function cathode. They combine to form an exciton in the light emission region, which subsequently recombines, resulting in light emission

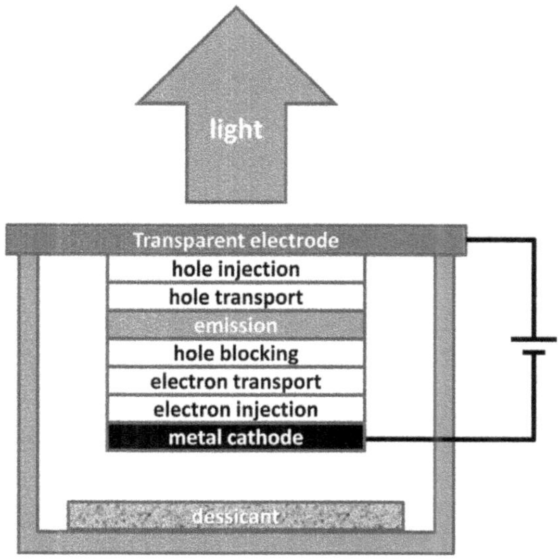

Figure 1.4 Multi-layered OLED structure with layers tailored for carrier injection, transport and blocking characteristics

impurity materials). Consequently, OLED devices have a multi-layered structure (Figure 1.4) incorporating layers that are tailored for electron and hole transport as well as hole-blocking layers to ensure equal electron and hole densities in the light emission layer.

In an OLED, the light emission region (or recombination zone) is typically designed to be in the middle of the emissive layer, thus requiring that equal fluxes of electrons and holes arrive in the centre of the OLED device. Consequently, balanced and efficient charge carrier injection and mobility need to be achieved since any unbalance results in a shifting in the position of the recombination zone. Unfortunately, neither balanced charge carrier injection nor mobility is usually obtained in single-layer OLEDs, due to non-radiative recombination of charges at the electrodes (dark current) or recombination of excitons in the near-electrode region, resulting in decreased quantum efficiencies [21]. In particular, the mobility of electrons in organic materials is typically much lower than holes (especially in semiconducting polymers), primarily due to the presence of trapping sites (such as oxygen-containing functional groups or other electronegative impurity materials). In order to address these issues, OLED devices have a multi-layered structure (Figure 1.4) incorporating layers that are tailored for electron and hole transport as well as hole-blocking layers to ensure equal electron and hole densities in the light emission layer. In a typical multi-layered OLED structure, the emitting layer is sandwiched between the hole and electron transport layers (Figure 1.4), which facilitate enhanced recombination of electrons and holes in the emissive layer, hence ensuring that light emission is maximised in the centre of the device.

As can be seen in Figure 1.3, the colour of the emitted light (wavelength) depends on the energy difference between the LUMO and HOMO levels (taking into account the exciton binding energy) and, thus, is controlled by the electronic structure of the polymer or molecule. At both electrodes a low barrier to charge injection is required. For the anode, indium tin oxide (ITO) is commonly used since its high work function (typically 4.8–5.0 eV) is well matched to the typical HOMO levels (5–6 eV) of most organic electronic materials and can be further enhanced by oxygen plasma treatment [22]. On the cathode side, a low barrier relative to the LUMO level of the organic material (typically 2–3 eV) is required for electron injection. OLED researchers pioneered the use of low work function metals such as calcium and magnesium as electron injection layers, but the low stability of these materials meant that they were superseded by more stable cathodes, such as magnesium/silver alloys [23], thin (~1–5 nm) lithium fluoride layers capped with aluminium [24], caesium fluoride [25] and magnesium oxide layers [26].

Another key criterion for OLED devices is a high solid state photoluminescent (PL) quantum yield. Quantum mechanically, an exciton is the bound state of an electron and a hole, forming a quasiparticle with zero net charge. Both electron and hole are particles with spin ½ and thus can be combined either symmetrically to form a triplet state (total spin quantum number, $S = 1$) or anti-symmetrically to form a singlet state ($S = 0$). Whereas the transition from the singlet state to the ground state (S = 0) occurs rapidly in nanoseconds leading to light emission, decay from the triplet state to the ground state is quantum-mechanically forbidden (since ΔS must be zero). The probability that a triplet or singlet state will form depends on the number of possible spin orbital configurations that are possible in each state, which for singlets is one and for triplets is threefold. Consequently, given the random nature of spin production in electroluminescent devices simple statistics predicts that only 25% of the injected charges should result in light emission from singlet states, whereas 75% of excitations will involve triplet states and thus, in principle, the internal quantum efficiency should be limited to 25%. However, the formation cross-section of singlet excitons can be much higher than triplet excitons in π-conjugated polymers [27] and it has been shown recently that singlets generated from triplet–triplet annihilation can reach 40% of all excitons generated by charge recombination [28]. Moreover, incorporating heavy atoms (such as platinum and iridium) into the emitter structure can allow triplet to singlet intersystem crossing (ISC), which can lead to highly efficient devices with potentially up to 100% of the excitons emitting light and very high PL quantum yields [29]. More recently, donor–acceptor complexes have been used to demonstrate highly efficient ISC without the use of heavy atoms [30].

Another important factor that governs the luminescence of OLEDs is the surface out-coupling efficiency, ξ, which relates the external electroluminescent (EL) efficiency, η_{EL}^{ext}, with the internal electroluminescent (EL) efficiency, η_{EL}^{int}, such that:

$$\eta_{EL}^{ext} = \xi \eta_{EL}^{int}. \tag{1.1}$$

Given that only the external EL efficiency can be measured, understanding the origin of ξ is therefore crucial to provide an estimate of the internal quantum efficiency and hence an insight into the device physics of OLEDs [31].

For isotropic emission, where there are no optical interference effects with the reflector, the external and internal EL efficiencies are simply related by the refractive index of the emitter layer, n, with $\xi \sim 1/2n^2$ for large n. In the case of in-plane dipoles (and no optical interference) the relationship is modified slightly so that $\xi \sim 3/4n^2$ [32]. However, for OLEDs, which consist of a thin organic light emitting layer sandwiched between a metal cathode and a transparent anode, the situation is far more complicated. Multiple optical reflections at the metal anode combined with complex refraction at the various multiple interfaces in the device result in optical interference effects that can significantly influence the value of ξ. For OLED devices where the emission location is optimal, optical modelling shows that $\xi \sim 3/4n^2$ and $\xi \sim 6/5n^2$ for isotropic and in-plane dipoles, respectively. More importantly, this analysis also revealed that the probability of forming an emissive singlet exciton from electrical injection was substantially larger than the value of 0.25 predicted from quantum mechanical spin statistics [33].

Table 1.1 *Summary of performance characteristics of various lighting technologies*

Light Source	Efficiency ($lm\ W^{-1}$)	CRI	Lifetime (hours)
Incandescent lamp	10 – 15	>90	1,000
Fluorescent tube	40 – 80	70	10,000
High-pressure sodium lamp	140	<10	10,000
Light-emitting diode (LED)	>80	80	>10,000
Organic light-emitting device (OLED)	>110[1]	>90	10,000

Table adapted from So et al. [37].
[1] Panasonic announced the world's highest efficiency white OLED for lighting on 24 May 2013.

1.2.2 White OLEDs and Application to Solid State Lighting

The development of the first white organic OLED was reported by Kido and co-workers in 1994 [34, 35]. The luminous efficacy of a source is a measure of the efficiency with which the source provides visible light and is the ratio of luminous flux to input power and is measured in lumens per watt ($lm\ W^{-1}$). Since this first report, the field developed rapidly with the white light efficacy improving from initial values of less than 1 $lm\ W^{-1}$, to figures of ~100 $lm\ W^{-1}$; comparable to that of a fluorescent tube [36]. White OLEDs offer a number of attractive features as a white light source. They are a lightweight, flexible and mercury-free illumination light source and already meet the requirements of the EU directives on Waste Electrical and Electronic Equipment (WEEE) and Restriction of Hazardous Substances (RoHS). As such, these devices are expected to usher a revolution in lighting design, including such developments as transparent lighting panels and luminescent wallpapers [37].

The requirements for general illumination lighting are: (a) sufficient luminous flux (where a typical office lighting fixture emits ~ 5000 lm), (b) a colour rendering index (CRI) greater than 80 (where the CRI measures the ability of a light source to render all of the colours that it illuminates) and (c) a lifetime of greater than 10,000 hours at high brightness. The comparative performance characteristics of various lighting technologies are listed in Table 1.1.

1.2.3 Conjugated Polymers for Lasing Applications

One of the holy grails of organic optoelectronics has been the development of the semiconducting polymer laser. The first report of the use of a conjugated polymer as the gain medium in a lasing system was in 1992, when MEH-PPV was used in solution as the dye in an optically pumped dye laser system using a Q-switched Nd:YAG laser [38]. Apart from their excellent material processing properties, conjugated polymers also have high photoluminescence yields, even in the solid state, which means that they have been identified as potentially attractive candidates for solid state lasing for a number of years [39]. Optically pumped lasing has been reported for a wide range of conjugated polymers, including in the polyphenylenevinylenes (PPVs) [40–42] and polyfluorenes [43, 44]. In addition, the lasing of host–guest systems composed of MEH-PPV (see Chapter 3) doped into F8BT (see Chapter 6) have been described [45].

The challenge faced in developing an electrically pumped organic solid state laser is, however, more significant [46]. In particular, the issues around overcoming the inherent losses, which lead to extremely high lasing thresholds, need to be overcome. These losses arise from optical leakage at the metal electrodes [47], polaron–exciton quenching in the active layer [48] and absorption by triplet excitons [49]. Thus, although the electrically pumped organic laser has yet to be achieved, despite more than 15 years of development since the first optically pumped organic laser, the underlying issues are now understood and are being quantitatively determined.

1.3 Developing Solar Paint: Organic Photovoltaics

The development of novel photovoltaic (PV) devices based on conducting polymers is one of the most rapidly developing areas in the field of organic electronics. The increasing demand for energy coupled with the rising world population provides a significant driver for the expansion of global energy resources. Moreover, the constraints of climate change and sustainability mean that meeting this so-called Terawatt Challenge will only be possible using non-CO_2 emitting energy generating technology. Indeed, ever since their first discovery, conducting polymers have been used to form the active elements of PV devices. The motivation is clear: of all the non-traditional renewable energy resources currently available, arguably only solar power has the potential to provide the world with the increased electrical generating capacity that it requires in a sustainable manner. Moreover, within the more specific area of solar photovoltaic devices, solar cells made from polymers can be printed at high speeds across large areas using roll-to-roll processing techniques, thus creating the tantalising vision of coating every roof and other suitable building surface with photovoltaic materials at extremely low cost.

1.3.1 Basic Operation

The photocurrent generation mechanism for inorganic semiconductors, such as silicon, is well established. In a conventional inorganic photovoltaic device, the photocurrent is generated across junctions between n-type and p-type semiconductors with the photoexcited electron and hole being separated by the strong internal electric fields that exist at the junction and thus becoming free to migrate to the opposite electrode, whereupon they can do useful work. In the case of semiconducting polymers, however, the photocurrent generation mechanism is more complex than for inorganic semiconductors. Although the action of an incident photon upon a conducting polymer excites an electron from the valence band into the conduction band, the low relative dielectric constant of organic materials (typically 2–4) means that the electron and hole that are produced are bound and their motion through the material is coupled. These coupled moieties are known as *excitons* and are responsible for many of the electronic properties found in the most common and efficient polymer-based electronic devices.

How then, can we obtain any useful work from a conducting polymer if the electron and hole are not separated? It turns out that the bound exciton can be split at interfaces with the simplest interface being created at the junction between the electrode and the conducting polymer. Under open circuit conditions, holes are collected at the high work function electrode (which is typically made of indium tin oxide (ITO), a transparent conductor) while electrons are collected at the low work function electrode (which is typically a metal electrode such as aluminium). Indeed, the open circuit voltage generated by these devices depends upon the work function difference between the two electrodes. Unfortunately, the exciton splitting process that occurs at a conducting polymer/electrode interface is not very efficient and was one of the causes of the low quality of early polymer photovoltaics. Another cause of the very low efficiencies of early devices was the effect of impurities, such as oxygen, which acted as traps for the migrating charge carriers.

Attempts to improve the efficiency of the exciton splitting process led to the development of new conducting polymer species that contained electron-donating and electron-accepting species. By creating interfaces between conducting polymer molecules of differing electron affinities it is possible to enhance the probability of electron transfer between molecules. This process (photoexcited charge transfer) causes the bound charges to separate and the junction formed at the donor–acceptor interface is analogous to a semiconductor heterojunction. These heterojunctions work very well at separating excitons that arrive at the junction. Unfortunately, the lifetime of excitons is short such that only excitons that are formed within about 10 nm of the junction will ever reach it. This short exciton range clearly limits the efficiency of these photovoltaic devices. In an attempt to develop a more efficient photovoltaic structure, interpenetrating networks of electron-accepting and electron-donating polymers were produced by blending electron-donating and electron-accepting materials together. With these materials, the number of heterojunctions within the polymer blend is greatly increased, thus so is the probability that an exciton will encounter a junction and be separated.

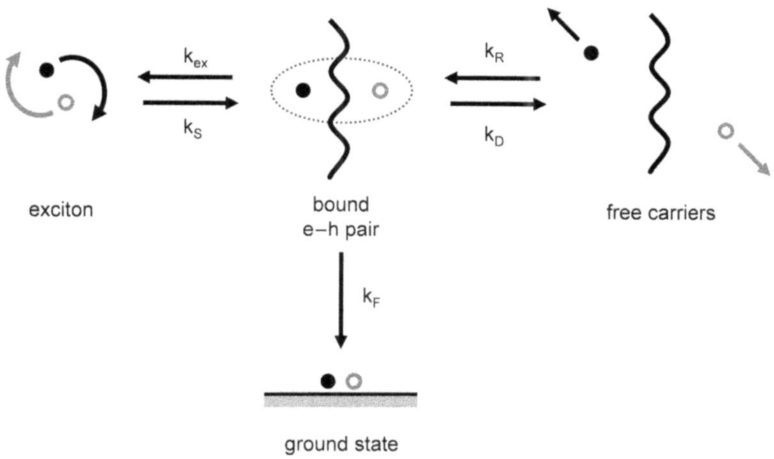

Figure 1.5 Schematic of free carrier generation in organic photovoltaic cells. The wavy line denotes an interface and the dotted ellipse denotes a Coulombic binding between an interface separated electron–hole pair

The model for exciton dissociation and recombination at these heterojunction interfaces is now well-established and is summarised in Figure 1.5. Dissociation of photogenerated excitons at donor–acceptor interfaces (with rate constant k_S) does not initially result in the production of free charges, but rather a bound electron-hole pair [50] (also known as a charge transfer state [51], bound radical pair (BRP) state [50] or bound polaron pair [52]) due to the Coulomb attraction across the interface between the charge carriers. Thus, a further dissociation step (with rate constant k_D) needs to occur, whereby the bound electron-hole pair separates into free charge carriers via a process that is dependent upon the temperature and electric field across the interface, before it decays to the ground state (with rate constant k_F). This bound pair is metastable and can also be regenerated via recombination (whether geminate or bimolecular) of free charge carriers (with rate constant k_R) [51]. Finally, it is thought that excitons can be regenerated at donor–acceptor interfaces (with rate constant k_{ex}) and can be transferred back into the bulk [52, 53].

The ratio of the dissociation, k_D, and recombination, k_R, rate constants can be expressed in terms of the charge separation, a, the temperature, T, and the binding energy, E_B, such that [51]:

$$\frac{k_D}{k_R} = \frac{3}{4\pi a^3} e^{-E_B/kT} \left[1 + b + \frac{b^2}{3} + \frac{b^3}{18} + \frac{b^4}{180} + \ldots, \right], \tag{1.2}$$

where, for a given electric field, E, relative permittivity, ε, and temperature, T, b is a constant defined by:

$$b = \frac{e^3 E}{8\pi\langle\varepsilon\rangle\varepsilon_0 k^2 T^2}, \tag{1.3}$$

where e is the electronic charge, k is, the Boltzmann constant and ε_0 is the permittivity of free space, and:

$$E_B = \frac{e^2}{4\pi\langle\varepsilon\rangle\varepsilon_0 a}. \tag{1.4}$$

Substituting Equation (1.3) into Equation (1.1) and rearranging:

$$\frac{k_D}{k_R} = A\left(\frac{1}{a^3}\right) exp\left\{-\left(\frac{B}{a}\right)\left(\frac{1}{kT}\right),\right\}, \tag{1.5}$$

where

$$A = \frac{3}{4\pi}\left[1 + b + \frac{b^2}{3} + \frac{b^3}{18} + \frac{b^4}{180} + \ldots\right] \tag{1.6}$$

and

$$B = \frac{e^2}{4\pi\langle\varepsilon\rangle\varepsilon_0}. \tag{1.7}$$

Thus, from Equation (1.4) we see that the ratio of the dissociation and recombination rates is a function of both the separation of the bound electron–hole pair (a) and the effective thermal energy of the bound state (kT) [54]. The overall probability that a bound electron–hole pair dissociates into free charge carriers is also a function of k_F [51], which is likely to have a similar dependence upon charge separation and effective thermal energy since a strong coupling between the bound electron and hole is also expected to increase k_F [52]. Therefore, the overall dissociation probability is a function of both initial charge separation (a) and effective thermal energy of the bound state (kT).

1.3.2 Progress towards Organic Solar Coatings

Photovoltaic devices made from organic semiconductors have come a long way since the first devices, using organic crystals such as anthracene and perylene, were reported in 1959 [55]. These early organic electronic devices were typically sandwich-type structures, with films or single crystals of thickness between 1 and 100 μm arranged between two conducting electrodes. Unfortunately, these molecular semiconductor-based devices tended to show low photo-voltages (typically less than 500 mV) and poor energy conversion efficiencies. These poor efficiencies were attributed to difficulties in achieving efficient ionisation of the photo-generated excitons together with inherently poor electron or hole mobility in these materials. However, different device architectures were devised to circumvent some of these limitations. In 1986, Tang demonstrated that exciton dissociation could be made more efficient in molecular photovoltaic devices by creating a heterojunction of two organic layers with different electronegativities (namely CuPc (copper phthalocyanine) and PV, a perylene tetracarboxylic derivative) [56]. This was the first heterojunction structure and showed that efficiencies could be improved by bringing the electron donor (D) closer to the electron acceptor (A) material. In 1992, Sariciftci et al. showed that, by combining a semiconducting polymer (D) with C_{60} (A), ultrafast D–A electron transfer occurs and they demonstrated for the first time the potential for fullerenes (and their derivatives) to act as highly efficient electron acceptors in these devices [57, 58]. In 1995, Yu et al. showed how the bulk heterojunction (BHJ) concept could be used to massively increase the number of excitons that reach an interface by intimately blending the donor and acceptor components [59]. Moreover, the soluble fullerene derivative phenyl-C_{61}-butyric acid methyl ester (PCBM) material that was developed as part of this work has gone on to become the mainstay electron acceptor material in OPV devices for more than a decade. In 2001, Shaheen et al. showed how surface morphology was a critical determinant for device performance and that the choice of solvent system could have a profound effect upon OPV efficiency [60]. Since then, OPV devices based on polythiophene derivatives have gone on to achieve routine efficiencies of around 4%. Further efforts to improve device efficiency have focussed upon the development of tandem cells; combining two or more OPV cells with complimentary absorption characteristics in series [61]. This architecture increases the open circuit voltage of the overall device and enables broadband absorption over the solar spectrum. The importance of controlling the light distribution in these devices has been recognised. Combining this architecture with the use of a TiO_2 spacer layer which can enhance device performance by ~50% through maximisation of the light field within the very thin active layer, has led to the most efficient OPV devices so far, with efficiencies as high as 6.5% reported in 2007 by Heeger's group for the first tandem cells involving two polythiophene derivatives combined with PCBM [62], which was increased in 2013 to over 10% for devices using alternating donor–acceptor copolymers [63, 64].

More recently, as will be expanded upon in Chapters 7 and 8, many more high-performance polymers have been developed [65]. The majority of these improvements have been based on extending the spectral response of the materials to longer wavelengths to produce so-called low band-gap materials. Materials such as poly [2,6-(4,4-bis-(2-ethylhexyl)-4H-cyclopenta[2,1-b;3,4-b']dithiophene)-alt-4,7-(2,1,

3-benzothiadiazole)] (PCPDTBT) have extended the wavelength response of these materials up to 900 nm, and have produced device efficiencies of up to 5.5% with appropriate alkanethiol additives [66]. Further synthetic developments led to the introduction of donor–acceptor (D–A) systems such as the thieno[3,4-b]-thiophene (TT) and benzodithiophene (BDT) D–A systems [67–69], capable of delivering power conversion efficiencies approaching 20%, as will be discussed in Chapter 8.

1.4 Developing Disposable Integrated Circuits: Organic Transistors

The first solid state organic field effect transistor (OFET) was demonstrated by Tsumura and co-workers in 1986 using polythiophene as the semi-conducting layer [70], with a similar device being reported by Burroughes et al. in 1988 [71]. In an OFET (Figure 1.6) the current between the source and drain electrodes (I_{SD}) is controlled by a voltage (V_G) applied to a third electrode known as the gate. The gate electrode is separated from the source–drain region by a thin insulating dielectric region and thus is capacitively coupled to the semiconductor. By altering the bias voltage applied to the gate region, the source–drain region can be altered from conducting to insulating and, hence, the device can be turned on or off. Importantly, the presence of a relatively small number of charges on the gate electrode (I_G) alters the flow of a great many charges (I_{SD}) controlled by the

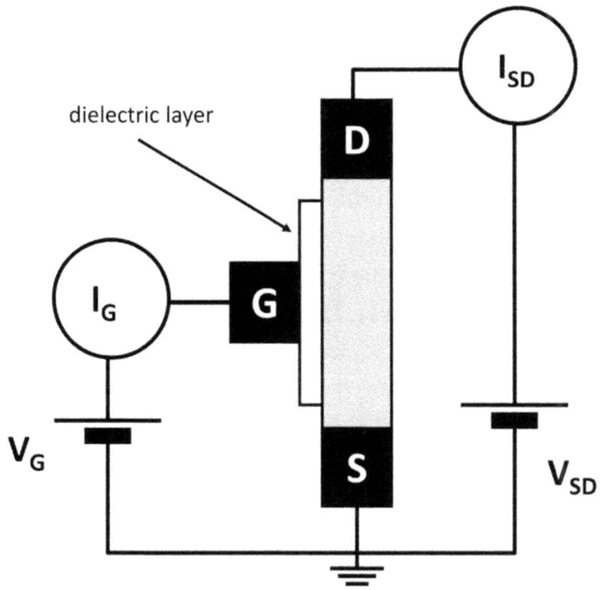

Figure 1.6 Schematic of free carrier generation in organic field–effect transistor. The letters **D**, **G** and **S** denote the drain, gate and source electrodes, respectively

voltage (V_{SD}) between the source and drain electrodes and, thus, as well as acting as a *switch*, the FET acts as an *amplifier*.

A common feature of these early organic thin film transistors (OTFTs) was the use of silicon as the substrate material. These hybrid devices were not truly all-polymer electronic devices and thus did not offer all of the advantages offered by organic materials such as flexibility. However, patterning, conducting and insulating regions on silicon is a well-established technology and thus the fabrication of these devices was relatively easy to implement. Subsequently, the first all-polymer FET was reported by Garnier et al. in 1994 and was fabricated by a printing technique [72].

As discussed previously in this chapter, the mobility of the charge carriers is a key parameter in determining the performance of conjugated polymers in organic electronic devices and this criterion is particularly relevant in the case of organic transistors. As such, a great deal of OFET research has focussed upon developing and optimising the charge carrier mobilities of existing and new organic materials. Traditionally, hole carriers (p-type materials) dominated the organic electronic landscape but advances in material design and synthesis have led to an increasing number of n-type molecules where electrons are the dominant charge carrier. A variety of semiconducting organic materials are now commonly available, ranging from small molecule systems based around the acenes to conjugated polymer systems (such as the polythiophenes) to larger macrocycles, with further intense effort aimed at improving morphology, processing and reliability of the materials through modification of side chain functionality, etc.

A much-discussed topic in the OFET field has been whether high polymer crystallinity is an essential determinant of high-mobility devices. Previous advances in polymer mobility appeared to be predicated on the development of highly crystalline polymers and this polymer engineering approach had resulted in OFET mobilities in excess of 1 $cm^2\ V^{-1}\ s^{-1}$ for solution-processed polymer semiconductors [73]. Recently, however, as will be discussed in Chapter 8, mobilities an order of magnitude higher have been reported, despite the fact that these new polymers are actually less ordered [74]. Interestingly, it appears that, although these high-mobility materials are not highly crystalline, they do gain longer-range structural order by having a more rigid polymer backbone [75]. This increased long-range order compensates for the lack of crystallinity by providing high-mobility pathways (which extend for up to 1 μm), which allow the charge carriers to more effectively bypass disordered regions in the material [76]. Excitingly, these discoveries may point towards the development of solution-processible polymers with mobilities as high as their small crystal counterparts [77].

One of the most active areas of current OFET research is focussed on the application of these organic electronic devices as sensors and detectors. Research into OFET-based sensors encompasses a number of areas, including large area detectors, transducers and biosensors. In the case of large area detectors such as pressure sensors, the OFETs are typically used to form active-matrix structures or other electronic circuits for reading data out from each sensing unit. As such, the OFET structures are simply used as electronic switches and are actually designed to be insensitive to their environment. Indeed, research at the University of Tokyo has shown how robots of the future could be endowed with the sense of touch using sensors based on organic electronics. Takeo Someya's group has built high-mobility organic transistors, with pentacene as the channel layer, that were fabricated into a flexible sheet made of carbon and rubber [78, 79]. Upon flexing, the electrical resistance of the carbon–rubber sheet is changed locally and this change in resistance operates on the nearest OFET. Using this so-called e-skin (Figure 1.4), the robot's control system could be made to 'feel', with a sensitivity of about 10 g cm^{-2} that is independent of the OFET array. The initial 16×16 sensor devices built by the University of Tokyo team were limited by the response time of the pressure-sensitive rubber (~500 ms) but for larger arrays the overall frequency response is currently limited by the response time of the OFET devices (~30 ms).

OFET devices are also attractive as candidates for sensors in the own right. In particular, the fact that OFETs are fabricated from a material system that can be made liquid soluble and is carbon-based means that they are much more compatible with other organic systems, such as biomolecules, than transistors built from conventional inorganic materials. Research at the University of Newcastle is focussed upon developing new biosensors based on integrating biomolecules (such as enzymes) directly into the OFET structure to create highly specific and highly sensitive detectors that can be printed at low-cost using conventional inkjet printing technology. The ultimate vision for this work, which is at an advanced stage, is the development of a sensor that responds to the much lower salivary glucose levels instead of the usual blood glucose levels [15, 80–83]. Such a development would open the door for a non-invasive glucose meter for diabetes sufferers that would only require a very small amount of saliva to determine glucose level as opposed to current devices that require the patient to provide blood samples via needle stick.

1.4.1 The Organic Electronic Age

Although the fabrication of devices and machines based on organic electronics is still in its infancy, this exciting new field is developing rapidly and this development is accelerating. It is widely anticipated that electronic devices based on organic materials will gradually replace those based on conventional inorganic semiconductors and metals, driven primarily by the material properties of plastics, which allow for the production of arrays of devices on flexible substrates using extremely low-cost printing and coating techniques. Indeed, rapid advances are already being made, with organic electronic materials offering enormous potential across many technologies, ranging from new sources of low-cost renewable energy to arrays of biosensors for medical applications. Optical pumping of organic lasers has also been demonstrated but electrically pumped organic lasers remain out of reach at present. These materials also open up the prospect for new technologies that are not accessible by current materials by providing the interface between biological systems and electronic systems. It is possible to envisage bionic devices based on organic electronics that could be readily interfaced with biological systems and potentially directly linked to a patient's muscles and nerves. Perhaps we will see, in our own lifetime, a world where organic electronics provides both the photovoltaic coating on our roof and the glucose sensor that we printed at home on our inkjet printer.

2

Polyacetylenes

It seems appropriate to begin our review of conjugated polymers with polyacetylene, as this is not only the simplest fully conjugated polymer but also the one whose properties first alerted scientists and engineers to the possibilities such polymers could offer for functional electronic devices.

2.1 Polyacetylene

Polyacetylene (PA, Figure 2.1, **1**) is the archetypal conjugated polymer consisting of alternating carbon–carbon single and double bonds. It receives its trivial name from it formally being a polymer of acetylene (ethyne) with formula $(C_2H_2)_n$. The double bonds have two possible isomers – *cis* and *trans*. Most syntheses of polyacetylene initially produce all-*cis* material which then isomerises upon heating to the thermodynamically more stable all-*trans* form – a process which can be monitored using infrared (IR) absorption spectroscopy by the disappearance of the *cis* C–H out-of-plane deformation band at 740 cm^{-1} and its replacement by the corresponding band for the *trans* isomer at 1,015 cm^{-1} [84].

The synthesis of PA illustrates an important recurring theme in conjugated polymer research – the final product is insoluble and infusible, but a proper study of its properties, let alone its use in a working device, requires high-quality thin films, possibly of quite large area. The earliest attempts to prepare PA by polymerisation of acetylene using Ziegler–Natta catalysts produced intractable powders. Eventually films of PA were obtained by the group of Shirakawa using a highly concentrated solution of catalyst (Figure 2.1) [84]. These films were good enough to enable the conductivity of pristine and doped PA to be measured, which led to the award of the Nobel Prize in Chemistry to Shirakawa, Heeger and MacDiarmid for the discovery that the doped films were conductive [85–87]. However, such films are not really further processable, and so are unsuitable for

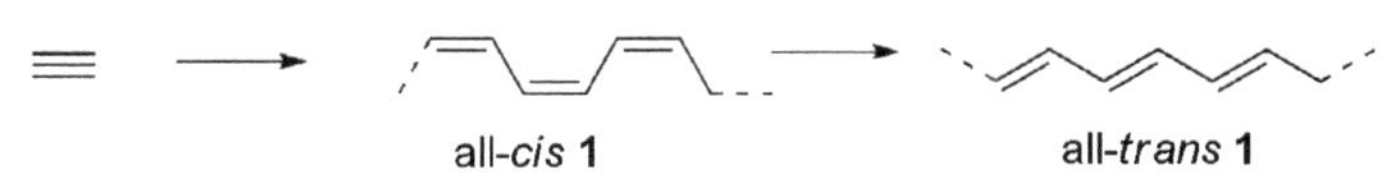

Figure 2.1 Shirakawa synthesis of polyacetylene

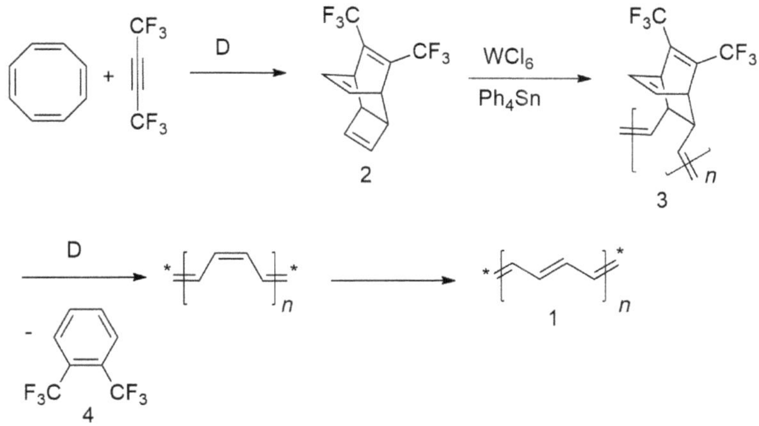

Figure 2.2 Durham precursor route to polyacetylene

devices as they require the ability to make films of any desired thickness, shape, etc. To achieve that one ideally needs to use a *precursor polymer.*

A precursor polymer is a polymer which can be processed (usually from solution) to form a film which can then be readily converted into a film of the (usually insoluble) active polymer by methods such as heating, irradiation or exposure to chemical vapours. The requirements for the precursor polymer are that the conversion process must be quantitative and occur without significant degradation of the polymer film and that the by-products should not cause damage to or adversely affect the optical or electronic properties of other materials (e.g. electrodes or charge-transporting layers) present during the conversion. For this reason, conversion reactions ideally should not form by-products that are corrosive (strongly acidic or basic) or non-volatile (i.e. persistent contaminants) or which can dope the active or charge-transporting layers within devices.

In the case of PA, a suitable precursor route was developed by the group of Feast at Durham – the so-called Durham Polyacetylene route (Figure 2.2) [88]. Here a ring-opening polymerisation of monomer **2** produces the soluble precursor polymer **3**. Upon heating, this undergoes a polymer analogous retro-Diels–Alder reaction to produce PA. The key point of the conversion is that the by-product 1,2-di(trifluoromethyl)benzene (**4**) is highly stable, making the reaction thermodynamically favourable and thus high-yielding, volatile and non-corrosive.

The conversion step in this synthesis occurs at 150 °C. To lower the required conversion temperature, an improved precursor route was developed at Durham

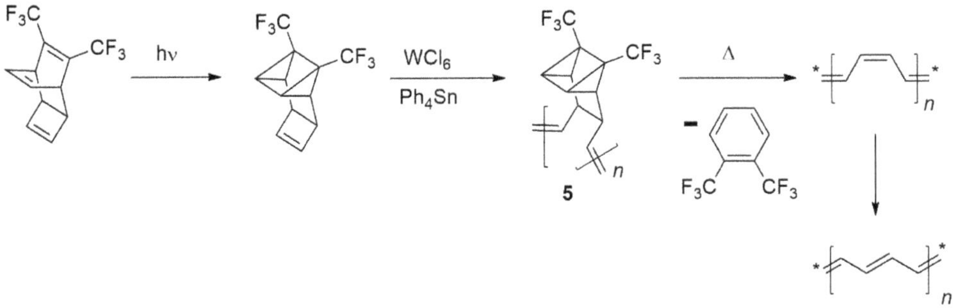

Figure 2.3 Revised Durham precursor route

(Figure 2.3) which required only 75°C for the conversion of the new precursor polymer **5** [89]. This lowering in the conversion temperature is due to the more exothermic nature of the reaction, as the opening of the cyclopropane rings relieves their high strain energy. The disadvantage of this is that the reaction can become explosive when performed on bulk samples and so is only safe for conversion of thin films.

As so often in materials science, one needs to find a balance – the conversion step must not be either too difficult or too facile from a thermodynamic viewpoint. If the conversion step is too energetically unfavourable, the conversion will not occur quantitatively, thus introducing defects into the polymer chain which will affect (in most cases adversely) the electrical and optical properties. If the reaction is too favourable then, as here, the conversion may become problematic to perform; indeed the precursor may become unstable at or near room temperature and start to spontaneously (and possibly uncontrollably) degenerate to the final (intractable) polymer before or during processing, which creates obvious problems for its use.

While the Durham precursor route has enabled the facile fabrication of films of polyacetylene, PA has to date not proven to be of any value as a material for practical functional devices. The charge carrier mobility of undoped PA in OFETs is too low (10^{-5} cm^2 Vs^{-1}) for practical applications, while conductive (i.e. doped) PA is unstable in air, making it unsuitable for devices outside an inert atmosphere (such as a glovebox). This problem with stability arises because of the nature of the charged species in the polymer. Doped PA is conductive because such doping produces delocalised carbocations or carbanions (Figure 2.4) but such species, unless stabilised by suitable substituents, are highly reactive towards molecules such as oxygen or water. As a result the only conducting polymer that is stable in its conductive form is polyaniline (PANi, **5**), as here the charges are produced by protonation of the nitrogen atoms to form ammonium ions which are stable towards air and water.

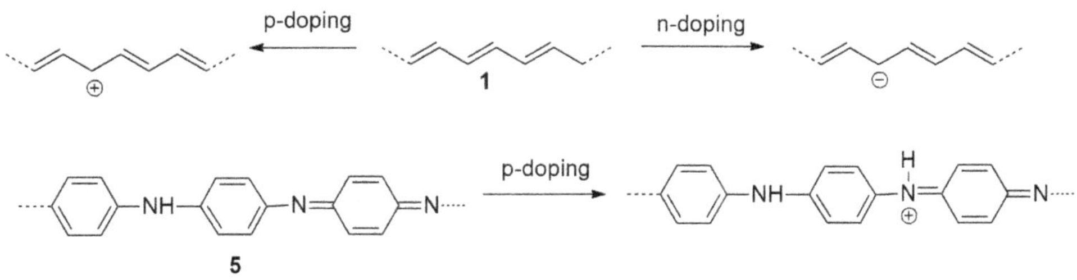

Figure 2.4 Comparison of charged species in doped PA and PANi

2.2 Electrical Properties of Polyacetylene

PA is the quintessential model system for 1-D conduction in an infinite polyene chain and has played a major role in the development of electrical models for conduction in the field of conducting polymers [90]. In the classic view of conjugated polymers, hybridisation of the sp^2p_z orbitals results in one unpaired electron per carbon atom. As such, the electrical nature of the polymer system depends upon the details of monomer repeat unit and can range from semiconducting to metallic behaviour. In PA, strong intra-chain (and weak inter-chain) interactions result in delocalisation of the π electrons along the polymer chain and a 1-D electronic system.

It has long been recognised that quasi-1-D metals tend to spontaneously reorganise according to the well-known Peierls distortion, whereby the spacing between successive atoms along the polymer chain undergoes a periodic modulation with periodicity π/k_F, where k_F is the Fermi wavevector [90]. In the case of PA, where the π band is half-filled, the energetic driver for symmetry breaking is highly favourable and results in dimerisation of the polymer chain. Despite the increased elastic energy caused by the distortion, the overall lowering of the electronic energy of the system means that modulation of bond length (which in the case of PA is 0.03–0.04 Å) is still energetically favourable.

PA is probably best known as providing the classic example of soliton-based electron transport in conducting polymer materials. The key energetic feature of PA is that it exhibits a two-fold-degenerate ground state. As such, PA can support non-linear electronic excitations which manifest as moving domain walls that separate regions having different ground states; an A phase and a B phase. The domain walls preserve their shape and actually alter the medium as they pass through it; thus, they can be considered to be *topological solitons*. The resulting soliton has a lower energy than either an electron or a hole and, thus, is spontaneously created when electrons or holes are injected into the system via doping or photoexcitation or are thermally generated.

High conductivity and dopant-mediated conductivity over several orders of magnitude were the two key drivers of interest in conducting polymers in the late 1970s and 1980s. However, progress towards understanding the mechanisms of conductivity was hindered by their complex morphology and semi-crystalline nature. Indeed, the challenging issue of understanding and manipulating the structure and morphology of semiconducting polymers remains as true today as it did 30 years ago. In the case of PA, its morphology is not well suited to device formation since it does not readily form films but rather inhomogeneous networks of low density [91]. These structures consist of colloidal-like clusters on the surface, which macroscopically resemble a metallic compact film [84].

Thus, while PA has proven interesting as a material in which to study the modes of conduction in conjugated polymers, it has proven of little practical benefit due to its poor morphology and the instability of the conductive form. In addition, it is not an emissive material so has no utility in LEDs. Substituted polyacetylenes, however, can be emissive and so have been used in PLEDs.

2.3 Substituted Polyacetylenes

Such materials can be made by the transition metal catalysed polymerisation of substituted alkynes, as illustrated in Figure 2.5. Polymerisation of diphenylethyne **6** gives the polymer **7**, which displays green luminescence but is unsuitable for use in LEDs as it has only limited solubility [92]. Partial perfluopropylation produces a soluble copolymer **8** with one perfluoroalkyl group per two monomer units, which can be processed from solution to make a green-emitting PLED [93].

Such a post-polymerisation reaction to increase solubility of a polymer is not a commonly used procedure as it is hard to control and produces random functionalisation with uncertain and possibly unreproducible effects upon the properties of the material. A better method to improve the solubility of conjugated polymers is to use monomers containing suitable solubilising groups, most commonly long-chain alkyl or alkoxy substituents. A range of soluble

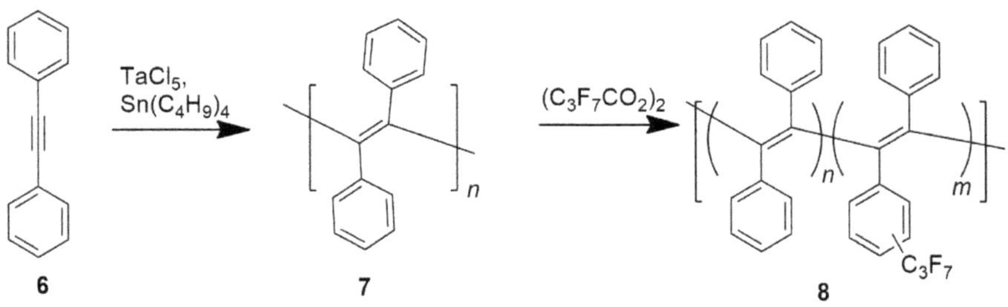

Figure 2.5 Synthesis of a soluble poly(diphenylacetylene) copolymer

polyacetylenes containing alkyl and aryl substituents have now been prepared by polymerisation of the corresponding alkynes [94–98].

The colour of the emission from these polymers depends upon the nature of the substituents – a theme we find recurring throughout this book. Poly(diarylacetylene)s such as the diphenylacetylene polymer **7** and its solubilised derivatives, whether random copolymers such as **8** or homopolymers such as **9–11** (Figure 2.6) [99–104], are typically green emitters with emission maxima around 520–540 nm. The luminescence intensity is reported to be enhanced by using bulky or long alkyl substituents. This may well reflect greater inter-chain distances, reducing non-radiative decay [101]. The suppression of inter-chain interactions is often desirable in LED applications as many such interactions act to decrease luminescence by permitting excitons to migrate easily to sites where decay of the excited states can occur by non-radiative mechanisms. Too little interaction between the chains on the other hand diminishes charge migration, which reduces the efficiency of the devices. The maximisation of device efficiency may therefore require a fine balance to be maintained between too much and too little chain interaction, as both charge migration and emission efficiency contribute to overall device efficiency. The incorporation of charge transporting units is commonly used as a way to boost device efficiency. An example of this in polyacetylenes is the polymer **12** bearing a hole-transporting carbazole substituent which reportedly displays green emission ($\lambda_{max} = 540$ nm), with much better efficiency and brightness than for the alkyl-substituted polymer **11** [92, 105]. By contrast, polymers of phenylacetylenes such as **13** or **14** are blue emitters with emission maxima around 470 nm, while halogenated polymers such as **15** are blue-green emitters ($\lambda_{max} = 500$ nm) [99, 102–104, 106]. Blue emission is also seen from similar polymers bearing other aryl groups such as anthracene [107] or fluorene [108]. The EL intensity of alkyl substituted polymers like **14** is reportedly enhanced by using longer alkyl chains [104]. As with the diarylacetylene polymers just discussed, this is presumably due to reduced interactions between the polymer chains.

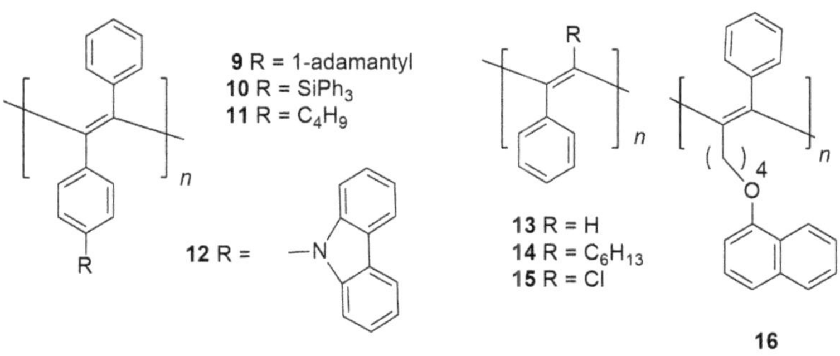

Figure 2.6 Some emissive soluble poly(arylacetylene)s

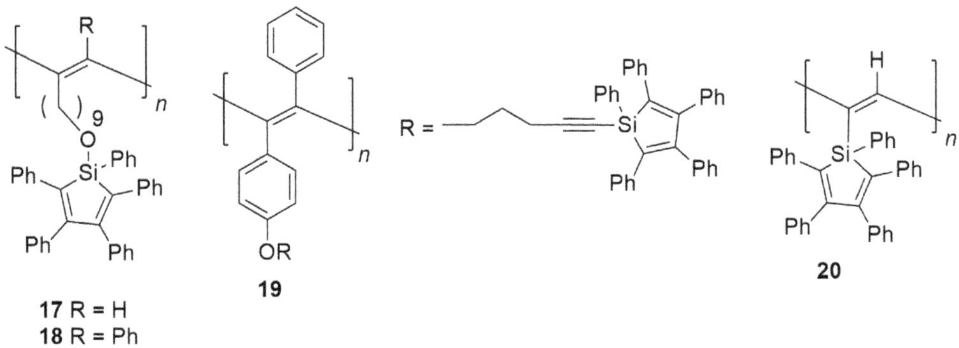

Figure 2.7 Emissive polyacetylenes bearing silole groups

It is reported that addition of the fullerene C_{60} to the reaction mixture increases the rate of polymerisation of 1-phenyl-1-alkynes, and that the resulting materials show stronger PL [109, 110], which is interesting as C_{60} usually suppresses PL by acting as an electron acceptor. Stretching of a film of **14** is reported to cause polarisation of the PL parallel to the stretching direction [104]. A wide range of polymers bearing alkyl chains ω-functionalised with chromophoric groups such as biphenyl, fluorene, carbazole or naphthalene have been prepared [94, 95, 111–113]. The emission is generally blue with high PL but only moderate EL efficiencies (e.g. 0.85%) were reported for polymer **16** (λ_{max} = 460 nm) [94, 114]. By contrast to the increase in luminance reported with longer alkyl chains in the polymers described previously in this chapter, in these polymers longer alkyl chains reduce the EL efficiency [94]. Attachment of mesogenic groups in the side-chains induces liquid crystallinity and thus the potential to produce polarised emission [94, 95, 111, 113, 115].

Interesting variance in emission colour is seen for the silole-containing polyacetylenes **17–20** (Figure 2.7). Whereas blue-green emission is seen from polymers **17–18** and green emission from polymer **19** (λ_{max} = 496 nm and 512 nm, respectively) [94, 116–118], polymer **20** with no linker displays red EL (λ_{max} = 664 nm), as here the silole group can directly influence the π-system in the main chain [116, 117]. A further interesting feature of polymers **17–19** is that they are only emissive in the solid state as energy transfer occurs from the polymer backbone to the siloles and the latter display *aggregate-induced emission* (AIE). Whereas most conjugated molecules and polymers tend to show some quenching of emission when the chromophores interact in aggregates, in AIE the opposite occurs so that siloles and polymers containing them as side-groups are not emissive in solution where the chromophores are isolated, but strong emission is seen from aggregates such as are found in the solid state [98].

Another example of colour modification is seen in polymers **21–23** (Figure 2.8) with ethynyl substituents. Polymer **21** with a terminal ethynyl substituent shows

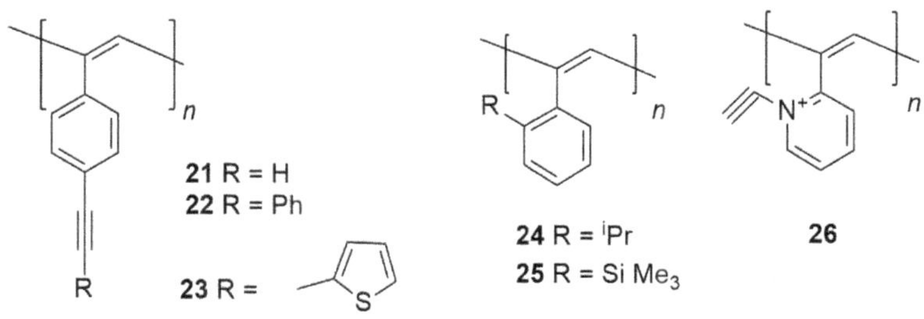

Figure 2.8 Other emissive poly(arylacetylene)s

green emission (λ_{max} = 521 nm), while the arylethynyl-substituted polymers **22** and **23** show violet (λ_{max} = 368 nm) and blue (λ_{max} = 405 nm) EL, respectively [119]. This marked blue-shift in emission colour is almost certainly the result of a severe decrease in conjugation length due to twisting of the polymer backbone arising from interactions between the side-chains. Usually such interactions are steric in nature – that is, the side-chains are repelling each other – but in this case it is possible that they are instead attractive π–π interactions between the side-chains. Such use of side-chain interactions to tune the emission colour of a conjugated polymer is a motif we will see again and is a reminder that one must consider not just the molecular but also the supramolecular nature of materials. However, bulky substituents do not always induce blue shifts in emission. Thus, polymer **24** with a bulky *ortho*-group on the phenyl ring shows orange-red EL (λ_{max} = 600 nm) [103], while weak red emission (λ_{max} = 680 nm) has been reported for polymer **25**. A similar pyridine containing polymer **26** has been reported to show red PL (λ_{max} = 710 nm), whose intensity decreased with increasing temperature [120]. In these cases, the most obvious explanation is that there is strong aggregation between the polymer chains as such aggregates often display red-shifted and weaker emission. A blend of polymer **25** with polymer **11** shows strong red emission, which is attributed to energy or charge transfer from polymer **11** to polymer **25** [104]. Similarly blending polymer **15** with polymer **11** leads to strong green emission from the latter due to energy transfer [121]. The use of such blends of emissive materials has the obvious problem that the blends may not be stable, leading to colour instability as the components start to phase separate. However, as we shall see, this approach can be useful in some cases. For example, it is a way to obtain white emission by mixing together materials with complementary emission colours, to enhance emission efficiency by using a wide-bandgap material as a charge transporting unit, or to suppress aggregation of an emissive species.

Poly(1-alkylacetylene)s (e.g. **27–28**) (Figure 2.9) [115, 122–124] are violet- or blue-emitters whose solid-state PL is more efficient than the poly(arylacetylene)s **13–14** . Longer alkyl chains generally produce higher PL intensity in these materials. We can thus see a clear substituent effect in emissive polyacetylenes – more aryl

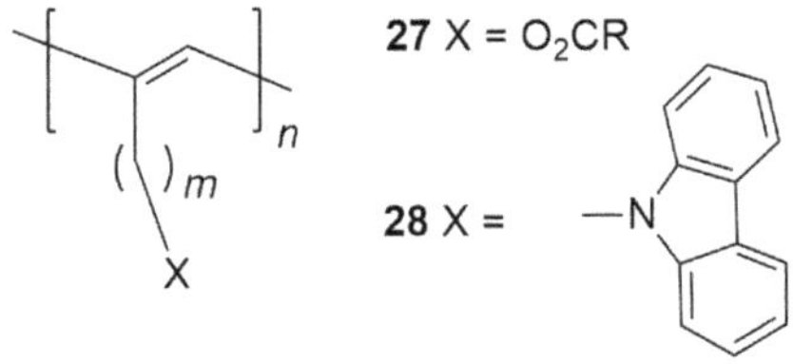

Figure 2.9 Poly(alkylacetylene)s

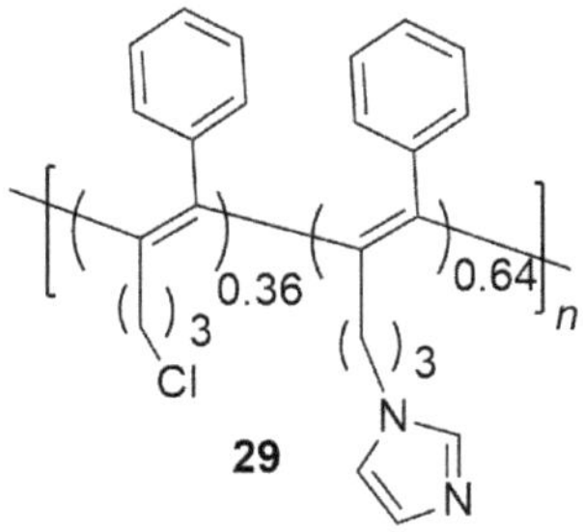

Figure 2.10 Emissive polyacetylene used in sensor

substituents produce more red-shifted emission, with the emission colour being generally blue or green. Steric or other non-electronic intra- or inter-chain effects can induce further changes in the emission colour, with red- or blue-shifts occurring depending upon the nature of the interactions.

Emissive materials which are capable of interacting with other species may be useful as sensors if the interaction induces changes in emission colour or intensity due to steric or electronic effects. The wide variety of side-chains that have been attached to emissive polyacetylenes suggests that they may have some potential in this regard, but to date there has been only one report of such an application [98]. This used the imidazole-containing copolymer **29** (Figure 2.10). The imidazoles can form a strong complex with copper ions, which quenches the luminescence, enabling detection of the metal ions at levels as low as 1.5 ppm. The emission is regenerated by addition of cyanide ions so that the metallated complex can itself act as a detector for those.

While polyacetylenes have been useful in illustrating many key themes in the design and synthesis of conjugated polymers, it must be admitted they have so far not proven to be of any great practical value. The conjugated polymers that first succeeded in demonstrating the real potential for organic electronics were the poly (arylene vinylene)s which may be thought of as alternating copolymers of polyacetylenes with polyarylenes, and it is to these that we now turn our attention.

3

Poly(arylene vinylene)s

Poly(arylene vinylene)s (PAVs) are, strictly speaking, alternating copolymers with aromatic units joined together by vinylene linkers. They are probably the most studied of all the classes of conjugated polymers, largely because they were the first such materials to be used successfully in polymer LEDs and relatively efficient (1% or above) polymer-based organic photovoltaic cells (OPVs). From the scientific point of view, they also provide one of the best examples of how the optoelectronic properties of conjugated polymers have been controlled by structural modification and/or choice of synthetic methodology. The investigations of the mechanisms of their synthesis have led to considerable advances in our understanding of how defects in conjugated polymers can have major effects upon their performance in devices and more especially of how the device performance can be improved by using our understanding of the fundamental chemical processes to minimise the formation of such defects.

3.1 Quinodimethane Routes to PAVs

Our account of the PAVs begins with PPV (**1**), which was the active material in the first conjugated polymer LED [19]. Due to its lack of solubilising substituents, PPV is insoluble and so films of it must be prepared by quinodimethane precursor routes (Figure 3.1). In these an α,α'-disubstituted *para*-xylene **2** is treated with one equivalent of base to produce a quinodimethane intermediate **3**, which undergoes spontaneous polymerisation to form the soluble precursor polymer **4**. This is isolated and is then converted by heating under vacuum to form the desired PPV. These routes can also be used to make substituted soluble PAVs in which isolation of the precursor polymers is no longer necessary, though may still be performed.

A number of these methods have been developed which differ in the choice of leaving groups X and Y. The first of these was the sulphonium precursor route first developed by Wessling and Zimmerman [125, 126], and later modified by others

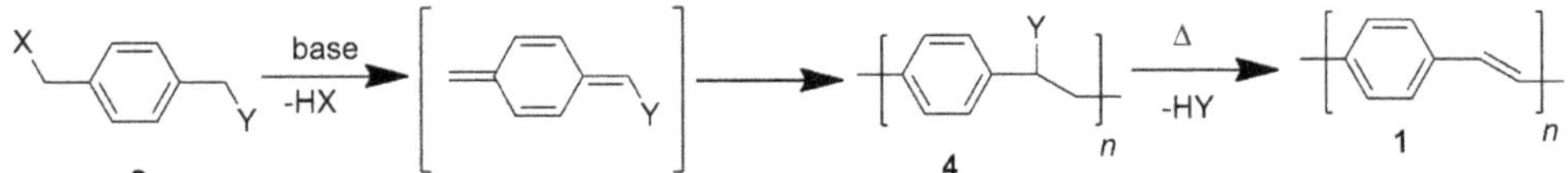

Figure 3.1 Quinodimethane precursor routes to PPV

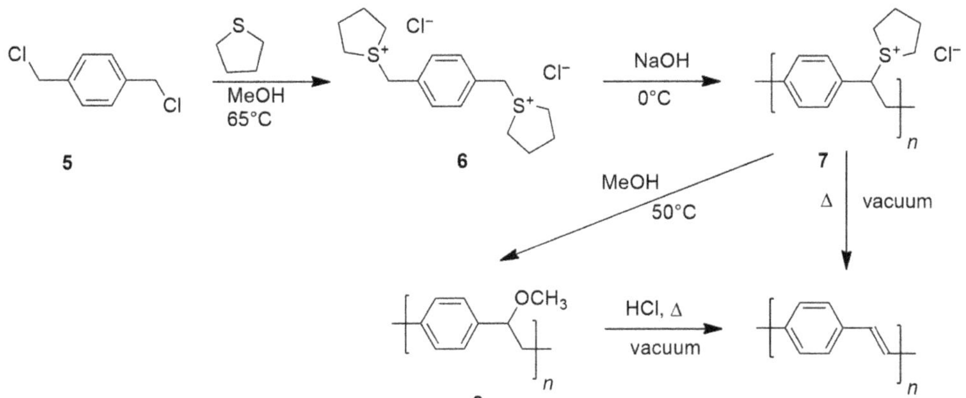

Figure 3.2 Wessling–Zimmerman route to PPV

[127–129], which is shown in Figure 3.2. The commercially available dihalide **5** is reacted with a sulphide to produce a disulphonium salt which acts as monomer. The choice of sulphide determines how readily the final precursor can be eliminated to form PPV **1**. Initially dimethyl sulphide was used but later the tetrahydrothiophene salt **6** was adopted as the standard monomer, as a tetrahydrothiophene is not as easily displaced as a dimethylsulphonium group, so one gets much fewer side reactions and the reagent is not as expensive as some other possible sulphides. It is also less hazardous and malodorous than dimethyl sulphide, though, as anyone who has been in a lab where Wessling PPV is being prepared or converted can testify, it is certainly not pleasant smelling. Treatment of **6** with just under 1 equivalent (typically about 0.9 equivalents are used) of base produces the soluble sulphonium precursor **7**, which is relatively stable in solution but can be readily converted into PPV by heating under reduced pressure. If exactly one equivalent of base is used, partial elimination of the sulphonium groups will occur, producing a highly coloured material with reduced solubility. The sulphonium groups are readily replaced by nucleophiles, and so the sulphonium precursor **7** can be converted into the methoxy-precursor **8** by treatment with methanol. This precursor can also be converted into PPV but requires the presence of acid as well as heat and vacuum to obtain complete elimination.

This ready nucleophilic displacement of the sulphide is a potential disadvantage of the Wessling route as it can produce defects, for example, reaction with water

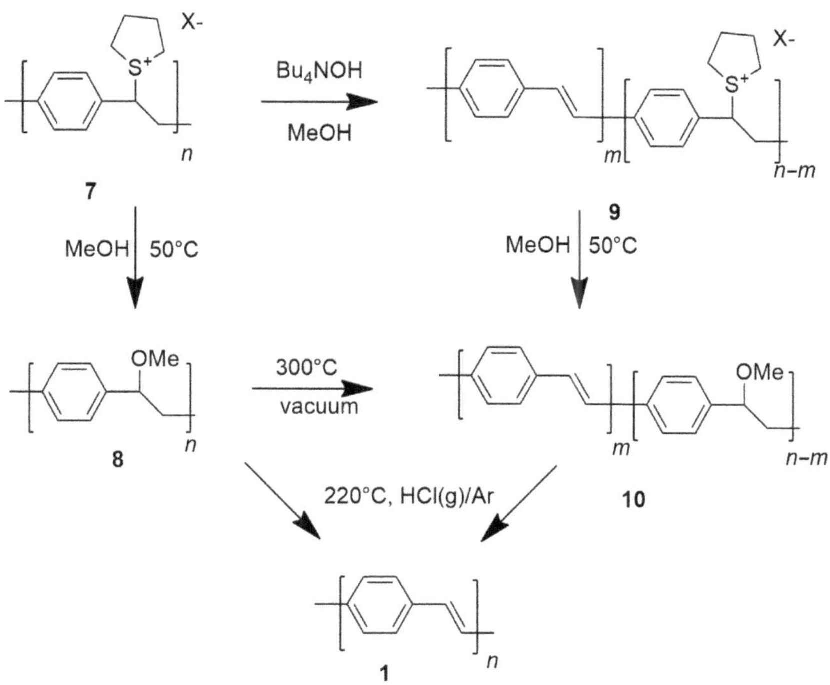

Figure 3.3 Synthesis of partially conjugated PPV copolymers

followed by aerial oxidation is a way for ketone defects to be formed. On the other hand it also offers a way to tune the conjugation length by replacing some of the sulphonium groups with other groups such as methoxy which are less easily eliminated – as so often in this field one has a two-edged sword with potential benefits and problems, and it is the task of the materials scientist to try to balance them so as to optimise the material's performance by maximising the former and minimising the latter. An example of how the conjugation length of PPV may be tuned is shown in Figure 3.3.

Treatment of the sulphonium precursor **7** with less than an equivalent of base induces partial elimination of the sulphonium groups to form the copolymer **9** which undergoes displacement of the remaining sulphides to produce the copolymer **10** [130, 131]. Such a copolymer is also obtainable by heating the methoxy-precursor polymer **8** in vacuum in the absence of acid [132, 133]. Because of the interrupted conjugation, the copolymer has blue-shifted emission compared to fully conjugated PPV **1**.

The conjugation length of PPV has also been controlled by partial displacement of sulphonium groups with acetate, which is less readily thermally eliminated than sulphide [134, 135]. This approach has been extended to substituted PPV derivatives bearing alkoxy solubilising groups where, by partial substitution of

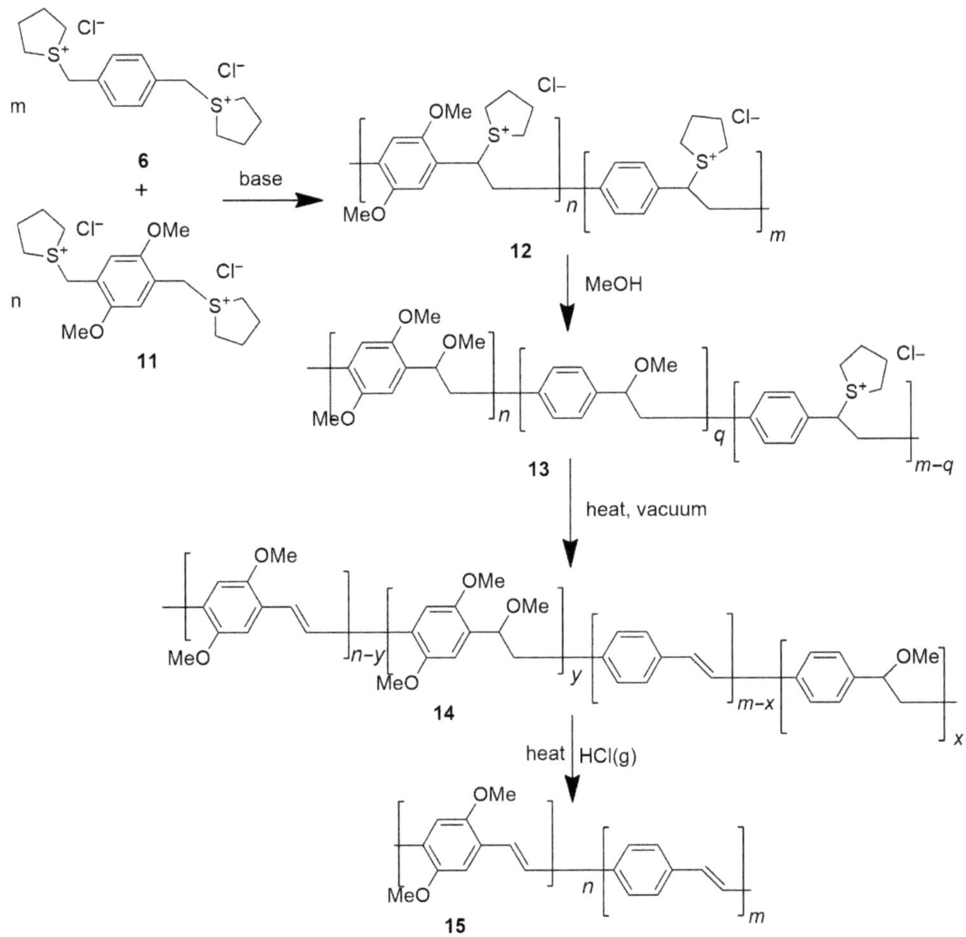

Figure 3.4 Colour tuning by selective displacement of sulphide groups in a PAV copolymer

sulphonium groups, the emission colour can be tuned between yellow and red [136, 137]. An extra point here is that the alkoxy groups assist the displacement of the sulphides upon the benzylic positions, which has been taken advantage of to pattern a conjugated polymer film by the method shown in Figure 3.4.

Copolymerisation of monomers **6** and **11** produces a statistical copolymer **12**. When this is treated with methanol, all of the sulphides adjacent to dimethoxyphenylene units get displaced but only some of those adjacent to unsubstituted phenylene units. When the resulting copolymer **13** is heated under vacuum all the sulphonium units are eliminated but merely some of the methoxy units, so copolymer **14** has a markedly blue-shifted emission. Treatment of copolymer **13** with HCl produces complete elimination to form the fully conjugated polymer **15** which displays red-shifted emission compared to either

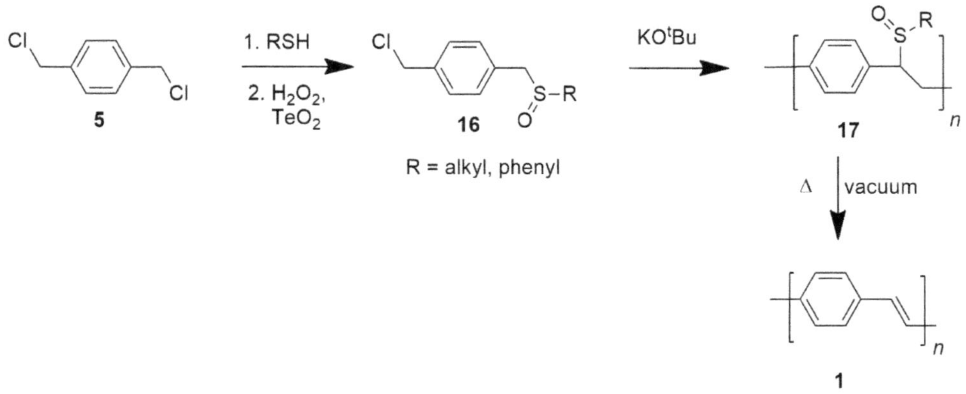

Figure 3.5 Sulphinyl-precursor route to PPV

PPV or copolymer **14**. By depositing a lithographically patterned mask of aluminium onto the surface of a film of copolymer **13** the film can be patterned as thermal elimination produces HCl which is trapped under the metal mask so that the areas under the mask are converted to polymer **15** while the uncovered polymer is converted only to copolymer **14**. The two areas thus possess very different colours. As the partially converted copolymer **14** is a much stronger emitter than the fully converted polymer **15**, due to exciton confinement, the emission from these patterned films in an LED comes from the former [138].

The Wessling route is still the most convenient route for making PPV but its utility for making other PAVs is limited; in particular it offers no advantage for the synthesis of soluble polymers which would outweigh its obvious disadvantages of toxic and malodorous by-products. The ready displacement of the sulphide units also offers too facile a way to introduce defects into the final polymers. Accordingly, it will probably continue to be used to make PPV but it is unlikely to be much used for preparing other polymers. As research into PPV currently seems to be inactive, for the immediately foreseeable future the Wessling route will be largely of historical interest and of value for the insights it has given into reaction mechanisms, into methods for control of defects and of conjugation length and into methods for film patterning in conjugated polymers.

The Vanderzande sulphinyl precursor method is a modification of the Wessling approach that has proved successful not only for the synthesis of PPV, for which it was originally developed [139, 140] but also continues to have some utility for the synthesis of other PAVs [141]. As shown in Figure 3.5 for the synthesis of PPV, this method starts from same starting material **5** as for the Wessling route. However, only one of the halides is displaced with a sulphide which is then oxidised to form the sulphinyl monomer **16**. Treatment of this with base then produces a precursor

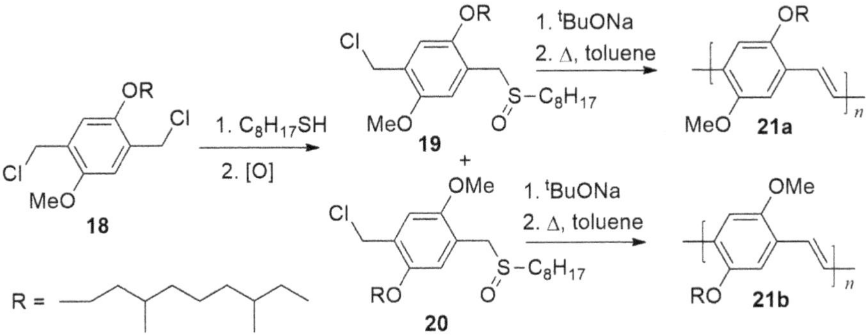

Figure 3.6 Regioregular PPV derivatives made using unsymmetrical sulphinyl monomers

polymer **17** which is soluble in organic solvents. Conversion to PPV (**1**) can then be done under similar conditions to those used in the Wessling route.

The major advantages of this method compared to the Wessling route are that the by-products are less volatile and toxic and that the precursor is much more stable than the sulphonium precursor towards nucleophiles. As a result, the PPV is of higher quality than standard Wessling–PPV. The main disadvantage of this method for making PPV is that the synthesis of monomer **16** is more complex and lower yielding than for the sulphonium salt **6**, though with improved synthetic procedures this is less of a problem than in the initial synthesis [141]. One point to note about this method is that the monomers are unsymmetrical and, due to the greater acidity of the proton α to the sulphinyl compared to the one vicinal to the halide, only one quinodimethane is produced. This can be taken advantage of to make regioregular polymers, as shown in Figure 3.6 [142]. Treatment of the dihalo compound **18** with octanethiol followed by oxidation produces a mixture of the two monomers **19** and **20**, which can be separated by chromatography. Treatment of either of these with base then gives the regioregular polymers **21a** and **21b**, respectively. Due to the long alkoxy substituent, these polymers are soluble and so can be prepared by conversion of the precursors in refluxing toluene.

While the sulphinyl precursor route has its advantages, the main route for synthesis of substituted PAVs is the Gilch method [143]. This is exemplified in Figure 3.7 for the synthesis of MEH-PPV (**22**), which is the most widely studied soluble PPV derivative [144]. This uses as a monomer a dibenzylhalide **23**, (either chloro or bromo monomers can be used) which is easily prepared by halomethylation of the activated aromatic molecule obtained by alkylation of 4-methoxyphenol **24**. For soluble polymers such as MEH-PPV, the final conjugated polymer is usually prepared directly by using two or more equivalents of base but, where the conjugated polymer is of limited solubility, a halo-precursor (Figure 3.7, **4**,

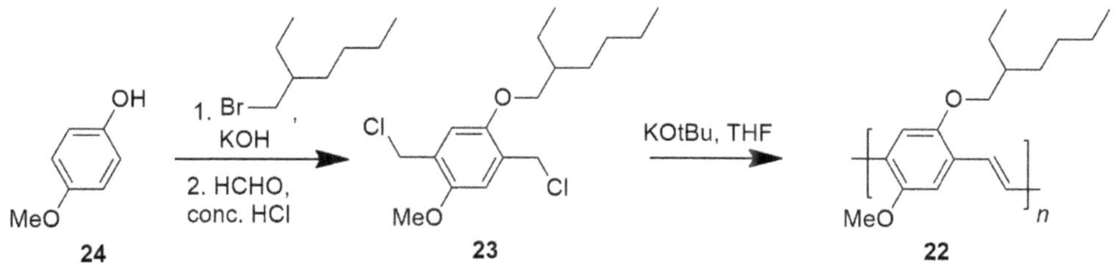

Figure 3.7 Synthesis of MEH-PPV by Gilch route

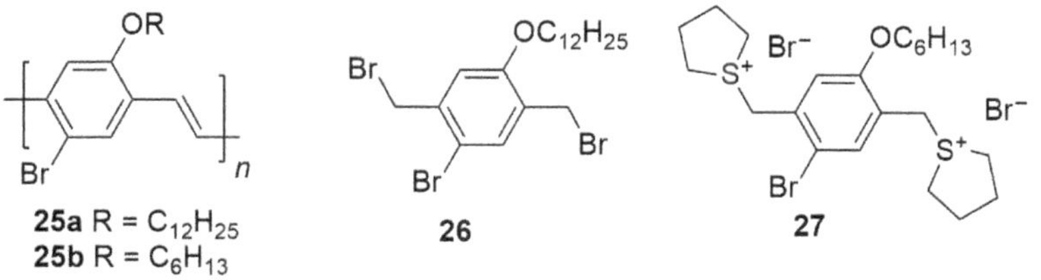

Figure 3.8 Bromo-substituted PPVs and the monomers used to make them

Y = Br or Cl) can be isolated by using one equivalent of base and then converted to the final polymer by heating in vacuum. Such monomers are also accessible by radical halogenations of xylene derivatives, though the yields are often significantly lower in such cases due to side reactions such as ring halogenation.

Since the dihalide monomers used in the Gilch route are intermediates in the synthesis of the monomers used in the Wessling and Vanderzande routes, this route is shorter and more efficient than the latter. As a result, this method is generally the best method for preparing PAVs, except for PPV (the chloro-precursor to PPV is much less soluble than the sulphonium or sulphinyl precursors). Regioregular polymers are sometimes accessible if the substituents on the aromatic ring are so different in electronic properties as to produce a significant difference in the acidity of the protons on the two benzylic groups. For example it was reported that, based upon NMR studies, polymer **25a** (Figure 3.8) made by Gilch route was regioregular, suggesting that the dibromo monomer **26** produces one quinodimethane due to regioselective deprotonation [145]. By contrast, when a similar polymer **25b** was made by the Wessling route, NMR analysis of the sulphonium precursor showed it contained an equimolar mixture of the two regioisomers, suggesting the sulphonium monomer **27** produces an equimolar mixture of both possible quinodimethanes [146]. This difference has been attributed to lesser steric bulk of the alkyl chain in **27** than **26**, but the greater acidity of the protons in the former making the difference in relative acidity between the two benzylic sites less significant than in the latter is a more obviously plausible explanation.

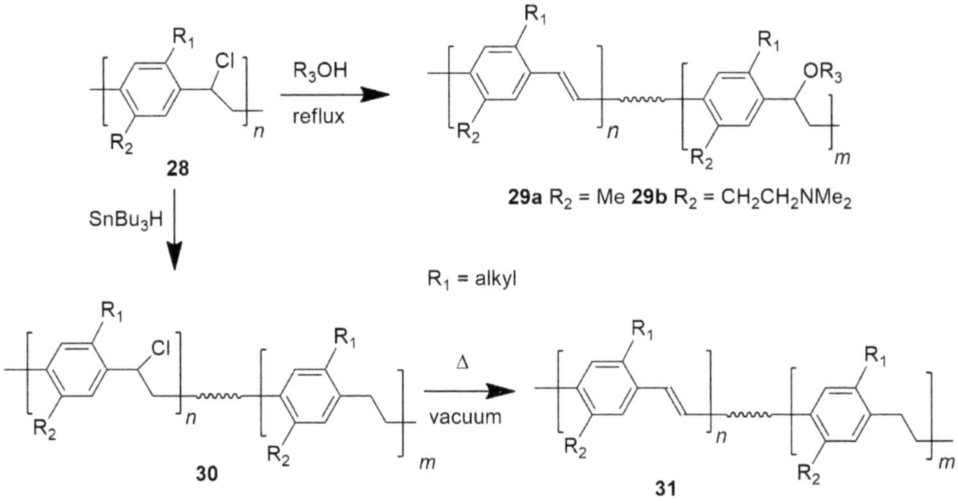

Figure 3.9 Control of conjugation length in PAVs made by Gilch route

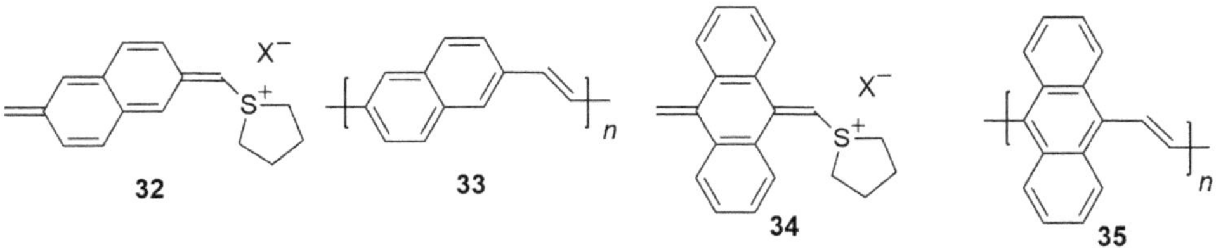

Figure 3.10 Examples of PAVs inaccessible by quinodimethane routes and their intermediates

The halide units in the halo-precursor polymers **28** are much less susceptible to nucleophilic displacement than the sulphide or sulphinyl groups used in the Wessling or Vanderzande routes, but the conjugation length in them has been controlled by treating the halo-precursor with alcohols, as shown in Figure 3.9 – initially methanol was used but later 2-dimethylaminoethanol was found to give better results [147]. Thermal conversion of the resulting materials in the same pot eliminates the halides but not the alcohols, producing partially conjugated polymers **29**. Partial reduction of chloro-precursor polymers **28** with trialkyltinhydride produces copolymers **30** containing saturated units. These copolymers undergo thermal conversion to **31** under reduced pressure [148]. The degree of conjugation can be controlled very closely by varying the amount of reducing agent used.

While the quinodimethane precursor routes are extremely useful for preparing PAVs, they suffer some limitations and not all PAVs are accessible by these methods. Firstly, only homopolymers and random copolymers are possible and, secondly, not all monomers are reactive under these conditions. Thus, the loss of aromaticity in forming the quinodimethane **32** (Figure 3.10) makes its formation

Figure 3.11 Initiation step for quinodimethane polymerisations

energetically unfavourable, inhibiting the synthesis of poly(2,6-dinaphthalene vinylene) (**33**), while the quinodimethane **34** is too stable for it to react to form poly(9,10-anthracene vinylene) (**35**). It has been shown that the ability to polymerise a bis-sulphonium salt (and presumably other monomer types) depends upon the enthalpy of formation of the quinodimethane, enabling one to predict whether or not a given PAV is accessible by these methods [149].

The mechanisms for these precursor routes to PAVs have been the subject of some lively debate in the literature [150]. The existence of the quinodimethane intermediate was rapidly proven spectroscopically, but there was some uncertainty as to whether the intermediate underwent radical or anionic polymerisation as both are mechanistically feasible under the reaction conditions. The suppression of the reaction by radical trapping agents was rapidly agreed to establish the radical nature of the Wessling process [151] and of the sulphinyl precursor route [152] but support remained for some time for a non-radical mechanism for the Gilch reaction [153]. This arose from the ability to control the molecular weight of soluble PAVs by adding reagents which enhance anionic polymerisation [154] of a known anionic initiator [155] and of benzyl bromide chain stoppers [156]. However, more recent studies [157, 158] have established that the Gilch reaction also proceeds primarily via a radical pathway, with anionic polymerisation being responsible at most only for the formation of low molar mass material. The initiation step is proposed to be the coupling of two quinodimethanes to form a diradical (Figure 3.11) [158]. One effect of this is to introduce a non-conjugated defect into the polymer chain.

A key issue in the synthesis of any conjugated polymer is the minimisation of defects, as they have a significant, usually negative, impact upon the optical and electronic properties of conjugated polymers, by acting variously as charge traps which reduce charge mobility and conductivity, as sites for non-radiative decay of excited states which reduce luminescence efficiencies and/or as low band-gap emissive sites which affect emission colour in LEDs. The formation of defects during the Gilch synthesis of PAVs has been shown to arise from non-standard coupling of the quinodimethane intermediates, as shown in Figure 3.12. It is likely that similar processes occur during other quinodimethane-based syntheses.

Whereas the precursor polymer **36** is formed by coupling of the quinodimethane **37** in the head-to-tail fashion, coupling in head-to-head and tail-to-tail fashions

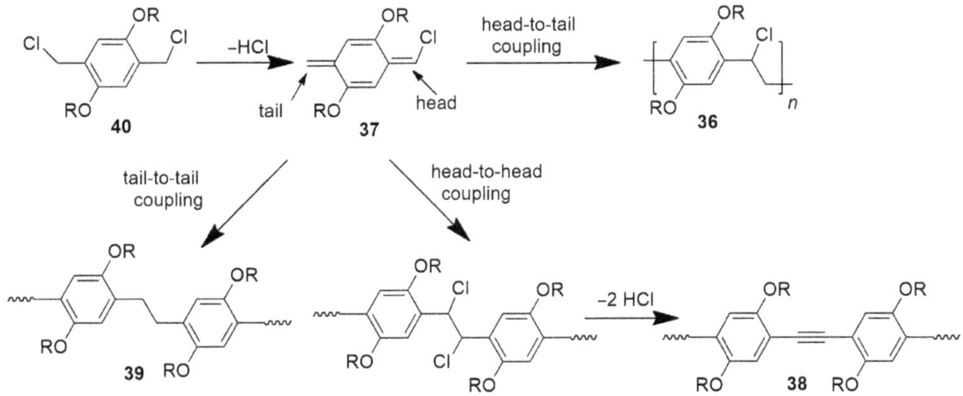

Figure 3.12 Formation of defects in PAVs during Gilch synthesis

produce, respectively, diarylethyne (tolane, **38**) and bisbenzyl defects (**39**) whose presence was proven by ^{13}C NMR studies on polymers produced by polymerisation of monomers **40** with ^{13}C labelling of the benzyl carbons [159–161]. Residual halide defects may also be present as a result of incomplete dehydrohalogenation during formation of **38**, or incomplete elimination of **36**, and it has been suggested that these may be even more important than the tolane or bisbenzyl defects in reducing LED device efficiencies and lifetimes [162]. As the initiation step introduces a bisbenzyl defect, all polymers made by the Gilch and related routes must, therefore, contain at least one defect. As will be discussed in Section 3.5, this understanding of the mechanism of defect formation has enabled new materials to be made with much lower defect levels and hence much better performance in light-emitting diodes (LEDs) [163].

One other source of defects in PAVs is the possibility of *cis* versus *trans* double bonds in the vinylene units. The final step in the formation of PAVs by quinodimethane routes is an elimination which proceeds via an E2 mechanism which favours formation of *trans*-double bonds [164]. However, some *cis*-double bonds are formed which change the polymer chain conformation and affect the packing of the chains. One study has shown that the presence of *cis*-bonds is higher in low molecular weight fractions of MEH-PPV made by the Gilch route, suggesting that these defects also act to limit chain growth [165].

3.2 Other Routes to PAVs

There are a number of direct routes for preparing soluble PAVs, of which the most important are Heck coupling (Figure 3.13) and the Wittig and Horner polycondensations (Figure 3.14). The two biggest advantages of these routes is that they avoid the formation of the tolane and other defects seen from the

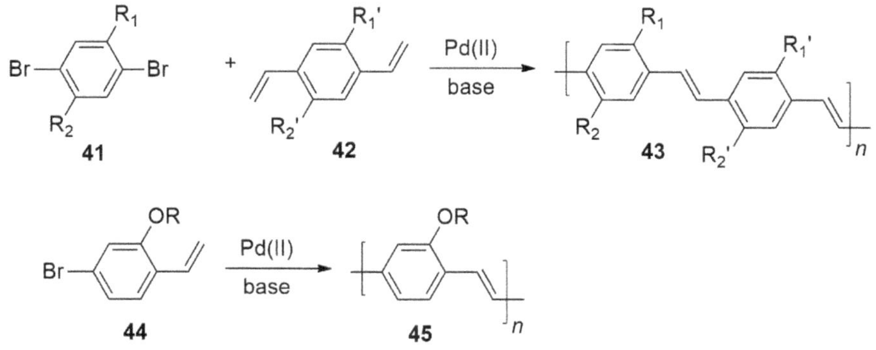

Figure 3.13 Heck coupling as a route to PAVs

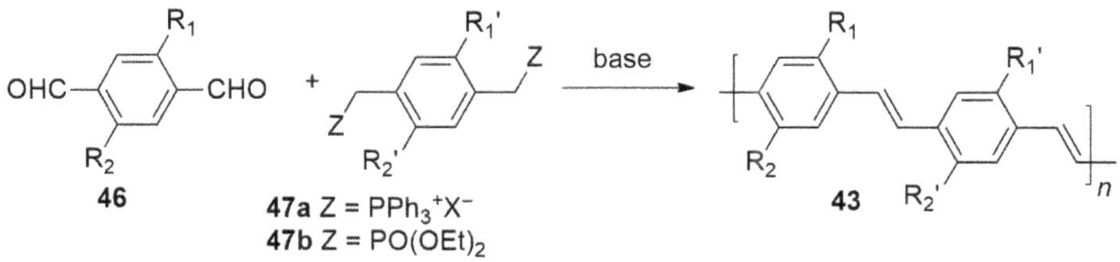

Figure 3.14 Wittig and Horner polycondensations as routes to PAVs

precursor routes, and also that they allow the synthesis of alternating copolymers with two different arylene moieties, whereas the methods described in Section 3.1 can only be used for making homopolymers or random copolymers. A major disadvantage is that the molecular weights obtained are generally significantly lower than those obtainable by the quinodimethane routes. The first synthesis of PPV and its derivatives by Heck coupling involved reacting dihalobenzenes **41** with gaseous ethene [166], but it was soon found to be much more synthetically convenient to use a divinylbenzene **42** as the alkene component [167, 168]. This latter method has since been used by many groups to make PAVs, especially alternating copolymers **43** (R_1, $R_2 \neq R_1'$, R_2') [150]. Homocoupling of a 2-alkoxy-4-bromostyrene **44** has been used to make a regioregular PPV derivative **45** [169], which would not be possible via the Gilch procedure.

The Wittig and Horner procedures involve coupling of a dialdehyde **46** with, respectively, a bisphophonium salt **47a** or a diphosphonate **47b**. The Horner route has been used to make homopolymers (**43** R_1, $R_2 = R_1'$, R_2') such as MEH-PPV [170–172], but more usually it and the Wittig method have been used to prepare alternating copolymers (**43**, R_1, $R_2 \neq R_1'$, R_2') [150]. The molecular masses from the Wittig method tend to be slightly lower than from the Horner and the latter produces much lower levels of *cis*-bonds [173]. The *cis*-bonds can be converted to *trans* bonds by using iodine but this lowers the overall yield as the material has to

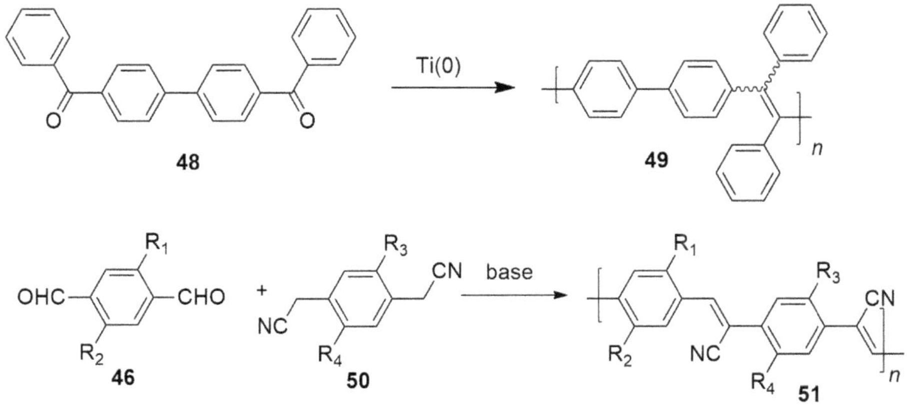

Figure 3.15 McMurry and Knoevenagel routes to PAVs

be rigorously de-doped afterwards [174], so that the Horner method is generally to be preferred.

Two other direct routes that have proved useful for making PAVs are McMurry coupling and Knoevenagel polycondensation (Figure 3.15). McMurry coupling of dialdehydes **46** produces substituted PAVs [175], while coupling of diketones such as **48** produces polymers like **49** bearing groups on the vinylene units [176, 177]. Knoevenagel polycondensation of dialdehydes **46** with dinitriles **50** produces CN-PPVs 51 which bear electron-withdrawing cyano substituents on the vinylene units [178]. The main advantage of these methods is that one can make polymers with substituents on the vinylene not readily available by other methods, but the McMurry route produces a large amount of *cis*-bonds in some cases [177] and the Knoevenagel is only useful for making polymers with strongly electron-withdrawing groups such as nitriles on the vinylene. As we shall see later, such materials are of interest due to their superior electron-accepting properties compared to standard PAVs.

Metal-mediated cross-coupling reactions such as Stille coupling of dihaloarenes with vinylbisstannanes [179–181], Suzuki coupling of arene bisboronates and vinyl dihalides [182] or a cascade Suzuki–Heck process in which dihaloarenes are coupled using a vinyltrifluoroborate [183] have also occasionally been used to make PAVs, but seem unlikely to be more widely used as the polymers made so far usually have low molar mass.

Metathesis is a synthetic method that has become widely used in synthetic chemistry in recent years but so far has shown only limited utility for the synthesis of PAVs. Metathesis of dialkyl- or dialkoxy-substituted divinylbenzenes **42** produces soluble PAVs **52** directly (Figure 3.16) [184–186], but to date the products reported are largely oligomers: with improved catalyst systems the molar masses could probably be improved significantly. Ring-opening metathesis polymerisation

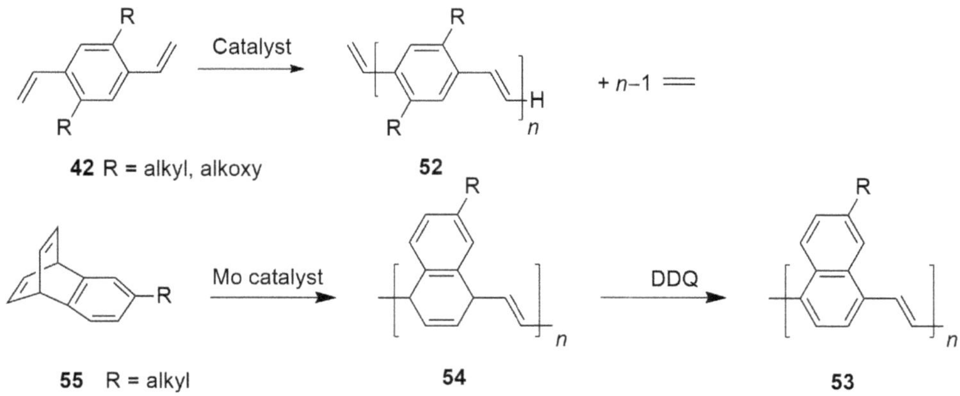

Figure 3.16 Metathesis routes to PAVs

(ROMP) is a polymerisation method of great potential due to its living nature permitting excellent control of molecular weights. ROMP has been used to make precursor polymers which on elimination or oxidation give PAVs [150], for example, poly(naphthalene vinylene)s like **53** have been made via precursors **54** [187], but the difficulty of making the monomers such as **55** severely restricts the utility of this method.

3.3 Structure–Property Relationships in Optical Properties of PAVs

Attaching substituents to PAVs, as well as enhancing solubility, is a simple way to tune the bandgap, by means of both electronic and steric effects. Alkoxy groups are the most widely used solubilising substituents on PAVs, especially PPVs, as the requisite monomers are readily prepared from phenols (see Figure 3.7). The effect of alkoxy groups on the ring is to enrich the electron density on the phenylene rings, which significantly reduces the bandgap and thus red-shifts the absorption and emission spectra, by raising the HOMO energy. As a result, whereas PPV (**1**) is a yellow-green emitter (λ_{max} = 520, 551 nm) [19], 2-alkoxy-PPVs (**45**) are yellow emitters (λ_{max} = 550 nm) [188] and the emission from 2,5-dialkoxy-PPVs like MEH-PPV (**22**) is mainly red (λ_{max} = 603, 650 nm), though the presence of some yellow emission, to which the eye is much more sensitive than it is to red light, makes the luminescence appear distinctly orange-red in colour [189]. It is reported that, as the length of the alkoxy substituents increases, the EL of the materials in devices first increases then reaches a maximum and starts to decrease again [190]. These effects can be explained by a decrease in non-radiative decay arising from decreased chain interactions due to increasing steric bulk of the side-chains, with extremely large side-chains leading to a dilution of the semiconducting properties of the materials due to the insulating nature of the alkyl chains. Alkyl and silyl groups by contrast do

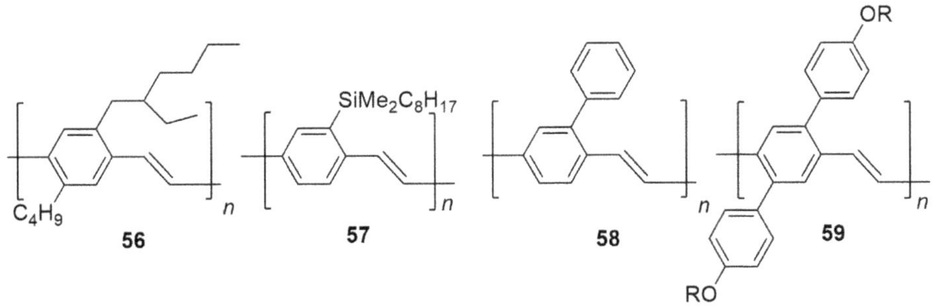

Figure 3.17 Some representative soluble PPV derivatives

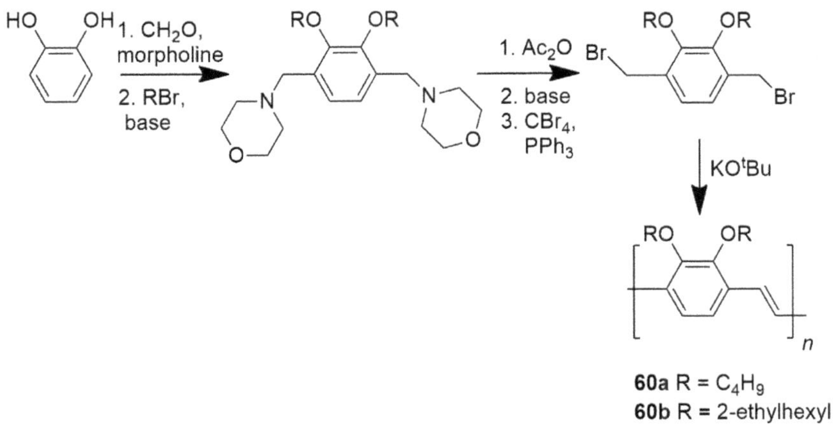

Figure 3.18 Synthesis of poly(2,3-alkoxy-1,4-phenylene vinylene)s

not red-shift the emission of PPVs so that polymers such as BuEH-PPV (**56**, Figure 3.17) [191] and DMOS-PPV (**57**) [192] are green emitters ($\lambda_{max} = 520$ nm). These polymers also show significantly higher solid-state PL efficiencies than PPV or MEH-PPV, which means their potential EL efficiencies are also higher. On the other hand, the monomers for these are more difficult to make than those for the dialkoxy-PPVs. Phenyl substituents also produce no marked red-shift due to electronic effects, so that polymers such PPPV (**58**) [193] or the 2,5-diphenyl-PPV (**59**) [194] are green emitters ($\lambda_{max} = 530$ nm).

The position of the substituents can be as important as their nature. For example, whereas 2,5-dialkoxy-PPVs such as MEH-PPV (**22**) display orange-red emission, the 2,3-dialkoxy-PPVs **60** are green emitters ($\lambda_{max} = 520$ and 505 nm, respectively) due to steric repulsions between the substituents which twist the backbone so as to reduce the degree of conjugation and thus blue-shift the emission [195, 196]. This ability to alter the emission colour by changing the substituent positions is useful for obtaining blue-shifted emission but, as shown in Figure 3.18, the synthesis of the monomers is a little longer than for the 2,5-dialkoxy-PPVs.

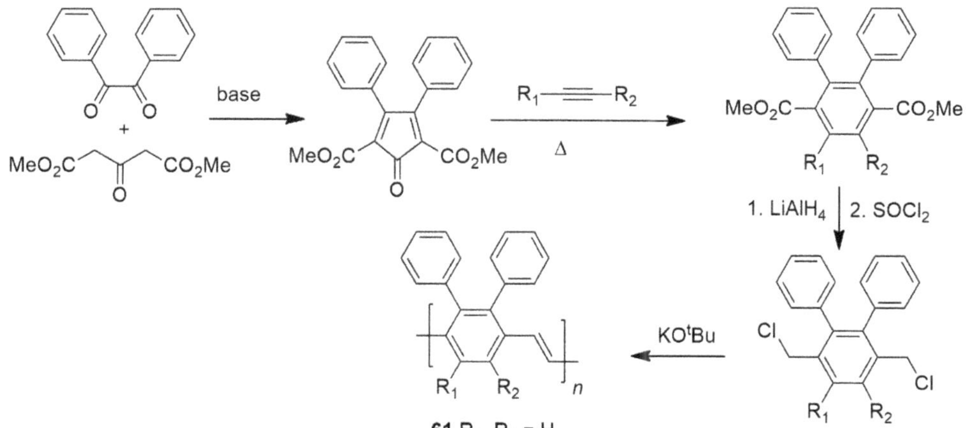

61 R_1, R_2 = H
62 R_1 = H, R_2 = alkyl

Figure 3.19 Synthesis of PPVs with 2,3-diphenyl substituents

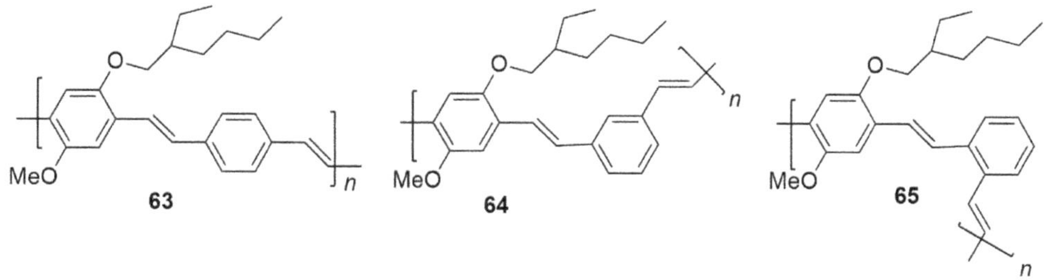

Figure 3.20 PPV copolymers with *para-*, *meta-* and *ortho*-linkages

A similar effect is seen with 2,3-diphenyl-PPVs; so that the 2,3-diphenyl-PPV **61** is blue-shifted (λ_{max} = 500 nm) compared to the 2,5-isomer **59** [197]. Further aryl substituents seem to have no effect on the emission colour, but alkyl-substituted polymer **62** has the most blue-shifted emission reported from a PPV (λ_{max} = 500 nm) with all *para*-linkages [198]. Here there exists an efficient synthetic route to the necessary monomers, involving both a Knoevenagel condensation and a Diels–Alder reaction (Figure 3.19), which illustrates the rich variety of synthetic methods available for synthesis of PAVs [198].

Conjugation along the backbone can also be reduced by introducing *meta-* or *ortho*-phenylene units. For example, the emission maxima of the polymers **63–65** (Figure 3.20) (which are readily made by Wittig or Horner routes) occur, respectively, at 550, 490, and 500 nm [199]. Whereas the blue-shift in the emission from polymer **64** compared to **63** reflects a disruption in the through-conjugation due to the *meta*-linkages, that in the emission from polymer **65** arises from a twisting of the polymer backbone which is induced by the steric repulsion between the adjacent groups on the *ortho*-substituted phenylenes.

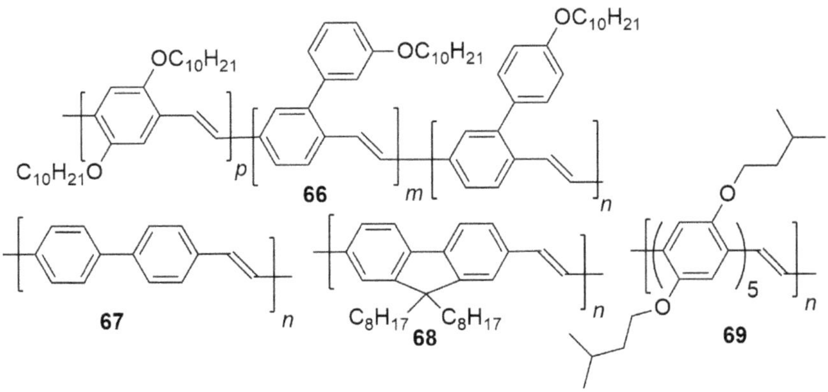

Figure 3.21 Phenyl–PPV copolymers and PAVs with oligophenylene units

An obvious way to tune the emission colour of PAVs is by copolymerisation of different units, For example, copolymer **63** is an alternating copolymer of PPV and MEH-PPV which shows emission intermediate between that of the homopolymers. This effect is also seen in random copolymers, for example, **66** (Figure 3.21), whose emission can be tuned between green ($\lambda_{max} = 515$ nm) and orange ($\lambda_{max} = 567$ nm) by varying the amount of the dialkoxyphenylene units [163, 200]. As mentioned, PPV is an alternating copolymer of polyphenylene and polyacetylene. Since polyphenylenes are well-known as blue-emitting polymers (see Chapter 5), increasing the proportion of phenylene units can be expected to blue-shift the emission compared to PPV. Thus, both poly(biphenylene vinylene) (**67**) [201] and poly(fluorenylene vinylene) (**68**) [202] are blue-green emitters ($\lambda_{max} = 467$, 497 nm), while the poly(pentaphenylene vinylene) **69** is a blue emitter ($\lambda_{max} = 446$ nm) [203].

Incorporating units other than phenylenes into PAVs is another obvious way to tune the emission colour. Replacing benzene with polycyclic arylenes has varying effects on the emission, for example poly(1,4-naphthalene vinylene) **70** is orange-red emitting ($\lambda_{max} = 605$ nm) [204], while poly(3,6-phenanthrene vinylene) **71** displays green fluorescence ($\lambda_{max} = 515$ nm) [205]. By contrast, electron-rich heteroaromatic units such as 2,5-pyridines or 2,5-thiophenes provide a reliable way to red-shift the emission of PAVs (Figure 3.22) compared to PPVs. The lower symmetry of the pyridine compared to the phenylene ring means that poly(pyridine vinylene) (PPyV) can be either a random polymer (**72**) or in two regioregular forms – head-to-tail (**73**) and head-to-head (**74**) [179]. All of these forms have been made by Stille coupling of halopyridines with vinylstannanes, and the EL emission of the regioregular forms was found to be red-shifted (emission maxima at 584 nm (**73**) and 605 nm (**74**)) compared with the regiorandom form ($\lambda_{max} = 575$ nm), which reflects better packing for the regular forms in the solid state. The

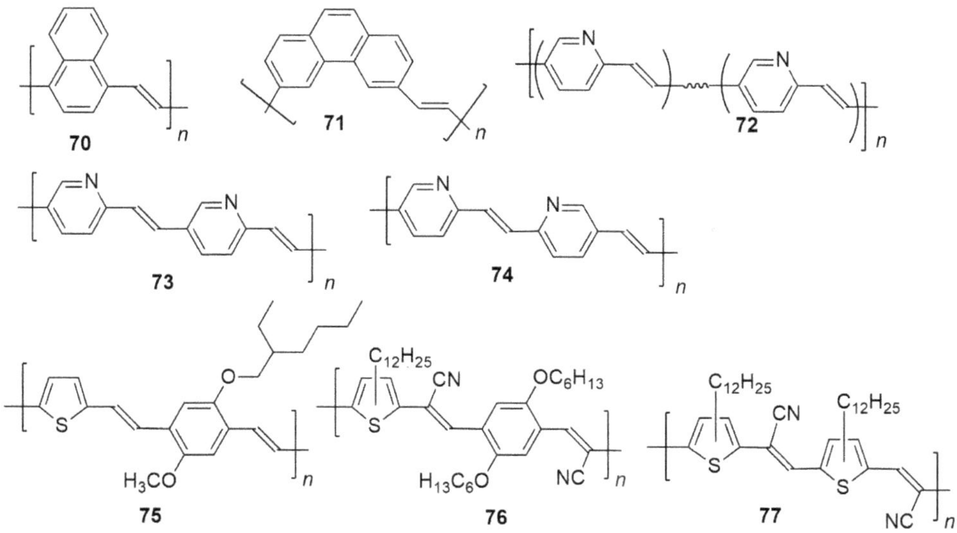

Figure 3.22 PAVs with aromatic units other than phenylenes

thiophene-containing copolymer **75** has even more red-shifted emission (λ_{max} = 620 nm) [206]. The most red-shifted emission yet reported from a PAV is from polymers **76** (λ_{max} = 740 nm) [207] and **77** (λ_{max} = 800 nm) [208], which emit in the near-infrared. A wide range of other heteroaromatic units have been incorporated into PAVs, with emission colours ranging from green to red [150], but the examples given illustrate some of the variations possible.

3.4 Applications of PAVs: FETs

Charge generation and dynamics are very different in PPV-based polymers compared with the polyacetylene (PA) materials discussed in Chapter 2. As discussed in Chapter 2, whereas charge transport is governed by soliton transport in PA, in PAVs the situation is very different, with the dominant charge carriers being charged polarons. In LED devices, it was established early on that light emission arises mainly from recombination of the singlet exciton and that the charge carriers are actually charged polarons in these polymers [209]. Muon studies probing the diffusion of negative polarons in two soluble PPV derivatives (poly(2, 3-dibutoxy-1, 4-phenylene vinylene) and poly(2, 5-bis(dimethyloctylsilyl)-1, 4-phenylene vinylene)) showed that scattering from vibrational modes hinders intra-chain transport but assists inter-chain transport [210]. Space-charge limited conduction dominates for negative polarons in PPV and related materials, with the occupation of traps having a strong influence on the observed transport behaviour [211]. For OPVs from PPV the situation is more complex. Moses et al.

showed that photoexcitation creates charged polaron pairs directly [212], which appeared to contradict the established model whereby charge carriers are regarded as a result of the field-induced dissociation of excitons (exciton model) [213]. In reality, both excitons and charged polarons are formed simultaneously on ultrafast timescales, with charged polarons generating about 25% of the total charge carriers [209].

Undoped PAVs are semiconductors – the conductivity of undoped PPV for example being a mere 10^{-13} S cm^{-1} [214]. This conductivity has been increased to as high as 100 S cm^{-1} upon acid doping [215], but this is still well below the values seen for doped polyacetylene. PAVs have not proven very successful as OFET materials either, with reported charge carrier mobilities being below 10^{-5} cm^2 Vs^{-1} [216, 217]. This means that PAVs are only useful for applications that do not require very high charge carrier mobilities or conductivities such as LEDs or OPVs. To date, PAVs have been one of the most widely studied classes of polymers for LED applications and also initially received considerable attention as donor materials in OPVs, though here they have been superseded by other classes of polymers. We will look first at how the emission efficiency of PAVs in LEDs can be optimised by synthetic and/or device design, and then at the use of PAVs in OPVs.

3.5 Increasing Emission Efficiency of PAVs by Minimising Defects

As mentioned in Section 3.1, the understanding of the mechanism of defect formation in the Gilch reaction has allowed the design and synthesis of relatively defect-free polymers which show particularly good performance in LEDs (high efficiency and long lifetimes) [163]. This is one of the best illustrations of how chemists can make use of their understanding of reaction pathways in order to deliberately create superior materials.

Each of the three unsymmetrical monomers **78–80** can form two isomeric quinodimethanes upon reaction with base (Figure 3.23). There is no electronic reason why one of these forms should be favoured for **78** and so the two intermediates **81a** and **81b** are probably formed in equal amount. The steric and electronic effects of the chlorine atoms, however, would strongly favour head-to-tail coupling of either of these, so the level of defects in polymer **82** is typically only 3%. While there probably is some electronic effect affecting the ratio of the intermediates **83a–b** derived from **79**, it is not so strong that only one form is produced and steric repulsion between the phenyl groups means that the coupling of **83a** and **83b** can only proceed in a head-to-head fashion, so that the resulting polymer **84** has a high defect level. The electronic effects of the methoxy group in **80** induce a sharp difference in the acidity of the various benzylic protons so that

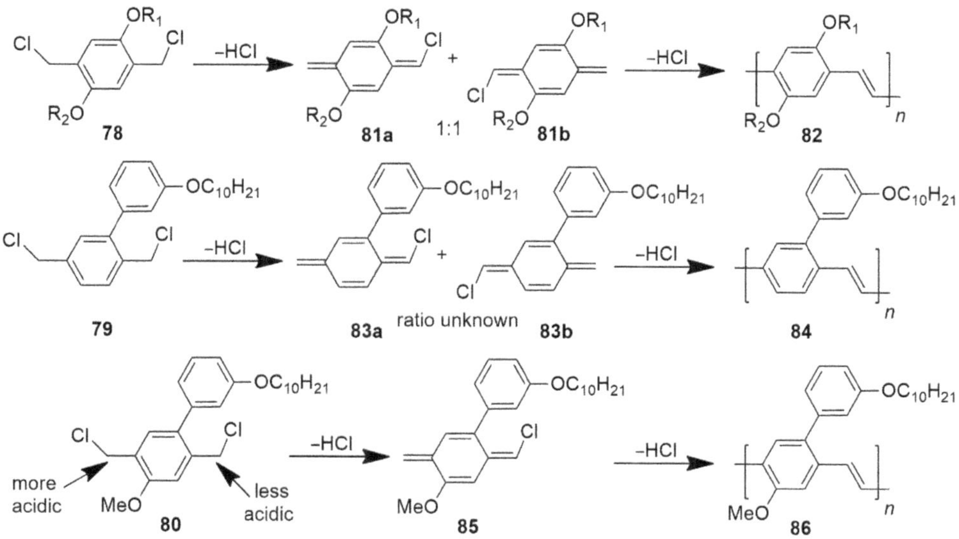

Figure 3.23 Mechanistic explanation for low defect levels in certain PAVs

formation of the quinodimethane **85** is strongly favoured. Steric repulsion between the aryl substituents hinders head-to-head coupling of **85**, so that the defect levels in the resulting polymer **86** are very low (<0.5%). Unfortunately, this polymer tends to form gels, which hinders processing, so more processable copolymers **66** were prepared. Such copolymers have now been optimised and marketed as Super Yellow™, which comprised the active material in the first commercial polymer LED devices.

3.6 Increasing Emission Efficiency by Attaching Charge Transporting Units

A second approach towards optimising the performance of PPVs in LEDs has been the improving of charge injection into the emissive polymers. Apart from device engineering approaches such as use of charge-transporting layers [150], this can be achieved by attaching charge-accepting groups onto the polymers. The limiting factor for efficiency of LEDs using PPV has been shown to be due to the injection of electrons as opposed to holes [150], as shown by the much greater EL efficiency when using calcium than aluminium cathodes. Most effort has been into the attachment of electron-accepting groups to PAVs (Figure 3.24) either on the phenylene (e.g. in the fluoro-PPV (**87**)) [218, 219] or on the vinylene as in CN-PPVs (**51**) [178]. In both cases the polymers display much higher electron affinities than PPV, and as a result produce significantly greater efficiencies than PPV in LEDs using air-stable electrodes such as aluminium. In particular, double-

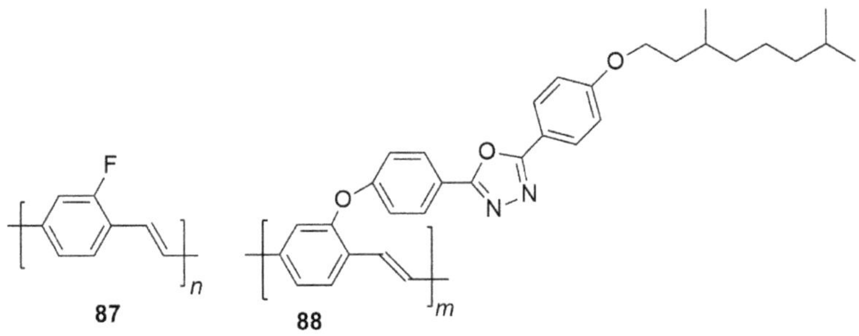

Figure 3.24 PAVs with higher electron affinity than PPV

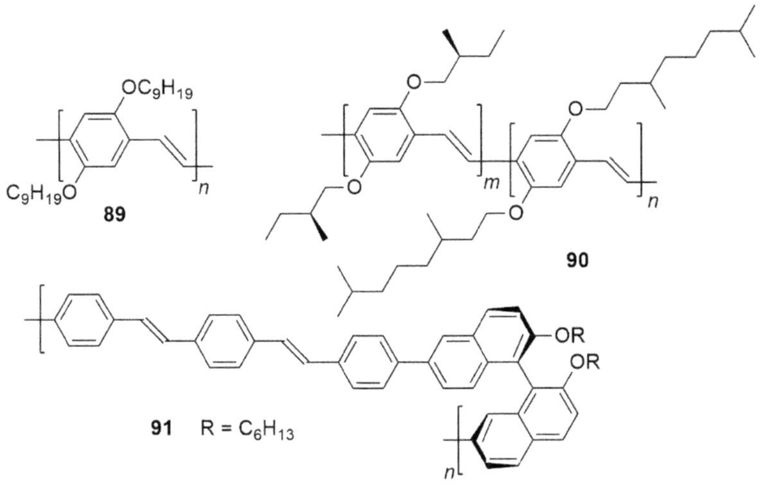

Figure 3.25 PAVs from which polarised or circularly polarised emission might be obtained

layer devices utilising **51** (R_1-$R_4 = C_6H_{13}$) as an electron-accepting emissive layer with PPV as a hole-transporting layer reached efficiencies of up to 4%, although their lifetime was somewhat limited [178, 220]. The highest efficiency yet reported for an LED using a PPV-derivative is 21.1 cd A^{-1} using the polymer **88** bearing oxadiazole units [221]. The key to obtaining such high efficiencies seems to be to achieve balanced injection of holes and electrons.

Polarised emission from LEDs is desirable for certain applications (e.g. as backlights for liquid crystalline displays) [222]. This is most easily obtained by alignment of the emissive layer, for which the ability of the polymer to form an ordered liquid crystalline phase is advantageous [223]. An example of a PAV with such a phase is the dinonyloxy-PPV **89** (Figure 3.25), which forms a nematic phase [224]. Cooling of the melted material to room temperature produces an oriented film with a lower energy gap (2.08 eV, 597 nm) than for an unoriented film (2.21 eV, 562 nm) formed by spin-casting of a chloroform solution of the polymer. Rubbing of a film of this material induces molecular orientation giving rise to polarised electroluminescence [225]. Some attempts have been made to induce liquid crystallinity and, thus, polarised emission by attaching mesogens as side-chains onto PAVs, but with only limited success [226, 227], Ordered films of insoluble PPV have been obtained by conversion of ordered films of precursors, for example obtained by the Langmuir–Blodgett technique [228–230] or from chemically modified precursors [130, 131]. Chiral polymers can display circularly polarised emission. To date, the only PAVs for which this has been observed are the copolymers **90**, bearing chiral substituents [231]. The best results were obtained for a copolymer with an *m*:*n* ratio of 8:1. A chiral PAV **91** has also been made containing chiral binaphthyl units in the main chain [232], but the EL properties have not been reported.

PAV-based OLEDs have found applications in other device areas. For example, an integrated PPV-based PLED excitation source was used for microscale fluorescence detection with application as a fully disposable, diagnostic microchip for determination of urinary human serum albumin [233].

3.7 PAVs as Electron Donors and Acceptors in OPVs

PAVs were one of the first extensively studied classes of organic polymers for OPV applications, and many seminal discoveries in the field were made using them. Although the development of other better performing materials has led to the main research focus being directed elsewhere, they are still the subject of ongoing investigations. PPV is photoconductive but single-layer devices containing PPV films display an efficiency of only about 0.1% [234, 235]. Bilayer devices using soluble PAVs are more efficient, with efficiencies of up to 0.46% obtained using layers of MEH-PPV as the donor layer and C_{60} (**92**) [236, 237] or the substituted fullerene PCBM (Figure 3.26, **93**) as acceptor layers [238], but the highest reported efficiency with a molecular electron accepting layer is only 0.9%, with aluminium tris(8-hydroxyquinolate) (Alq_3, **94**) used as an acceptor [239].

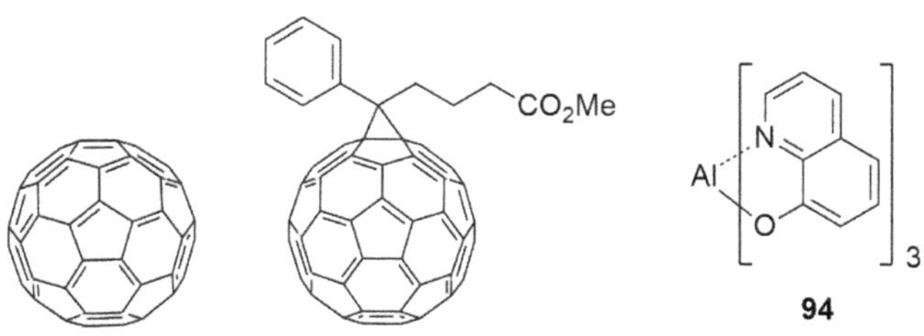

92 **93** **94**

Figure 3.26 Molecular additives used as electron acceptors with PAV donors

As mentioned in Chapter 1, blending of a donor polymer with an electron-accepting material to form a bulk heterojunction (BHJ) has generally proved to be a better way to obtain high efficiency, as the interface area is much higher and a significantly higher proportion of the excitons generated produce photocharges. It was discovered that the photoconductivity of soluble PPV derivatives was markedly enhanced by the addition of C_{60} (**92**) which acted as a highly efficient electron acceptor [240–242]. For example, the photovoltaic efficiency of a 1:1 blend of MEH-PPV and C_{60} is two orders of magnitude higher than for neat MEH-PPV [236, 243]. Subsequently, the more soluble PCBM generally replaced the parent fullerene as the electron-accepting component in blends with conjugated polymers including PAVs [244]. In addition, early work on PAVs revealed the importance of inter-layers in determining device characteristics. An OPV cell model consisting of a MDMO-PPV PCBM cell was fabricated with and without an organic inter-layer. Marks et al. found that a $TPDSi_2TFB$ cross-linked electron-blocking layer (EBL) chemisorbed between the active layer and the anode produces a record $V_{oc} = 1.08$ V for a materials system having a theoretical maximum $V_{oc} \sim 1.2$ V [245].

Considerable progress has been made towards making organic solar cells based on blends of PAVs with PCBM more efficient. Efficiencies of up to 3% have being reported from blends of MDMO-PPV (Figure 3.6, **21**) with PCBM [142]. Here the regioregular polymer gave higher efficiencies than the regiorandom material. It was found that the solubility of the fully regioregular polymers is not as high as for copolymers containing a portion of the other isomer and so in order to optimise processability a 70:30 copolymer was used for device fabrication. While these results are interesting, the lifetimes reported to date for OPVs using PAVs are still short [246] and much higher efficiencies have since been obtained from other types of polymers (*vide infra*), so, as mentioned in Section 3.4, PAVs are currently receiving only minor attention as potential OPV materials. The main reasons for the fading interest in the PAV materials system are their relatively large bandgaps combined with their low charge transport mobilities [247]. A key parameter for obtaining high efficiency from such blends is the phase separation of the two components, which can be influenced by factors such as the choice of solvent, the solvent evaporation time, surface interactions, and annealing, and several seminal studies into the effects of such parameters upon device performance have been made on blends of **21** and PCBM [248]. The insights gained from these studies have since been used to help optimise the performance of OPVs using other conjugated polymers. The common view is that the PAV polymers are constrained as organic electronic device materials by their inherent processing and morphology characteristics. As discussed, the insolubility of PPV itself, combined with its amorphous nature, has limited its application to only the earliest OLED devices.

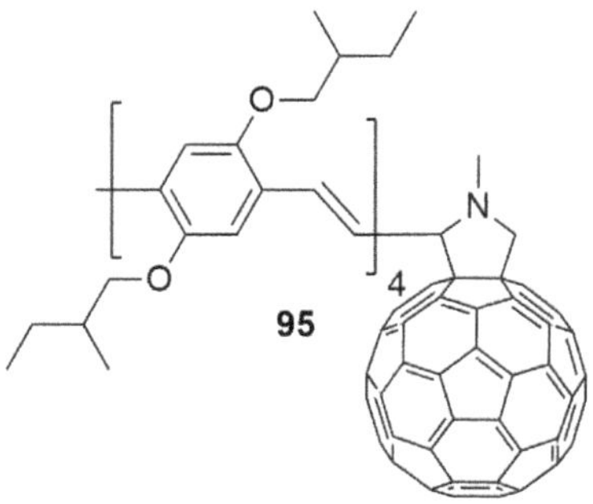

95

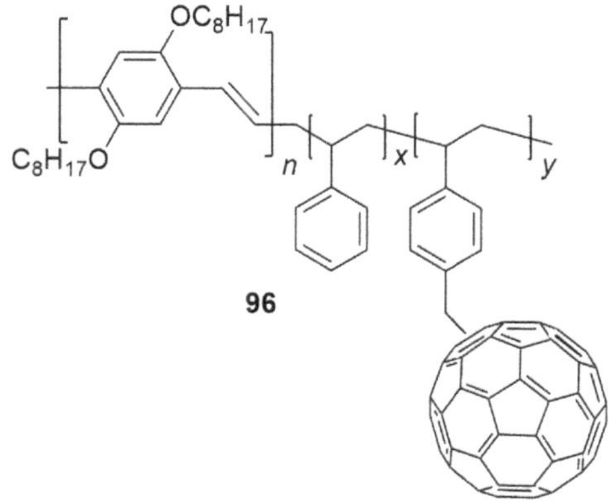

96

Figure 3.27 Oligo-PAV–fullerene dyad and block copolymer with pendant fullerene groups

The more soluble derivatives, such as MEH-PPV, have also been considered to be essentially amorphous polymers [249] and this lack of order has been one of the factors that has led to them being replaced by their more ordered counterparts such as the polythiophenes. However, work in the late Professor Fann's group has shown that the large directional anisotropy in charge mobility observed in amorphous MEH-PPV thin films is not due to differences in charge-carrier density, trapping, or electric-field effects but is actually a consequence of inherent structural anisotropy induced in the spin-coating process. These results indicate that, despite a material being nominally amorphous, processing and surface effects may introduce molecular alignment and hence further allow for the deliberate design of devices with tailored properties [250].

Thus, while at the moment PAVs do not feature as major players in the development of highly efficient OPVs, their study has contributed greatly to the field.

One potential way to avoid the problem of phase separation from polymer–fullerene blends is to prepare dyad materials with fullerenes attached to one end of a conjugated donor molecule [251]. There have been some dyad molecules made containing oligo-PAVs attached to fullerene, such as **95** (Figure 3.27) [252]. An extension of this concept are block copolymers such as **96** in which a fullerene-bearing polystyrene block is grafted onto a PAV [253]. To date, however, this approach has not borne fruit, though it offers fascinating possibilities for future development.

In order to improve the photoconductivity of PAVs, a number of other dopants have been tried, including viologens [254], fluorescent dyes [255–257], carbon nanotubes [258] and nanoparticles of inorganic semiconductors such as cadmium selenide [259] or zinc oxide [260]. None of these have shown any improvement over fullerenes but the search for new acceptors remains ongoing. In an intermediate approach between the blending and bilayer approaches, nanostructured titanium

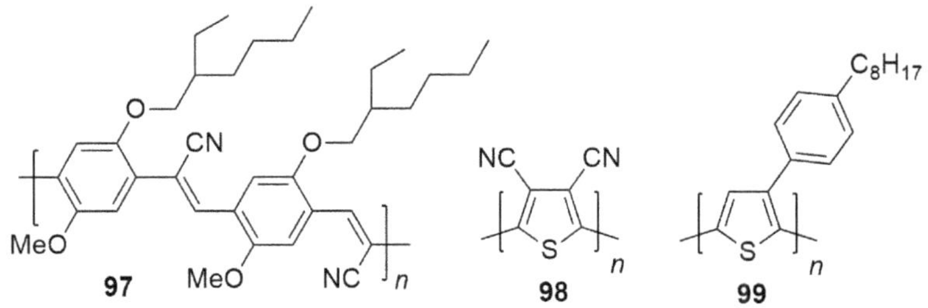

Figure 3.28 PAVs used in all-polymer BHJ OPVs

dioxide has been used as a charge-accepting layer in hybrid inorganic/organic solar cells using PAVs [261–263]. Such cells bear an obvious resemblance to those developed by Grätzel [264], though to date their efficiency remains considerably lower.

One is not of course restricted to small molecules for the electron acceptor. The concept of the all-polymer OPV in which an electron-accepting polymer is blended with an electron donating polymer has its attractions. In this regard cyano-substituted PPVs such as **50** have been used successfully as electron acceptors. The first examples of this were in 1995, when it was reported that blends of MEH-PPV with CN-PPV (**50**) [265] or with MEH-CN-PPV (Figure 3.28, **97**) [266] produce efficiencies of up to 0.9% in OPVs. This sparked further research efforts [267–270], but the maximum efficiency of devices using blends of PAVs with cyano-containing PPV derivatives remained around 1%. CN-PPV has also been used as the electron-accepting component in a blend with other donors such as polythiophenes [271], while poly(3,4-dicyanothiophene) (**98**) has been used as an electron-accepting polymer with MEH-PPV [272], though the efficiencies in both cases were not high. The best all-polymer cell to date using a PAV is a bilayer device using **97** as an electron accepting layer and a poly(3-arylthiophene) **99** as an electron donating layer which shows 2% efficiency [273]. Curiously, the efficiencies of devices using a blend of these polymers was lower than from the bilayer device. The reasons for this difference are as yet unelucidated.

3.8 Poly(thienylene vinylene)s as Low Bandgap Polymers for OFETs and OPVs

As mentioned in Section 3.3, replacing the phenylene unit in PPVs with thiophene is a good way to reduce the bandgap. There has been some interest in poly(2,5-thienylene vinylene)s (PTVs) as low bandgap materials for transistor and solar cell applications. Unsubstituted PTV (Figure 3.29, **100**) is, of course, insoluble but films of it can be prepared by the Wessling and other precursor routes. A film of this material made by the

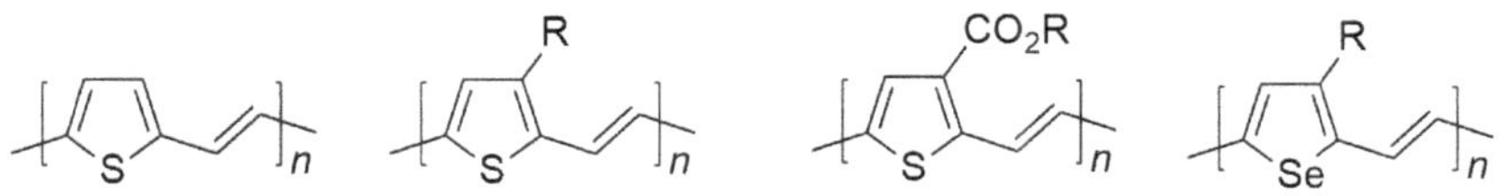

100 **101** R = alkyl **102** R = alkyl **103** R = alkyl

Figure 3.29 PTV and derivatives for use in OFETs and OPVs

Wessling route using the methoxy precursor has been reported to show a very good mobility of 0.22 $cm^2 Vs^{-1}$ in an FET [274]. While this value is promising, the insolubility of the material and the strong possibility of doping and/or damage to electrodes during the acid-mediated conversion of the precursor make it unsuitable for device applications. Accordingly, methods have been developed for the synthesis of soluble alkylated PTVs. Regiorandom poly(3-alkylthienylene vinylenes) (**101**) have been obtained by metathesis polymerisation of 2,5-dipropynylthiophenes [275], while regioregular polymers have been prepared by Heck coupling of a 5-vinyl-2-bromo-3-alkylthiophene [276]. The regioregular materials are reported to show better conductivity than their irregular isomers [276], but there are no reports of high charge carrier mobility from such materials. In view of the low bandgaps (1.5–1.8 eV) of these soluble PTVs, they have been the subject of some study as donor polymers in OPVs, but the best efficiency reported is under 1% [277]. Higher efficiencies (up to 2%) have been obtained using the carboxylated polymer **102**, which seems to be due to the lower HOMO and LUMO energy levels of **102** compared to **101**, which produces a lower loss in energy during electron transfer to the fullerene acceptor and a higher V_{oc} [278]. The carboxylated polymer also showed much stronger PL and a slightly higher (3.1×10^{-3} vs 1.6×10^{-3} $cm^2 Vs^{-1}$) charge carrier mobility than the corresponding polymer **101** (R = hexyl). The selenium analogues of **101**, poly(3-alkyl-2,5-selenylenevinylene)s (**103**), have recently been obtained with high regioregularity by Stille coupling of a dibromoselenophene with a distannylethene [279]. These show better charge carrier mobilities than their thienyl analogues **101** with values of up to 0.01 $cm^2 Vs^{-1}$ reported from OFETs, but there is to date no report on their use in OPVs. While PTVs show promise rather than actual demonstrated performance as electronic materials, they do illustrate that there is still much scope for design and development of new and improved PAVs.

PAVs, especially PPVs, have done much to establish organic polymers as materials for viable electronic devices, but in areas such as OFETs and OPVs the spotlight has since moved on to other classes of materials based on arylenes, but they remain an important group of compounds and they may well still have some surprises in store for us.

Before we move on to look at polyarylenes we will first consider polymers in which arylene units are linked not by double bonds but by triple bonds – the poly (arylene ethynylene)s.

4

Poly(arylene ethynylene)s

Poly(arylene ethynylene)s (PAEs) are structurally very similar to PAVs but the seemingly slight difference in their structures – the replacement of a double with a triple bond linking the aromatic rings – has profound effects upon their properties. One immediately obvious difference between PAEs and PAVs is that the former are inherently more linear and rigid than the latter and there is no issue of *cis* versus *trans* isomerisation. As a result, the chain packing of PAEs is often more regular than for PAVs, which has implications for their optical and electrical properties. They also tend to be intensely fluorescent – a property which has been made use of in their application as active materials in sensors (see Section 4.2).

4.1 Synthesis of PAEs

There exist two major synthetic routes to PAEs [280]. The Sonogashira polycoupling of aryldiiodides **1** (dibromides can also be used but are less reactive) with diethynylarenes **2** illustrated in Figure 4.1 for a poly(dialkoxyphenylene ethynylene) **3** is the most widely used method.

This method has the advantages of being relatively simple as the diethynyl unit can be readily produced from a dihalide by Sonogashira coupling with a silylacetylene followed by desilylation. It is also highly versatile, being compatible with most aromatic groups, and enables alternating copolymers to be made. The main disadvantages of this method are that it requires exact stoichiometry to obtain high molar masses (a consequence of the Carothers equation), that the nature of the end groups is indeterminate, and that alkyne coupling to produce diyne defects is a side-reaction which is hard to suppress. A reportedly defect-free polymer has been reported from a modified procedure [281] which, if generally applied, would overcome the problems arising from the diyne defects such as photo-induced cross-linking.

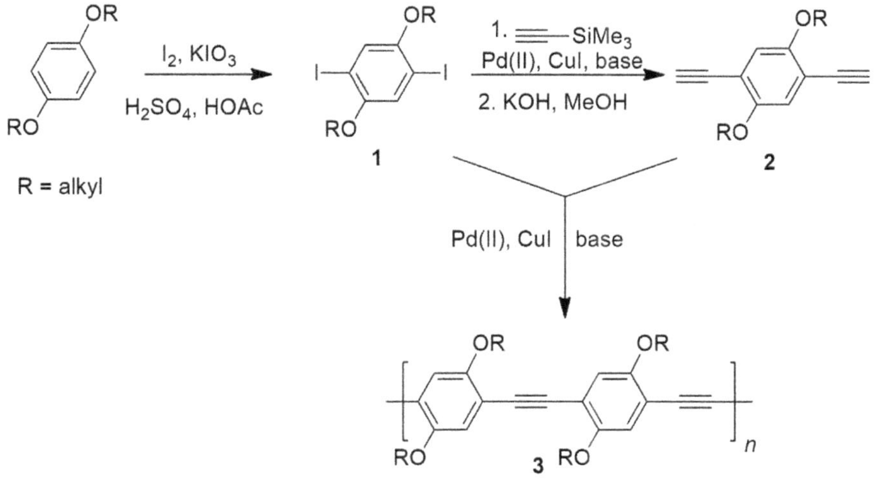

Figure 4.1 Sonogashira route to dialkoxy-PPEs

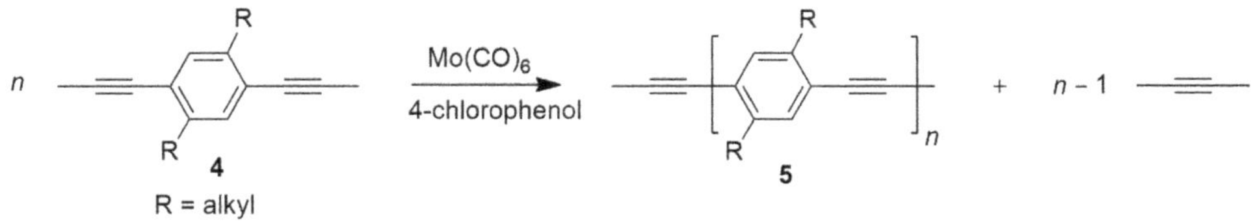

Figure 4.2 Synthesis of PPEs by alkyne metathesis

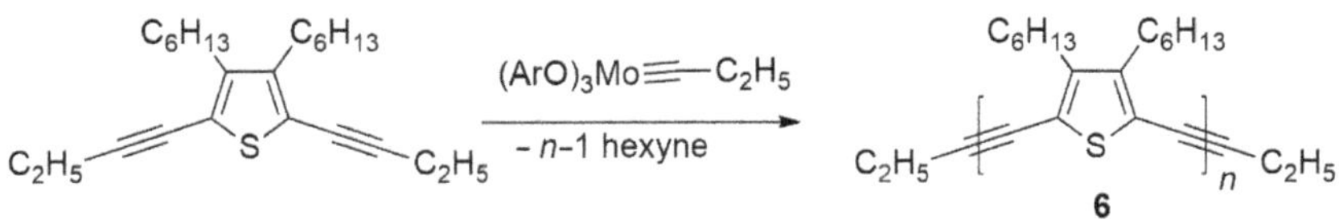

Figure 4.3 Synthesis of a poly(thiophene ethynylene) by metathesis

The second route to PAEs is alkyne metathesis; as shown in Figure 4.2, treatment of a dialkynylphenylene **4** with a carbyne complex generated *in situ* produces the polymer **5** with the only by-product being butyne, whose removal by evaporation from the refluxing reaction mixture drives the reaction forward.

The great advantages of this method are that there are no defects in the resulting polymers, whose end groups are exactly known. The main disadvantages are that alternating copolymers are not accessible and not all arenes are compatible with this procedure. For example, the synthesis of thiophene–ethynylene polymers **6** by metathesis (shown in Figure 4.3) required the development of much more reactive catalysts, which are less straight-forward to prepare and handle [282]. Most syntheses of PAE homopolymers reported in recent years have used metathesis with the Sonogashira method reserved for alternating copolymers.

4.2 Optical and Electronic Properties of PAEs

The effects of substituents upon the optical properties of PAEs are generally very similar to those observed in PAVs. Thus, for example, RO-PPEs **3** show red-shifted spectra compared to alkyl-substituted PAEs **5**, incorporation of thienyl units produces a red-shift in emission, while blue-shifts can be introduced by incorporation of *meta-* or *ortho*-linkages, use of bulky substituents, or 2,3-disubstitution in place of 2,5-substitution, just as was seen for PAVs [150, 280]. However, the emission colours from PAEs are often quite different from those of the corresponding PAVs, and there appear to be much greater effects of aggregation on the solid state PL and EL spectra [150]. Thus, the RO-PPE **7** (Figure 4.4) produces blue-green PL (λ_{max} = 478 nm), and yellow-green EL (λ_{max} = 535 nm) [283, 284], whereas, as seen in Chapter 3, the corresponding RO-PPV is a yellow-orange emitter. The carboxylated polymer **8** shows even more remarkable variations in emission colour as it produces blue PL in solution (λ_{max} = 426, 468 nm), green PL in the solid state (λ_{max} = 516 nm), and yellow-green (λ_{max} = 557 nm) EL [281]. Strong evidence has been produced that these effects for PPEs are due to formation of emissive aggregates or excimers [285]. Blue PL (λ_{max} = 424 nm) has been obtained from a PPE **9** containing a bulky iptycene unit which hinders inter-chain interactions and thus suppresses aggregation [286].

PAEs have not been used much as materials for PLEDs. The early experiments in this area used rather poor-quality RO-PPE materials made by the Sonogashira method, which produced red emission with low efficiency. Subsequently better results have been obtained using improved materials but the potential of the latest high-quality materials seems to have been largely ignored, possibly because industry already has other good green and yellow emitters available. One example of where their emissive properties have been used to advantage has been as sources of polarised emission. Blends of RO-PPEs with high molar mass polyethylene can be cast as aligned films containing molecular-oriented polymer chains which produce strongly polarised emission [287]. If dyes which can transfer energy to the emissive polymer are incorporated into the blend, the emission is still polarised even when the dye molecules are randomly oriented [288]. These films have been

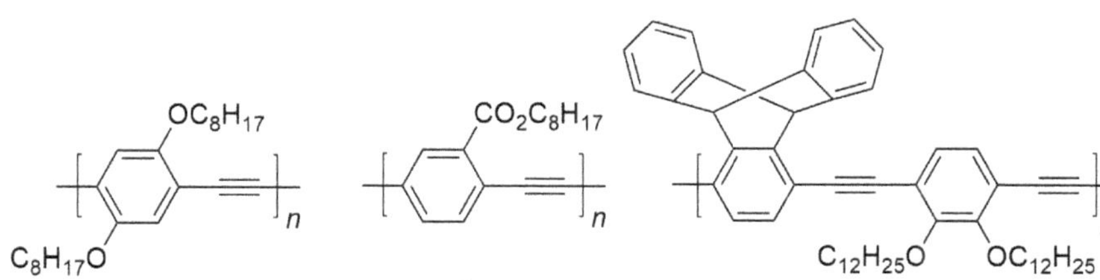

7 8 9

Figure 4.4 Some representative luminescent PAEs

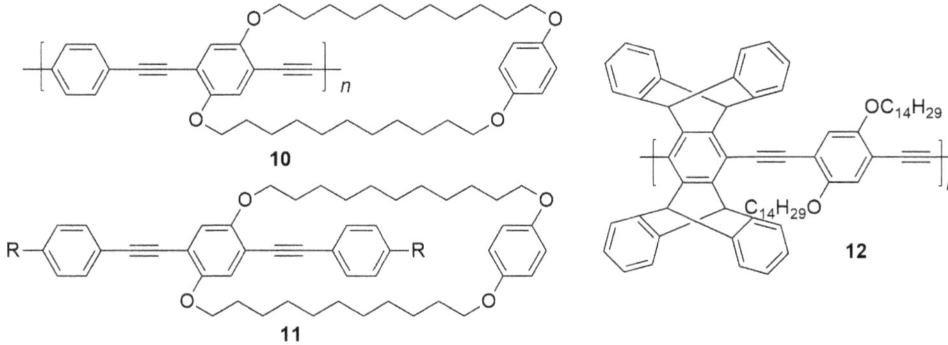

10

11

12

Figure 4.5 Examples of PAEs used for sensing

shown to be suitable for possible application as backlights for liquid-crystal displays (LCDs) [289].

A second possible application of PAEs is in chemo- and biosensors [290, 291]. Here the high fluorescence efficiency of RO-PPEs is the key feature of the proposed devices. These sensors work by the efficient quenching of fluorescence when the desired analyte interacts with suitable substituents in the polymer sidechains to form a complex which quenches the luminescence. Because of the extended conjugation within the polymer, excitons tend to migrate efficiently to the complex as this is the lowest energy site, so conjugated polymers are inherently more sensitive to luminescence quenching than small molecules, as complete quenching does not require binding to all of the receptor sites on the polymer. An early example of this was the discovery that polymer **10** (Figure 4.5) is 65 times more sensitive towards fluorescence quenching by paraquat than oligomer **11** [292]. Similar polymers bearing ethylene oxide or sugar sidechains have been found to be efficiently quenched by metal ions such as lead and mercury [293, 294]. Other polymers with sugar or biotin-containing sidechains have been used to detect proteins [295–297] or DNA [298]. Films of iptycene-containing PPEs such as **12** undergo strong fluorescence quenching in the presence of dinitrotoluene, which is exuded by explosive devices such as landmines – the active ingredients of which include related substances such as trinitrotoluene [286, 299–301]. This occurs because the electron-poor dinitrocompound forms a charge-transfer complex with the electron-rich PPE. This has been used as the basis for a prototype chemosensor for detection of landmines [302].

PAEs like PAVs have not generally proven of much use as transistor materials. However, recently it has been reported that relatively good charge carrier mobilities of up to 0.1 $cm^2\ Vs^{-1}$ have been obtained by using aligned nanowires of a thiol-endcapped RO-PPE (Figure 4.6, **13**) [303]. The values from spin-cast films were three-to-four orders of magnitude lower, indicating the value of order in

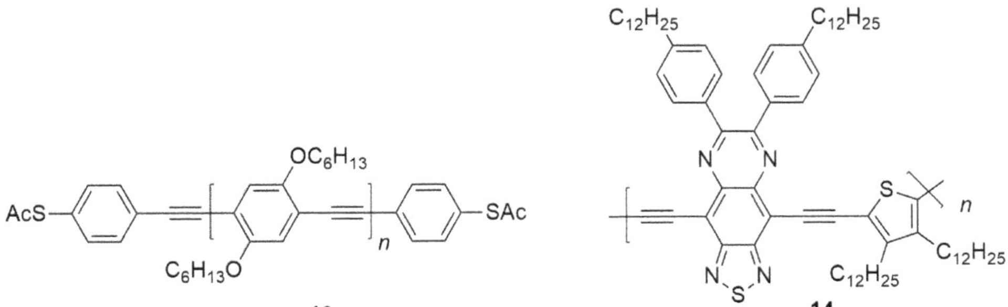

13 **14**

Figure 4.6 PAEs with high charge carrier mobilities

attaining good mobility and suggesting that charge transport along the polymer chain, which is aligned parallel to the wire long axis, is better than between chains. It is interesting to note that this polymer aggregates to form ordered fibres with the polymer aligned parallel to the fibre long axis as many other polymers, for example, polythiophenes (see Chapter 7), self-assemble with the polymer chain perpendicular to the fibre long axis. The highest electron mobility recorded for a PAE is 0.042 $cm^2\ Vs^{-1}$ obtained for the copolymer **14** [304]. The hole mobility was only slightly lower (0.028 $cm^2\ Vs^{-1}$), indicating that this polymer has ambipolar characteristics due to the combination of the electron accepting thiadiazoloquinoxaline and electron donating thiophene units. As we shall see in Chapter 8, such combinations of electron-accepting and donating units are widely seen as being the best choices for obtaining truly ambipolar charge transport.

4.3 PAE–PAV Mixed Polymers

Polymers have been made containing combinations of arylene vinylene and arylene ethynylene units for use in LEDs or OPVs [305]. The usual method for synthesising such materials is by Wittig coupling of an alkyne with aldehyde end groups **15** with an aryldiphosphonate, as illustrated in Figure 4.7. Knoevenagel condensation of the dialdehyde **15** with an arylbis(acetonitrile) produces polymers with cyano substituents on the vinylene units. These methods are compatible with a wide range of aromatic and heteroaromatic moieties, enabling a wide range of copolymers to be made, but to date only a relatively small number have been made and tested in devices.

The materials made to date have generally been green, yellow, or red emitters, with emission maxima varying between 510 and 600 nm. The highest efficiency LED yet reported using such polymers utilised the orange-emitting ($\lambda_{max} = 587$ nm) polymer **16** (Figure 4.8) [306]. The efficiency of this device reached 3.6 cd A^{-1} (2.15%). Replacement of the octadecyl substituents with ethylhexyl as in **17**

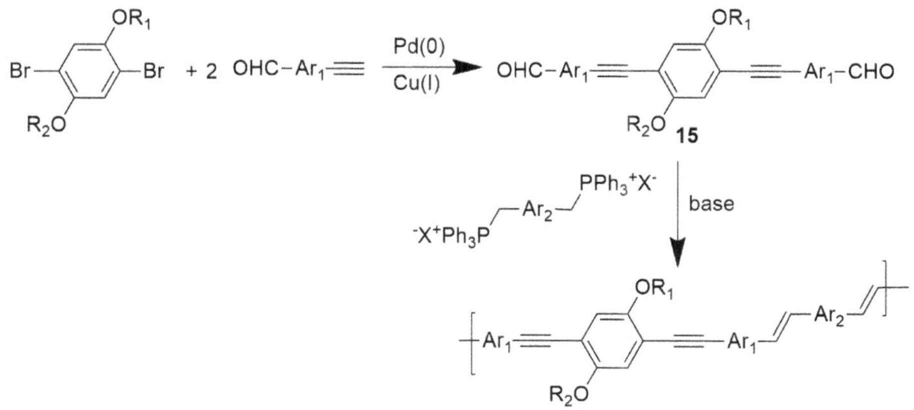

Figure 4.7 Wittig route to alternating PAV–PAE copolymers

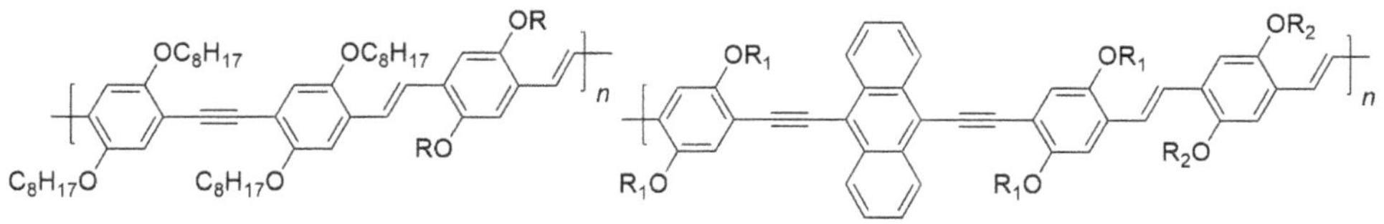

16 R = $C_{18}H_{37}$
17 R = 2-ethylhexyl

18 R_1, R_2 = C_8H_{17}, 2-ethylhexyl
19 R_1 = C_8H_{17}, R_2 = 2-ethylhexyl

Figure 4.8 PAV–PAE copolymers with especially good LED and OPV performance

produced a small drop in efficiency to 3.5 cd A^{-1} (1.82%) but other substituents reduced the emission efficiency markedly. This seems to reflect the ability of the polymer chains to pack closely and illustrates well how seemingly minor differences in polymer structure may have profound (and not always predictable) effects upon solid state properties and device performance. Interestingly, the efficiency of OPVs using **17** with PCBM as an acceptor is better than for **16** (1.7% vs 0.5%) [305], indicating that the parameters for optimal device performance vary between device types. The best performing OPV devices yet reported using this class of polymers were fabricated from anthracene-containing copolymers **18–19**. Efficiencies of nearly 4% have been obtained from copolymers **18** containing a statistical mixture of octyl and ethylhexyl sidechains (i.e. R_1, R_2 are randomly either octyl or ethylhexyl), while the regular polymer **19** produces a slightly lower efficiency of just over 3%, and polymers with other sidechains get up to only 1–2% efficiency [305]. More recently, work has focussed on controlling the electron and hole mobilities in PAV–PAE polymer:fullerene systems [307]. By using specialised fullerene derivatives it was shown that device efficiencies as high as 4.8% could be achieved [308]. Even with $PC_{60}BM$, by fine-tuning the donor and acceptor ratio and the morphology, it is possible to achieve 4.7% with

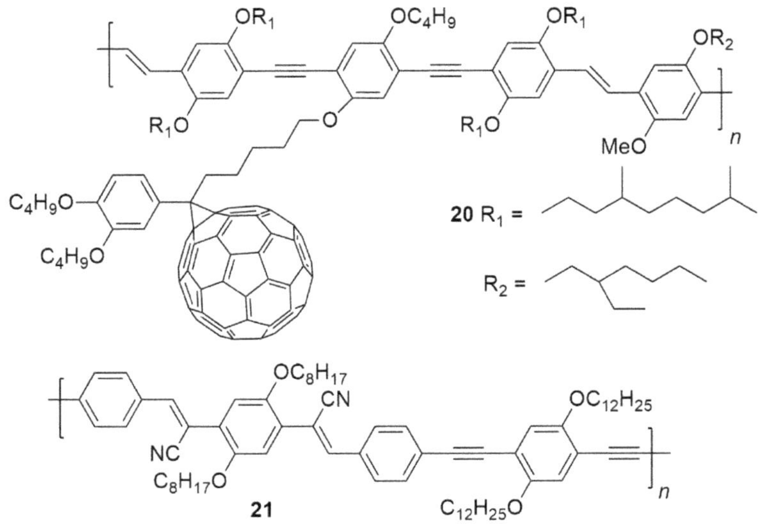

Figure 4.9 Other PAV–PAE copolymers for OPV applications

this system [309]. Again, these differences reflect the ability of the polymers to pack in the presence of a fullerene acceptor to form suitable nanomorphologies with large donor–acceptor interfaces and efficient charge transport of generated photocharges.

One interesting concept for obtaining high efficiency in OPVs is to make polymers such as **20** (Figure 4.9), bearing pendant fullerenes, but to date these have not produced good results [310]. Cyano-containing PPV–PPE derivatives such as **21**, prepared by Knoevenagel condensation, as mentioned at the start of this section, have been used as electron-accepting components in blends with PAVs for all-polymer OPV devices but to date the efficiency from such devices has not risen above 1% [305, 311].

As noted in Chapter 3, PAVs can be considered as copolymers of polyacetylenes and polyarylenes. While both PAVs and their close cousins the PAEs (and the mixed copolymers just discussed) have proved very effective as active materials in OPVs and LEDs producing light between the green and infrared parts of the spectrum, they have not been useful for the efficient production of blue light as their bandgaps are generally too low. To obtain efficient blue emission requires large bandgap materials and it is in this area that polyarylenes, especially polyphenylenes, have proven their worth, and so it is to them we turn in Chapter 5.

5

Poly(phenylene)s

Polyphenylenes are an important class of conjugated polymers as they have intrinsically wide band-gaps and so are blue-emitters. Blue emission has been the hardest colour to attain for commercially viable PLEDs, and phenylene-based polymers are currently the best candidates for achieving this, though their utility for other applications looks decidedly limited. Because of their large bandgaps they have generally not been considered as donor materials in solar cells, as their absorption covers only a small part of the solar spectrum, which severely limits their potential efficiency in energy conversion, and their frontier orbital energy levels make charge injection a problem from standard transistor electrode materials, thus making them poor candidates for transistor materials.

5.1 Poly(para-Phenylene) (PPP)

Any discussion of these polymers must logically start with the parent polymer – poly (*p*-phenylene) (PPP, **1**). As one would expect, this polymer is not soluble, and so, when made directly, as shown in Figure 5.1, by oxidation of benzene [312] – the so-called Kovacic-PPP – or by Yamamoto polycondensation of *p*-dihalobenzenes [313], one obtains an intractable powder. The polymer obtained by oxidation is a mixture of 1,4- and 1,2-linked phenylenes, whereas the Yamamoto method gives an all-*para* product. Films of oligomers can be obtained from these crude products by sublimation under high vacuum, which have been used to fabricate blue-emitting LEDs [150], but this is hardly a viable method for making LEDs industrially. Better results are achieved by using specifically synthesised oligomers – particularly the hexamer sexiphenyl [314] – but this involves lengthy synthesis followed again by expensive evaporation techniques. So, as with PPV, to obtain films of PPP one must use precursor routes.

The unusual feature of these precursor routes is that they start by a biological oxidation of benzene to produce a *cis*-diol **2**. In the first route developed at ICI

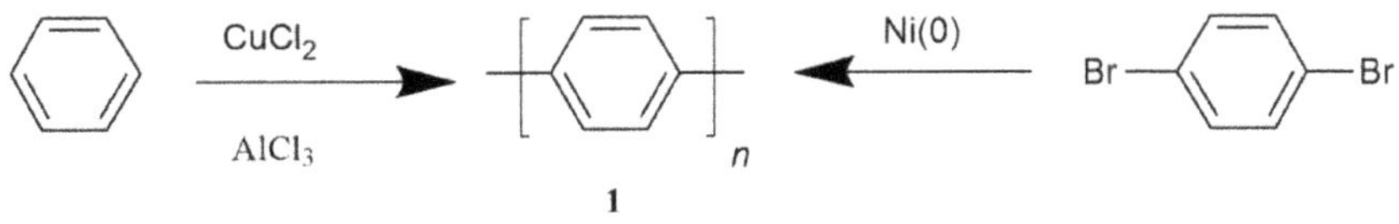

Figure 5.1 Direct routes to PPP

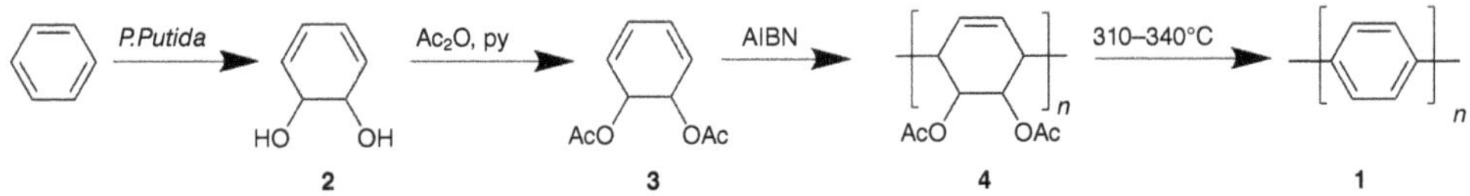

Figure 5.2 ICI route to PPP

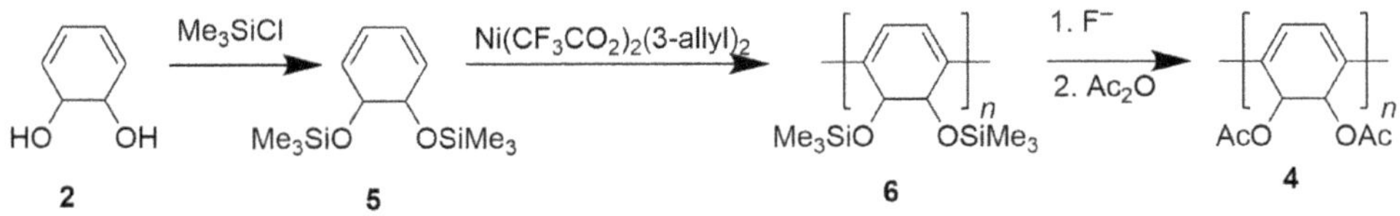

Figure 5.3 Grubbs precursor route to poly(*para*-phenylene)

(Figure 5.2) [315, 316], this diol is converted to the diacetate **3** which undergoes radical polymerisation to form the soluble precursor **4**, that can then be thermally converted to the desired PPP.

The disadvantage of this procedure is that it produces a polymer that is not completely *para*-linked, with about 15% of *ortho*-linkages. A completely *para*-polymer has been made by a variation of this approach developed by the group of Grubbs (Figure 5.3) [317–320]. Here the bissilylether **5** undergoes a regioregular step-growth polymerisation with a nickel(II) catalyst to produce the precursor **6** which is then converted to **4**. The regularity of the product arises from the catalyst binding to the chain end and moving to the new chain end upon insertion of the next monomer unit – this is an example of a living polymerisation involving catalyst transfer, which we will encounter again later with other polymer systems (e.g. regioregular polythiophenes). Such polymerisations not only occur with high regioregularity but also with excellent control of molecular mass and low polydispersities.

PPP is a blue emitter with a reported EL emission maximum around 459 nm [321]. The effective conjugation length of PPP has not been determined as larger oligomers are extremely insoluble, meaning their purification and processing is difficult. Estimates made from small, processable oligomers suggest the ECL is at least 10 phenylene units, but it may be as high as 20 units for absorption judging by the measurements done on related materials such as polyfluorenes (see Chapter 6). The precursor methods produce acidic by-products during the final conversion steps, which can dope the polymer and corrode electrodes, adversely affecting device performance. Therefore, since the initial work on blue-emitting LEDs using PPP, attention has switched to soluble derivatives as potential blue emitters for PLEDs.

5.2 Soluble PPPs

PPP derivatives **7** with one or two long (at least C_5) alkyl or alkoxy solubilising substituents on the phenylene units are soluble in standard organic solvents. A number of methods are available for their synthesis [150, 322, 323]. Both oxidative and Yamamoto polymerisations, as shown in Section 5.1, for making PPP can be used, but the most commonly used method for their synthesis is Suzuki polycondensation (Figure 5.4) [324, 325]. This has been reported to produce the highest degree of polymerisation in one comparative study of making a substituted PPP [326], but is experimentally more difficult to optimise than the Yamamoto method, which requires only the dihalomonomer **6**. From a practical point of view, the Yamamoto method is often best used for making a first sample of a new polymer for initial investigation of its optical properties and device performance, with the optimised Suzuki then being used to obtain higher molar mass samples of polymers which show promising properties and performance.

Generally the AA–BB coupling method involving the reaction of a dihalide **8** with a bisboronic acid **9** is used as the monomers are relatively easy to prepare and purify, but considerable care must be taken to use exactly equimolar amounts of the two monomers to obtain high molecular weights – a consequence of the well-known Carothers equation. The various, sometimes rather involved, methods used to optimise this procedure are discussed in the reviews by Schlüter [324, 325]. These include, for example, spectroscopic analysis of the purity of the boronic acid to be sure of using exactly stoichiometric amounts of it, due to problems with their

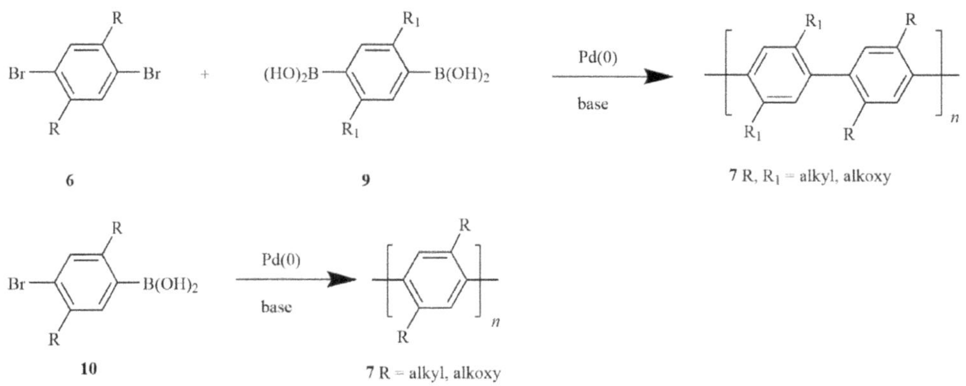

Figure 5.4 Synthesis of soluble PPPs by AA–BB and AB Suzuki polycondensation

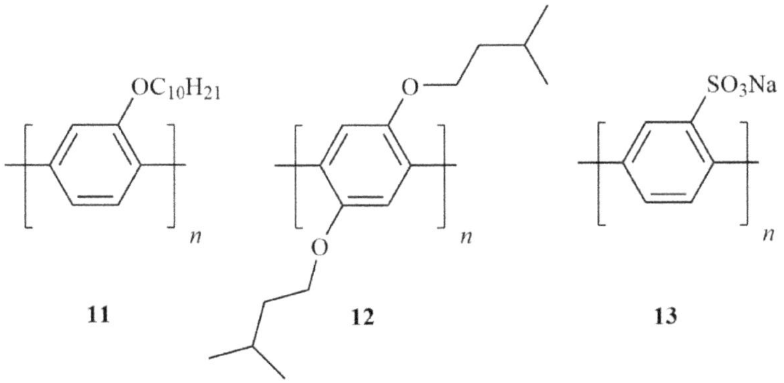

11 12 13

Figure 5.5 Some soluble PPPs used in PLEDs

being both hygroscopic and conversely susceptible to dehydration to form anhydrides. As a result, boronic esters are often used which are hydrolysed to the acids under the reaction conditions, but here partial hydrolysis of the ester during purification becomes an issue. This problem is not seen for the AB coupling of the monoboronates **10** but this method is used less often, due to the greater synthetic difficulty of making and purifying the monomers. A further obvious advantage of the AA–BB method is that it can be used to make alternating copolymers – in fact this is the most widely used method for making alternating copolymers containing arenes and/or heteroarenes. Recently microwaves have been used as heat sources for Suzuki polycondensations as a way to reduce the overall use of energy and solvent (this is an example of 'green chemistry') [327].

PLEDs showing violet-blue ('deep-blue') emission with maxima typically around 400 nm have been prepared from a variety of mono- or dialkoxy-substituted PPPs [150] (e.g. **11** [328] and **12** [329]) (Figure 5.5), with the highest efficiencies (3%) being reported for devices using **11**. Solubility in highly polar solvents such as water or ethanol is obtainable by incorporating charges, as in the polyelectrolyte **13**, whose EL efficiency is reportedly higher than that of **12** [330]. While water is not a suitable solvent for fabrication of electronic devices as it will corrode the electrodes, processing from ethanol or similar highly polar solvents may have advantages in constructing multilayer devices due to the low solubility of most organic electronic materials in such solvents. PLEDs using **13** are reported to display much higher EL efficiencies than those using **12**.

Three issues limit the potential of substituted PPPs as blue emitting EL materials. First, they have large bandgaps with low-lying HOMOs and high LUMOs, which make charge injection from standard electrode materials a problem, lowering efficiency. This can be overcome fairly simply by using hole injection layers and low work function calcium cathodes. Second, steric repulsion between the solubilising side-groups leads to greater twisting of the polymer backbone in substituted PPPs than in PPP with phenylene–phenylene torsion angles of 60–80°, compared to 23° in PPP [323]. As a result, their conjugation lengths are shorter than for PPP and so their emission is strongly blue-shifted and occurs mainly in the violet or far blue region of the spectrum between 400 nm and 420 nm, in which the eye is rather insensitive. Their effective conjugation lengths have been determined from measurements on oligomers to be only 11 phenyl rings for absorption and 7 for emission [331]. Introducing meta-linkages reduces the conjugation even more so that poly(*m*-phenylene)s emit in the ultraviolet ($\lambda_{max} = 346$ nm) [332]. Copolymers containing dialkyl- or dialkoxyphenylenes and phenylenes show more red-shifted emission with emission maxima tuneable between 415 and 450 nm by increasing the proportion of unsubstituted rings [150]. The final problem is that the EL emission can be red-shifted by the formation of aggregates which show longer wavelength, usually yellow, emission. For example, the EL emission from spin-cast films of polymer **12** contains the violet-blue band around 400 nm seen in the PL spectrum and a new yellow-green emission band around 530 nm [329], whose intensity increases with time during device operation. Films of **12** prepared by the Langmuir-Blodgett (LB) technique display almost entirely yellow emission, with the emission parallel to the dipping direction slightly red-shifted ($\lambda_{max} = 536$ nm *cf.* 524 nm) compared to the emission perpendicular to that direction. The emission is also polarised as the LB technique induces alignment of the polymer chains. Polarised emission is also obtainable from dialkoxy-PPPs by using rubbing-alignment, or by attaching mesogenic or chiral sidechains which induce order in the solid state [150]. That the yellow emission arises from aggregation is demonstrated by the fact that blends of **12** with a polysilane display only violet-blue emission around 400 nm [333]. The emission colour from the blend shows none of the red-shifting seen for pure **12**, and the EL efficiency is higher due to suppression of non-radiative inter-chain decay pathways; phase separation of the blend is a potential problem so that the emission colour may not be stable in the long-term.

While blending of soluble PPPs with other polymers can produce relatively stable emission, the emission colour is still too far into the violet to be truly acceptable for a display device, and so soluble polymers with more red-shifted emission have been sought. Since the problem arises from the large torsion angle between adjacent phenylenes, tethering adjacent rings together to form a planarised ladder polymer is an obvious solution.

5.3 Ladder-Type PPPs

Ladder-polymers are polymers in which adjacent units are connected by more than one covalent linkage [334]. They are ribbon-like materials intermediate in nature

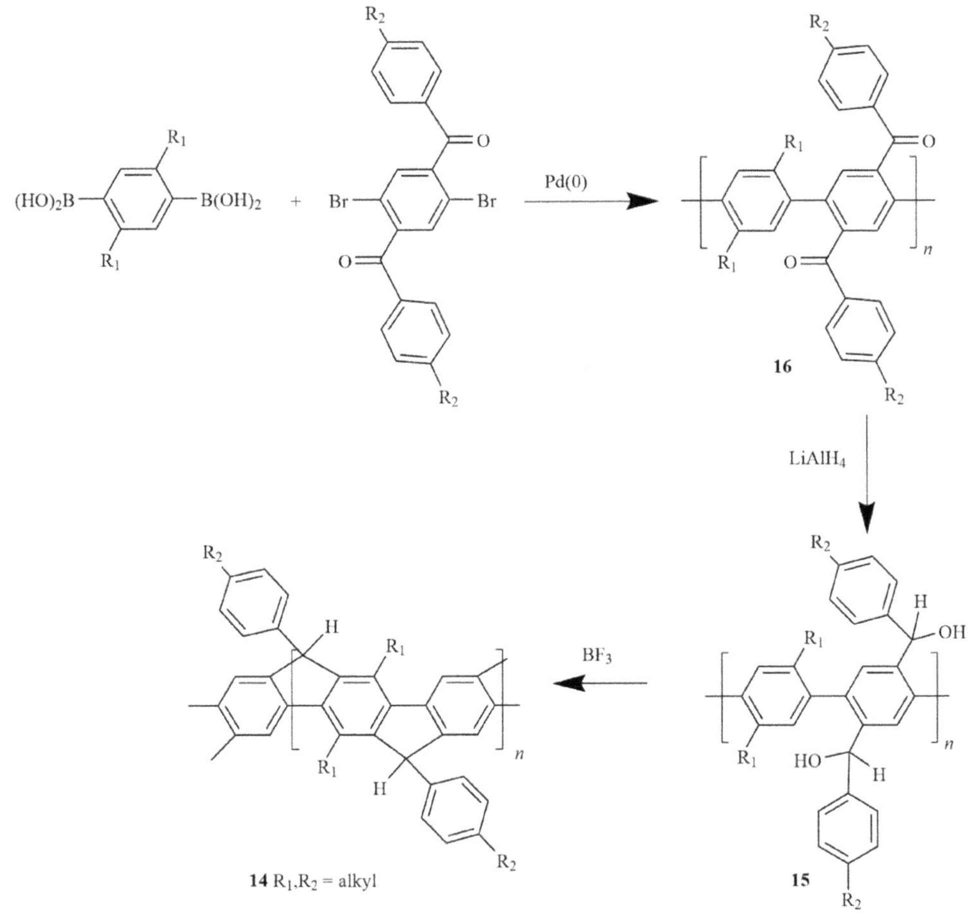

Figure 5.6 Synthesis of LPPP

between linear polymers and two-dimensional structures such as graphene. Ladder-type oligo- or polyphenylenes with single carbon bridges are planar as the torsion angle between adjacent phenylenes is 0°, while two carbon bridges induce a torsion angle of 20° [335] and longer bridges result in even larger angles with a loss of planarity [336].

Ladder-type PPP (LPPP, **14**) is prepared via a precursor polymer **15** (Figure 5.6) produced by an AA–BB Suzuki coupling followed by reduction of the resulting polyketone **16** [337–339]. The key step is the polymer analogous Friedel-Crafts ring closure of **15** which can be done quickly and smoothly by addition of the Lewis acid boron trifluoride to a solution of **15** in dichloromethane. LPPPs with number average molecular weights (M_n) of up to 50 kDa (corresponding to about 150 phenylene rings) have been obtained by this method. Spectroscopic analysis of the final product **14** suggests the presence of less than 1% of defects due to incomplete ring closure, but the results from X-ray and neutron scattering [340] and dynamic light scattering experiments [341] indicate the polymer has a worm-like structure with a persistence length of only 6.5 nm, which is inconsistent with the rigid ribbon-like structure expected for a fully ladder-type material. Indeed this persistence length is shorter than that obtained for some of the stepladder polymers such as polyfluorenes. discussed in Chapter 6 [342]. This suggests that the polymer **14** contains considerably more defects than are detectable by techniques, such as NMR, and is best thought of as a series of rigid ribbon-like segments connected by single bonds. This is consistent with the results of single molecule spectroscopy studies on a substituted LPPP, which revealed emission from multiple chromophores of various sizes on a single polymer chain [343].

The absorption (λ_{max} = 440–450 nm) and emission (λ_{max} = 450–460 nm) spectra of **14** are both considerably bathochromically shifted with regard to single-stranded PPP, so that the emission is blue rather than violet-blue. The planarisation of the PPP backbone in LPPP (**14**) also produces good vibrational resolution in both absorption and emission spectra and a small Stokes shift characteristic of a rigid ladder polymer, in which the rigidity of the polymer hinders deformation in transitions between the ground and excited states [344]. This rigidity also means that the PL quantum efficiencies in solutions of LPPPs are very high (up to 90%) as there are fewer possible non-radiative decay pathways involving vibrational modes [345].

While solutions of **14** produce blue PL, the emission from thin films is dominated by a broad, featureless band in the yellow (λ_{max} = 600 nm) [314, 346, 347]. The relative intensity of the blue and yellow bands is strongly dependent upon the method of preparation of the films, varying with the solvent used and the film thickness. The blue band disappears completely upon annealing a film of **14** at 150°C, and LEDs using **14** generally produce yellow EL [348]. While blue emission (λ_{max} = 450–460 nm) has sometimes initially been observed from LEDs using **14**, the yellow band always appeared rapidly after the device began operating [347].

The results of site-selective excitation experiments and time-resolved PL measurements [349], and of photovoltaic experiments [350], suggested that, like the yellow emission seen for polymer **12**, this yellow emission band was due to excimers from aggregates formed by π-stacking of the polymer chains. Supporting this view, pure blue EL was obtained from blends of **14** (1 wt%) in the wide bandgap polymer poly(N-vinyl carbazole) (PVK) [348]. The emission from the blends, however, became white after only a short period of device operation, which has been attributed to formation of excimers due to the heat produced by the passage of electricity through the device [351]. Accordingly, to obtain stable emission from LPPP, chemical modification is clearly necessary.

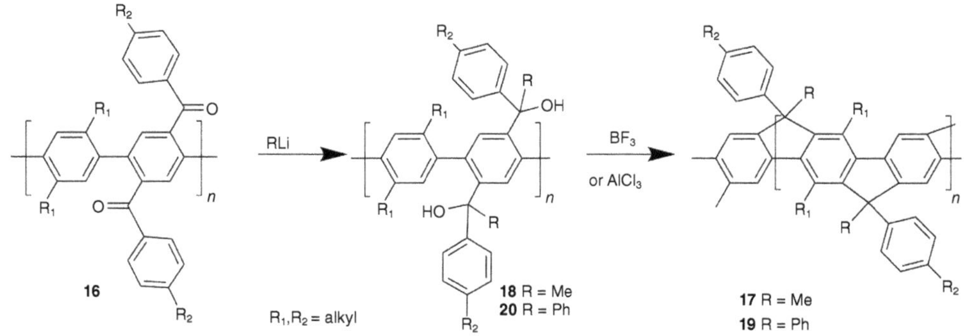

Figure 5.7 Synthesis of Me-LPPP and Ph-LPPP

The stability of emission from LPPPs can be substantially enhanced by replacement of the hydrogens on the bridges with a methyl group to give Me-LPPP (**17**), which is readily made by a slight modification of the procedure used to make LPPP **14** (Figure 5.7) [352]. Treatment of the precursor polymer **16** with methyl lithium, followed by ring-closure of the resulting polyalcohol **18** with borontrifluoride, gives **17**. If one uses phenyllithium instead of methyl lithium, Ph-LPPP (**19**) can be made, but one has to use $AlCl_3$ as the Lewis acid to achieve complete ring closure of its precursor **20** [334].

Me-LPPP (**17**) shows virtually identical emission from solutions and films, with very high PL quantum efficiencies (> 90% in solution and 60% in the solid state), but the emission from **17** is slightly red-shifted compared with **14**, with an emission maximum at $\lambda_{max} = 461$ nm and a secondary peak at $\lambda_{max} = 491$ nm, so that the emission colour is blue-green. This emission is stable and highly efficient (up to 4% efficiency) LEDs have been made using **17** [353]. The effective conjugation length for Me-LPPP was first estimated by extrapolation from the spectra of small oligomers to be about 12 phenyl rings [335], but comparison of the spectra of an undecamer with that of Me-LPPP suggests that it is slightly longer (13–14 phenyl rings) [343]. The excellent optical properties of MeLPPP make it an excellent material for photophysical studies and, as we shall discuss in Chapter 13, it is also an excellent candidate for fabricating polymer-based lasers.

The emission spectrum of MeLPPP also contains a band centred at 560 nm [354]. This band is much weaker than the corresponding yellow band from LPPP (**14**), and the emission does not change upon annealing. This band was initially attributed to emission from aggregates but was later observed from dilute solutions of Me-LPPP [355]. Since such solutions should not contain any aggregates, it is implausible that the emission comes from them. It was, therefore, proposed that the long wavelength emission band originates from defects on the polymer chains. As we shall see in Chapter 6, there is convincing evidence that similar long

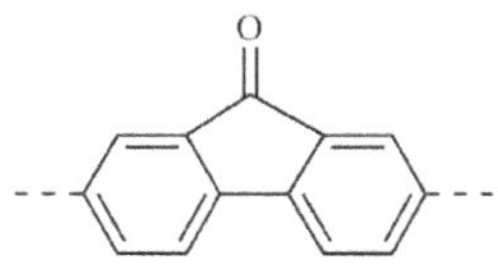

21

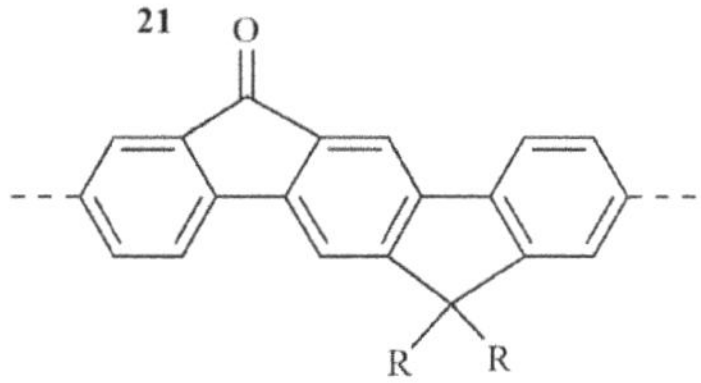

22 R = alkyl

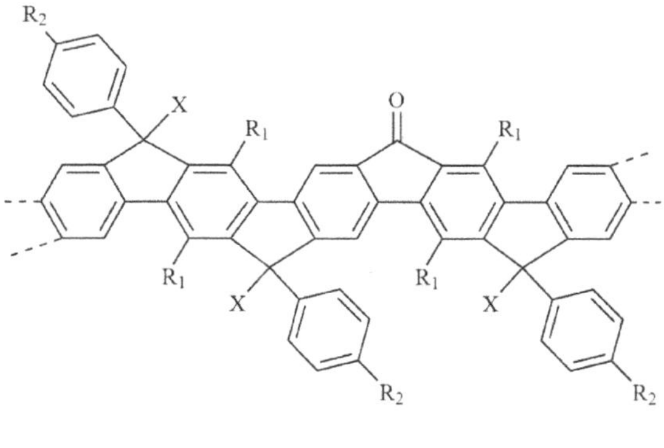

23 R_1,R_2 = alkyl, X = H, CH_3

Figure 5.8 Proposed ketone defects in ladder and stepladder-type polyphenylenes

wavelength emission bands observed around 530–540 nm from polyfluorenes and polyindenofluorenes, which are structurally related to LPPPs, is due to ketone defect units **21** and **22**, so it is probable that the defect in LPPPs is also a ketone like **23** arising from oxidation of bridgehead positions by aerial oxygen (Figure 5.8). As this ketone has more extended conjugation than fluorenone (**21**) or indenofluorenone (**22**), one would expect its emission to be bathochromically shifted, which is consistent with the long-wavelength emission from LPPPs occurring at 560–600 nm.

The difference between the emission spectra in solution and the solid state can be explained by the more efficient energy transfer to the defect sites in the latter due to the increased intermolecular interactions [355]. The much lower intensity of the defect band and the greater emission stability of Me-LPPP over LPPP can be explained by the greater difficulty in oxidising the methyl-substituted bridge, resulting in a much lower level of defects. As will be shown in Chapter 6, the presence of bridgehead hydrogens is also implicated in the formation of defects in other bridged phenylene materials such as polyfluorenes. The synthetic scheme used to make Me-LPPP precludes the presence of bridgehead hydrogens, thus enhancing its stability and giving an example of how the choice of synthetic methodology can directly affect polymer performance.

The PL emission from Ph-LPPP **19** is very similar to that from MeLPPP **17**, with maxima at 460 and 490 nm. The EL spectrum of **19** shows an additional long wavelength band, which is not a broad featureless band as seen for the defect emission from **14** or **17** but has well-resolved maxima at 600 and 650 nm. Photophysical investigation of this emission showed the feature at 600 nm to be emission from a triplet exciton (phosphorescence) with a vibronic shoulder at 650 nm [356]. Such strong phosphorescence requires the presence of a heavy atom, so, as elemental analysis of the polymer showed it contained much more

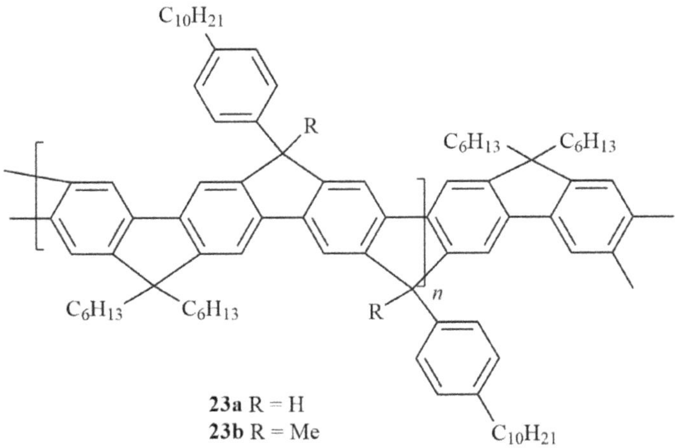

23a R = H
23b R = Me

Figure 5.9 Relatively low defect LPPPs made using a fluorenebisboronate

palladium than **17** (80 ppm *cf.* <2 ppm), it was proposed that residues of the palladium catalyst used to make the precursor polymer **19** reacted with the phenyl lithium and the polymer to introduce covalently bound palladium centres onto the polymer chain, which were sites for phosphorescent emission. That defects at such a low level could produce noticeable emission is indicative of just how sensitive the emission from conjugated polymers can be to the presence of emissive defects.

A way to reduce defects through incomplete ring-closure in LPPPs is to replace the benzene-bisboronic acid in Figure 5.6 with a fluorenebisboronate. LPPPs **24a, b** have been made by this method (Figure 5.9) and the emission from **24a** is reportedly more stable than for **14**, but the emission colour is blue-green ($\lambda_{max} = 460$ nm) rather than blue [357].

While substitution of the bridgeheads as in Me-LPPP does stabilise the emission colour, it also changes it from blue to blue-green. The only way yet found to obtain relatively stable blue emission from LPPPs is to produce statistical copolymers **25** called 'stepladder' copolymers containing phenylene spacers randomly distributed between oligo-LPPP units (Figure 5.10). These are made by addition of a dialkyldibromo comonomer (40–70 mol%) into the Suzuki copolymerisation used to make the LPPP precursor **16** (Figure 5.6), followed by reduction and ring-closure [347]. The emission from these copolymers **25** is blue-shifted with respect to LPPP (**3**) due to the shorter length of the LPPP segments and/or to out-of-plane twisting of the phenylene rings [346, 358]. Films of the copolymers with higher unbridged phenylene content ($m > 50\%$) remain blue-emitting, even after annealing [347], and blue-emitting LEDs with efficiencies of nearly 1% have been made using these materials [359]. This difference in behaviour from LPPP cannot be due to any greater resistance of the copolymers towards oxidation to form

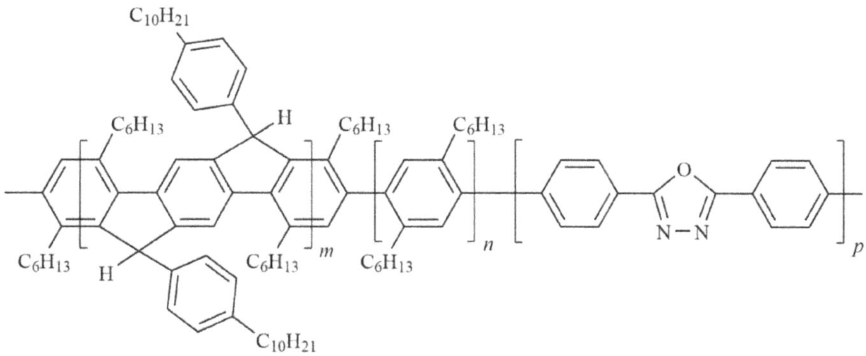

25 $p=0$ **26** $p>0$

Figure 5.10 Stepladder-type polyphenylenes

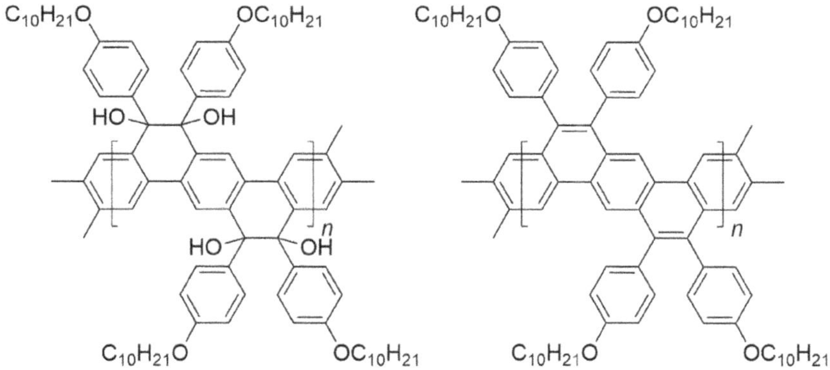

27 **28**

Figure 5.11 Ladder-type PPPs with two-carbon bridges

defects, but must reflect less efficient exciton diffusion to defect sites due to less close packing of the polymer chains brought about by the random introduction of the phenylene groups. The EL efficiency of these polymers has been further improved by incorporating charge-transporting oxadiazole units as in the copolymers **26** which produce blue emission ($\lambda_{max}=410$, 480 nm) with greater EL efficiency than the corresponding copolymers **25** without the charge-transporting units [360].

Ladder-type PPPs with two carbon bridges have also been reported. The polymer **27** (Figure 5.11) with ethylene bridges, which was made by samarium(II) iodide coupling of a poly(diacylphenylene) precursor, shows strong blue-green fluorescence in solution ($\lambda_{max}=459$ nm) and in the solid-state ($\lambda_{max}=482$ nm) [361]. The polymer **28** with ethene bridges is a polyacene and could be thought of as a very narrow substituted graphene ribbon. It was prepared by treatment of a poly(diacylphenylene) with boron sulphide [362, 363]. This polymer also shows blue-green emission (λ_{max} = 478 nm) with some long wavelength emission in the solid state which has been attributed to aggregates, but the EL efficiency is reported to be very low (<0.1%) [364].

5.4 Application of Poly(phenylene)s in Organic Electronic Devices

5.4.1 OPVs

The same issues that have limited the performance of linear chain PPPs in OLED devices have also meant that their application in OPV devices has been constrained. The fabrication of suitable films of unsubstituted PPPs is challenging, while the torsion in substituted PPPs means that optical absorption is limited to the UV region, resulting in poorly performing devices [365]. Ladder-type PPPs (which avoid the torsional defects seen in substituted PPPs) have been more successful in OPV devices, with V_{oc} values close to 1 V. However, their intrinsic photo absorption in the optical and IR regions is still relatively low, resulting in reduced current generation and efficiencies of the order of 1% [366].

5.4.2 OFETs

The low mobility of polyphenylene has limited its application in OFET devices [367]. Recently, it has been used as the gate material in gas sensors based on a FinFET architecture [368]. In these devices the polyphenylene gate forms a large active sensing area that is wrapped around the semiconducting junction. These sensors have been capable of selective detection of a range of organic vapours such as isopropanol, chloroform, dichloromethane and methanol [369].

5.4.3 OLEDs

The greatest application area for PPP materials has been in optoelectronic devices. However, as we have seen, neither single-stranded nor double-stranded polyphenylenes have been totally suitable as blue emitting materials. Relatively efficient violet-blue and blue-green emission have been obtained from them, but a good stable pure blue has proven elusive. In the case of unsubstituted PPPs, fabrication of defect-free films remains a challenge with, as has been discussed, aggregate formation, resulting in changes to the electro-optical properties. By contrast, substituted PPPs are more readily spin-coated. Moreover, substitution with bulky side groups can improve EQE and produce longer excited state lifetimes (narrower emission bandwidth) through a reduction of exciton migration and non-radiative losses, consistent with studies in other systems [370]. However, these device advantages tend to be outweighed by the negative effects of steric interactions between side groups, producing increased chain torsion and consequent blue-shifted emission (sometimes even into the UV) and lowering EQE.

Only the statistical stepladder copolymers display a good blue with reasonable stability in emission colour, but the random nature of these polymers makes it hard to control their properties reliably. As a result, there has been great interest in regular stepladder polymers in which the breaks in the ladder structure come at regular intervals along the chain, that is, in which the monomer units are bridged oligophenylenes. The simplest bridged phenylene is fluorene, a bridged biphenyl which is readily available. As we shall see in Chapter 6, the synthesis of polyfluorenes is also relatively straightforward, so that polyfluorenes are probably the most widely studied of all blue-emitting conjugated polymers, and it is to them that we now turn our attention.

6

Polyfluorenes and Related Polymers

Polyfluorenes are the most widely studied class of phenylene-based conjugated polymers. They are of interest primarily because of their great suitability to be the active materials in blue-emitting PLEDs. This importance is because of their ready synthetic accessibility, good film-forming properties and high fluorescence quantum efficiencies. They also represent one of the best examples of how the relative importance of defects and aggregates in the properties of conjugated polymers has been determined, and in how synthetic design can be used to attain superior device performance.

6.1 Polyfluorenes

Polyfluorenes (PFs) are the simplest regular stepladder-type polyphenylenes, in which only every second ring is bridged. Soluble polyfluorenes, especially poly (9,9-dialkylfluorene)s (PDAFs, Figure 6.1, **1**), have been much studied in recent years due to their excellent physical properties, including high PL quantum efficiencies, and excellent chemical and thermal stability [43, 323, 371–374]. A particularly attractive feature of PDAFs is the ready synthetic accessibility of the monomers. The bridgehead (C9) protons in fluorene (**2**) are relatively acidic so treatment of fluorene (**2**) with an alkyl halide and base, usually under phase transfer conditions (Figure 6.1), produces 9,9-dialkylfluorenes (**3**) in very high yields. Some care is required to exclude oxygen, since oxidation of the fluorenyl anion intermediate **4** can lead to formation of fluorenone (**5**) as a by-product. Another problem with this reaction is that the product may contain traces of the monoalkylfluorene **6**, which can be extremely difficult to detect and remove. Bromination of **3** efficiently produces **7**, which is also available by alkylation of 2,7-dibromofluorene. Both **3** and **7** can be used as monomers to make PDAFs.

The first reported synthesis of PDAFs was by oxidative coupling of the monomer **3** (R = hexyl) with iron(III) chloride, but the products had low molar

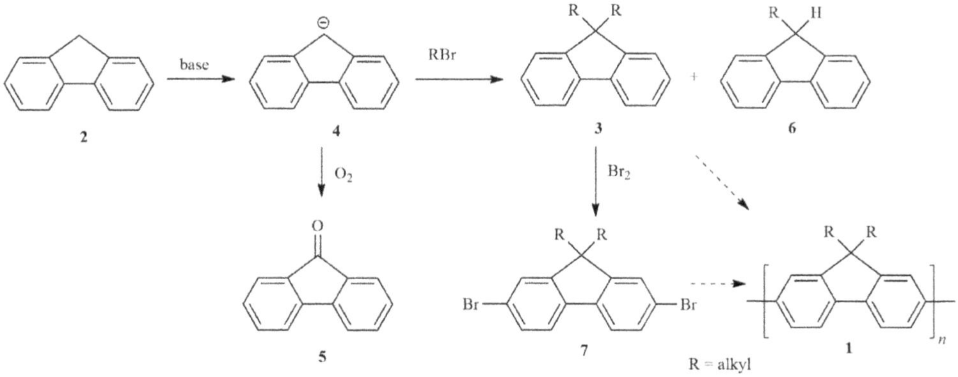

Figure 6.1 Alkylation of fluorene to produce monomers for PDAFs

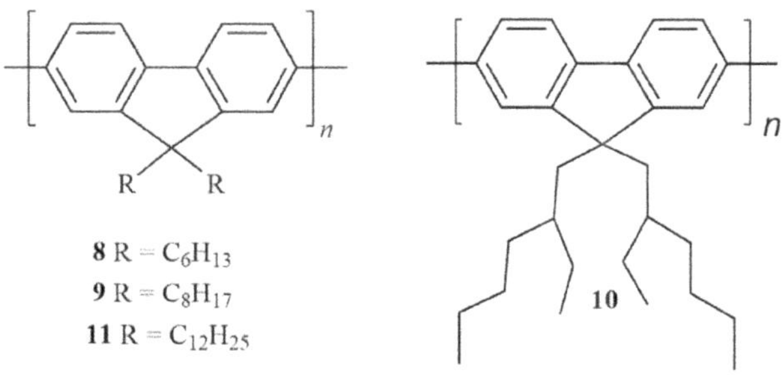

Figure 6.2 PDAFs with straight and branched chains

masses with high levels of defects due to non-selective coupling [375, 376]. Devices using poly(9,9-dihexyl-2,7-fluorene) (Figure 6.2, **8**) made in this way were among the first PLEDs to produce blue-emission, but with low emission efficiency [377–379]. Subsequently, better syntheses of PDAFs were developed using Suzuki polycondensations or Yamamoto couplings of dibromo-monomers **7**. In particular, a group at Dow [380–385] have developed a highly efficient AA–BB Suzuki cross-coupling (see Figure 5.4) route leading to high molecular weight PDAFs. For example, poly(9,9-dioctylfluorene) (**9**) with a $M_n > 100{,}000$ g mol^{-1} was obtained by this method after less than 24 hours reaction time. This material was used to fabricate the first relatively efficient (*ca.* 1%) blue ($\lambda_{max} = 436$ nm) PF-based PLED; a hole-transporting layer being used to improve the device efficiency [386]. Yamamoto-style polycondensations have also been used to prepare high molecular weight PDAFs. For example, polymer **10** has been prepared with $M_n >$ 100,000 g mol^{-1} by coupling the dibromofluorene monomer **7** (R = 2-ethylhexyl) with bis(cycloocta-1,7-dienyl)nickel(0) and bipyridine [387]. As was mentioned in Chapter 5 for PPP synthesis, the highest molecular weights are generally obtained from Suzuki couplings, while the Yamamoto method is experimentally simpler but impractical for large scale synthesis, as the expensive nickel reagent must be used stoichiometrically.

PDAFs produce violet-blue light ('deep blue') with an emission maximum at about 425 nm and a secondary peak around 445 nm. This is not a totally satisfactory blue colour as the eye is not very sensitive towards emission around 425 nm but is good enough for most display purposes. Companies working on PLED displays generally do not reveal the structures of the emissive materials used in their prototype devices, but it is likely that many of them are using fluorene-based polymers as their blue emitters. The ECL for PFs was initially estimated by extrapolation from the optical properties of short oligomers to be 6 units (12 phenyl rings) for emission and 12 units (24 phenyl rings) for absorption [388], but more recently following the synthesis of defined oligomers up to a 64-mer it has been re-estimated that it is actually 39 phenyl rings (19.5 fluorene units) for absorption [389]. This again illustrates the problem of accurately estimating an ECL by extrapolation from properties of oligomers.

Polarised emission is desired for some applications such as backlights for liquid crystal displays (LCDs), which requires aligned films. These can be facilely produced from PDAFs with liquid crystalline properties. PDAFs with unbranched alkyl substituents (e.g. **8**, **9**, and **11**) (Figure 6.2) show two thermotropic nematic liquid crystalline phases whose phase transition temperatures decrease with increasing sidechain length [390, 391]. By contrast, the polymer **10** with branched alkyl chains displays only one nematic phase while other polymers with branched chains show no liquid crystalline phases at all. The liquid crystalline order can be retained at room temperature by annealing films of the polymers at a temperature just above the (highest) liquid crystalline transition then cooling them rapidly [390, 391]. Polarised PL and EL has been obtained from such films deposited upon a rubbing aligned polyimide alignment layer [222, 392]. A rubbing-aligned layer of PPV has also been used as an alignment layer to obtain polarised EL from films of **9**, but the emission spectrum was slightly red-shifted due to some absorption by the PPV layer in the region around 433 nm [393, 394]. Circularly polarised PL and EL emission has been obtained from polyfluorenes bearing chiral side-chains or from a copolymer with a mixture of dioctylfluorene and chiral units, though the degree of dissymmetry was much reduced in the latter case due to the two units wanting to adopt different conformations [395, 396]. A combination of a helical backbone conformation (favoured by branched alkyl side-chains) and liquid crystallinity is thought essential for obtaining a high degree of circular polarisation.

Two problems have hindered the development of polyfluorenes as efficient blue-emitting materials. First there are problems in obtaining high efficiency long-lasting devices, because the low lying HOMO and high LUMO in these polymers mean efficient charge injection requires use of hole-transporting layers and of low

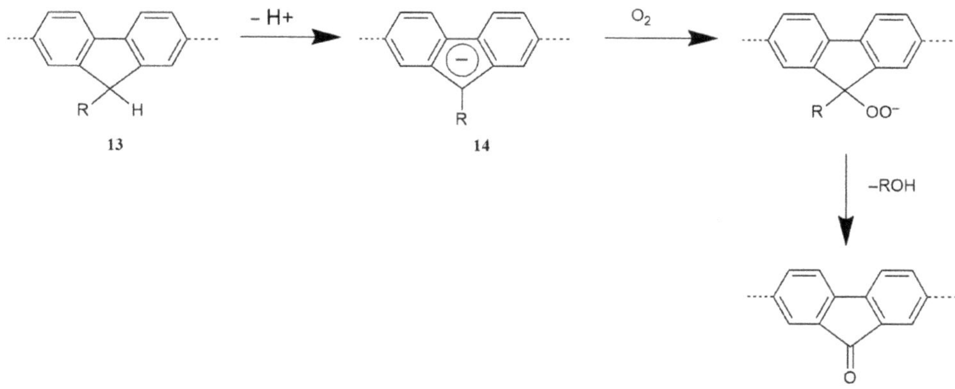

13 14 12

Figure 6.3 Proposed mechanism for formation of ketone defects **12**

work function electrodes such as calcium, which are air- and water-sensitive, requiring good encapsulation. This limitation has been overcome by device optimisation and, as we shall see later in this chapter (Figures 6.9–6.11), by incorporation of charge-transporting units into the polymers. But the second problem has been even more of an obstacle – the blue emission from PDAFs is unstable, with a strong emission band around 530 nm appearing rapidly after annealing or upon running an EL device [371, 372]. At first this long wavelength emission band was believed to arise from emissive aggregates or excimers, similar to those seen for substituted PPPs (see Chapter 5), but convincing evidence has been found that emission from fluorenone defect sites **12** (Figure 6.3) is a better explanation for this emission [372, 397]. The long wavelength emission thus arises from energy transfer to the ketone defect sites, with the greater strength of this band in EL compared to PL spectra being due to the defects acting as charge trapping centres. This picture is consistent with the results of time-delayed PL experiments [398]. As we saw in Chapter 5, similar defects are now thought to be responsible for the appearance of long wavelength emission in ladder-type PPPs. To account for the formation of the defects it has been proposed that monoalkylfluorene units **13** are deprotonated to form anions **14** which then react with oxygen to form the fluorenone units **12**.

The crucial evidence for fluorenone formation came from studies on the polymers **15** and **16** (Figure 6.4), which were prepared by Yamamoto polycondensation. The infrared (IR) absorption spectrum of even pristine monoalkylfluorene polymer **15** displayed a strong ketone carbonyl stretching band at 1721 cm^{-1}, and the solid-state PL and EL spectra were dominated by a band at 533 nm [399]. Exactly the same band was observed from the dialkylfluorene polymer **16** upon photo-oxidation or after running an LED for a

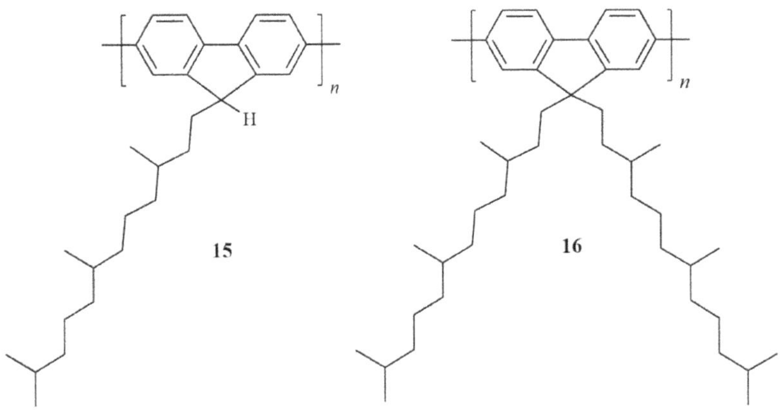

Figure 6.4 Mono- and dialkylated polyfluorenes used to prove the ketone defect hypothesis

short time. These absorption and emission bands match closely with those seen for fluorenone in the solid state [400]. The presence of the ketone in the pristine **15** can be explained by the low valent nickel catalyst reducing some of the monoalkylfluorene units **13** to the anions **14** – the anions would also be formed under Suzuki conditions due to the presence of base. These anions would then be oxidised by air during workup. The susceptibility of fluorenyl anions to oxidation is well known as fluorenones can be obtained in high yield by treating fluorenes with base in the presence of air [401]. The fact that no long wavelength emission was observed when polymer **16** was heated at 200 °C under illumination in a dynamic vacuum ($<10^{-4}$ mbar), but only upon heating in air, proves that oxygen must be involved [402]. That the green emission appeared when a film was heated under a nitrogen atmosphere in a glovebox, where the oxygen levels were very low, illustrates the extreme susceptibility of PDAFs to aerial oxidation [403]. There is evidence that reduction of fluorene by calcium cathodes is implicated in the formation of ketone defects during LED operation, which can be suppressed by introduction of a buffer layer between the polymer and the cathode – nicely illustrating how an understanding of polymer chemistry can lead to advances in device design [404]. It also proves a point often forgotten – materials still undergo chemical reactions inside devices, especially in the presence of current (as in LEDs or FETs) or strong light (LEDs and OPVs). Such reactions will particularly occur at interfaces such as electrode surfaces, where they have major effects upon the electrical and other properties of the devices.

It is now generally accepted that the green emission band is caused by fluorenone defects, but inter-chain interactions do play some part in the appearance of the long wavelength emission. This is not only because inter-chain interactions (aggregation) enhance exciton migration to the defect sites, but because evidence suggests that ketone defects on different chains tend to aggregate due to dipole–dipole interactions and that the defect emission actually arises from excimers formed between two ketone defects [405]. The details of the mechanism for the formation of the defects have also not yet been elucidated, so considerable work still has to be done to fully understand the origins of the ketone defects and the green emission band associated with them. Nonetheless the story of the identification of the fluorenone defect is an extremely good example of how a close collaboration between physicists and chemists can lead to major advances in understanding the behaviour of conjugated polymers and devices. The identification of the source of the green emission has been of immense value in developing effective ways to combat it.

Even before this discovery, a number of often simple but effective ways had been found which significantly enhanced the colour stability of emission from PFs. For example, fractionation of **9** to remove low molecular mass material was reported to reduce the long wavelength emission [406]. This result seems counterintuitive, as the probability of a polymer chain containing a defect increases with increasing chain length, but the oligomers in the low molecular mass fraction may assist in inter-chain charge and energy migration, so removing them might reduce exciton migration to defect sites. Also, fluorenone units on a chain reduce its solubility, thus hindering chain growth in solution; low molar mass fractions may therefore contain a higher proportion of defects. Blending a PDAF with hole-transporting molecules also produces more stable blue emission [407]. This improvement can be attributed to the dopants acting as hole traps, which reduce the amount of charge trapping at the defect sites (which are electron acceptors due to the oxygen) and thus the emission from them. Blending PDAFs with low molar mass poly(phenyl ether) has also been found to enhance the stability of the emission colour, which is attributed to the dopant hindering interactions between the polymer chains [408]. However, these blends of PDAFs with other materials are likely to undergo phase separation over time, so the long-term stability of emission from them is dubious.

The strength of the green emission, even at low levels of defects, is due to efficient energy transfer from the wide bandgap fluorene chromophores to the smaller bandgap ketones. The efficiency of such processes is illustrated by the fact that EL from random copolymers of fluorene with only 1–5 mol% of perylene dyes results in the EL coming only from the perylene chromophores [409]. If significantly lower amounts of comonomers are used, emission from both the fluorene-backbone and the dye units occurs which, as we will see in Chapter 12, is a promising method for obtaining white emission. Incorporation of blue-emitting comonomers has been used in order to stabilise the emission from PDAFs. For example, copolymers **17** (Figure 6.5) of dihexylfluorene with anthracene, made by

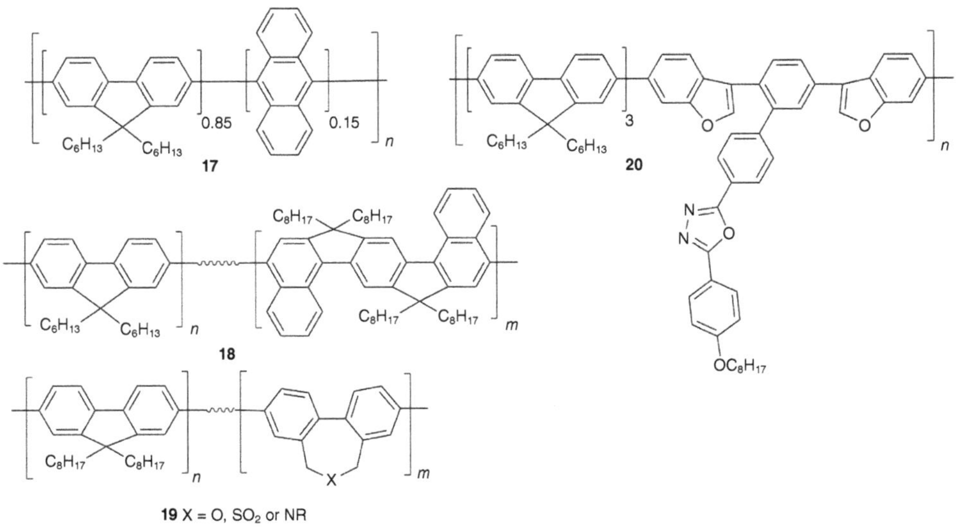

Figure 6.5 Copolymers of fluorene with stable blue emission

the Yamamoto method, show stable blue emission ($\lambda_{max} = 455$ nm) even after prolonged annealing due to trapping of the exciton at and subsequent emission from the anthracene moieties [410–412]. Another blue-emitting comonomer which has been used to produce stable blue emission is dinaphtho-s-indacene. Copolymers **18** containing 1–15 mol% of this unit display stable blue emission with emission maxima between 448 nm ($m = 1\%$) and 460 nm ($m = 15\%$) [413]. When chains longer than two atoms are used to bridge biphenyls, the biphenyl units become twisted, reducing the conjugation, so that polymers containing these units can become ultraviolet emitters [336, 414]. Introduction of 30–40 mol% of such units into fluorene copolymers such as **19** (Figure 6.5) has been shown to suppress the long wavelength emission seen for the fluorene homopolymers, at the expense of lowering the EL efficiencies [415]. These effects are attributed to disruption of the chain packing which not only hinders exciton migration to emissive defects, but also charge transport. This loss of charge transport can be overcome by incorporating in charge transporting substituents, for example the terfluorene copolymer **20** bearing electron-transporting oxadiazole side-chains combines stable blue EL ($\lambda_{max} = 434$ nm) with good efficiency due to the improved charge transport [416]. The stability of the emission from PDAFs has also been improved by incorporation into the main chain of hole-injecting 3,6-carbazole units [417, 418], or a combination of hole-transporting triphenylamine and electron-accepting diphenylsulphone units [412]. These units act by disrupting the packing and/or being sites for charge trapping, but the device results from these copolymers are not outstanding.

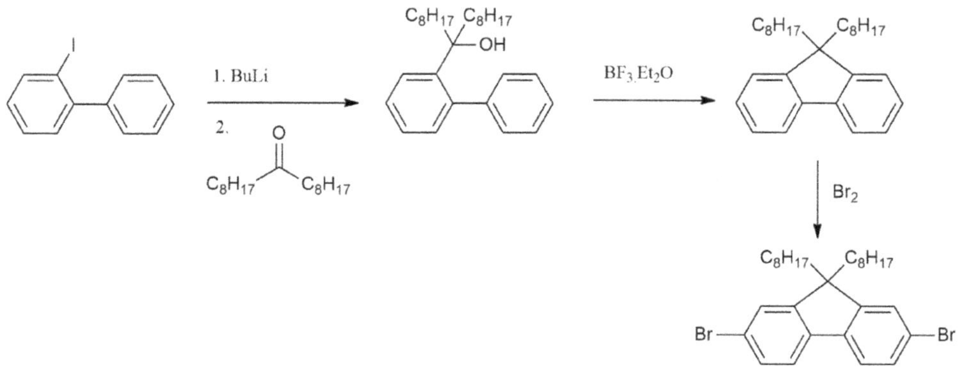

21

Figure 6.6 Route to dialkylfluorenes without monoalkylfluorene impurities

Other approaches towards obtaining stable blue emission from PDAFs through reduction of inter-chain interaction have included cross-linking polymers bearing vinyl end groups [419] or attaching bulky groups such as dendrimers [420, 421] or silsesquioxanes [422, 423] as end- or side groups. Again, the reduction in interactions produces a loss in device efficiency as charge transport is also hindered.

Obviously the best way to avoid ketone defects in PDAFs would be to ensure that the dialkylfluorene monomers contain no monoalkylfluorene impurities. It has been found that treating the monomers with base (to remove the monoalkyl derivatives as the anions) before polymerisation greatly enhanced the colour stability of the resulting polymers [424]. The impurity was not detectable either before or after this treatment, indicating it was present in very low concentration, so checking for its removal still meant making the polymer and testing it in a device. A new synthesis of 9,9-dialkylfluorenes has been developed (Figure 6.6), which precludes the formation of monoalkylfluorenes [425]. Polymers made from the 'defect-free' monomer **21** showed stable blue emission in initial tests, though lifetime tests to determine their absolute stability have not been reported. When as little as 0.06 mol% of a monoalkylfluorene comomoner was incorporated into the polymerisation mixture, a green emission band was observed from the resulting copolymers in an LED. This observation illustrates not only how sensitive the emission colour is to the presence of emissive defects, but also the immense practical problems that would arise from trying to produce defect-free monomers by other methods – attempting to detect the monoalkyl impurity at such low levels is presently impossible.

The long-term stability of blue emission from PDAF-based LEDs has been thrown into doubt by a recent study which showed that ketone defects can be produced by degradation of the alkyl side-chains in PDAFs under aerial oxidation

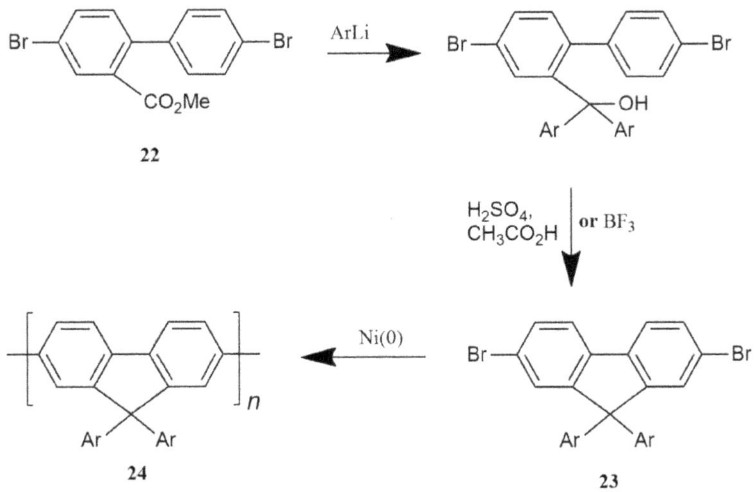

Figure 6.7 Synthetic route to poly(9,9-diarylfluorene)s

[426]. This observed degradation implies that even defect-free polymers must be rigorously protected from oxidation sources, though acceptable lifetimes should be possible for suitably encapsulated devices. Removal of possible oxygen sources such as ITO from the devices would be another possible option for enhancing lifetimes, though this runs into the problem of finding alternative electrode and charge transport materials which are acceptable from both a device performance and a cost perspective.

As aryl sidechains should be less susceptible to oxidation than alkyl chains, polymers based on 9,9-diarylfluorenes should display more stable blue emission than their dialkyl analogues. A synthesis of polyfluorenes with aryl substituents at the 9-position has been developed (Figure 6.7) [420, 427]. This synthetic route precludes formation of the bridgehead hydrogens which are so readily oxidised to form ketones and the bulk reduces inter-chain interactions, and hence exciton diffusion to any defect sites that may be formed. As aryl groups, unlike alkyl groups, generally cannot be directly substituted onto the 9-position of fluorenes, the monomers were prepared by addition of aryllithium reagents to the biphenyl-2-carboxylic acid methyl ester **22**, followed by ring closure using sulphuric acid in acetic acid [420], or Lewis acids such as BF_3 [427], to give the diarylfluorene monomers **23**. The polymers **24** were then made by Yamamoto polymerisation. This route is fairly general as it can be used to add any arene which is compatible with the presence of aryl lithium, though it is inefficient for adding bulky aryl groups due to steric congestion at the reaction site.

The diphenylfluorene polymer **25** (Figure 6.8) is insoluble, but the materials **26** and **27** with solubilising alkyl or dendrimers substituents are soluble in toluene

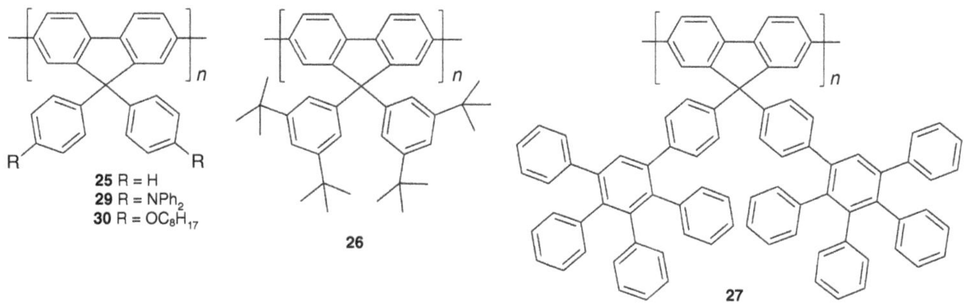

Figure 6.8 Poly(9,9-diarylfluorene)s

[420]. The degrees of polymerisation for these polymers were rather low, possible reflecting unoptimised reaction conditions. The dendronised polymer **27** in particular had a M_n of only 10,000 g mol^{-1}, corresponding to a degree of polymerisation of only 10 attributable to the bulkiness of the substituents. This polymer displays excellent thermal stability, with no significant mass loss being observed in Thermal Gravimetric Analysis (TGA) below 570°C.

Stable blue PL emission has been obtained from films of both **26** and **27**, with no sign of a green even after prolonged annealing in air at 100 °C, while blue EL with no significant defect emission has been obtained from **27** even after running the device for several minutes [428]. As already stated, a strong defect emission band would be expected from a PDAF under these conditions, so confirming the greater stability of the aryl polymers. The use of bulky dendrimers as in **27** is not to be recommended, as not only do they make the synthesis longer and less efficient, but they have also been found to affect the photophysics of the polymer [429–432]. One of these effects is to increase the lifetime of triplet excited states, which might reduce the efficiency of LEDs due to increasing singlet–triplet quenching [432]. The use of simple substituted phenyl substituents to suppress the long wavelength emission would thus appear to be advisable on the grounds of device performance as well as of synthetic simplicity and efficiency.

A second and synthetically simpler way to attach aryl groups at C9 is by acid-catalysed condensation of dibromofluorenone **28** with phenol or triphenylamine (Figure 6.9). While experimentally very simple and high-yielding, this method is restricted only to highly activated arenes with amine, hydroxyl or alkoxy substituents. This method has been used to make polymers **29** [433] and **30** [434], which both show stable blue EL. The hole-transporting triarylamine substituents in **29** mean that a hole transporting layer is not needed to increase the efficiency of LEDs containing it, but the solid-state PL efficiency of **29** is lower than for PDAFs such as **9**, so that the overall EL efficiency is similar for the two materials [433]. This problem can be mitigated by making copolymers containing

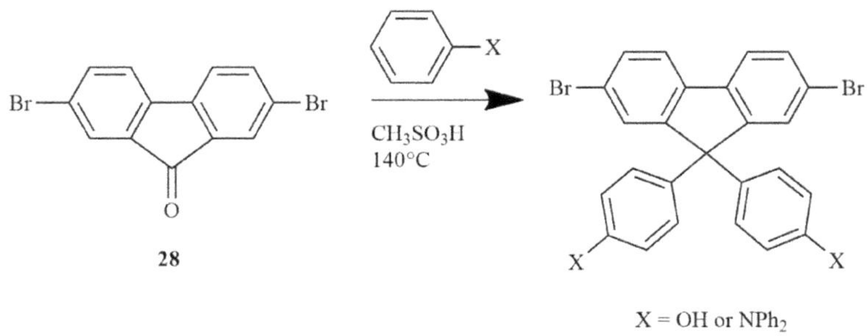

28

Figure 6.9 Addition of activated arenes to fluorenone

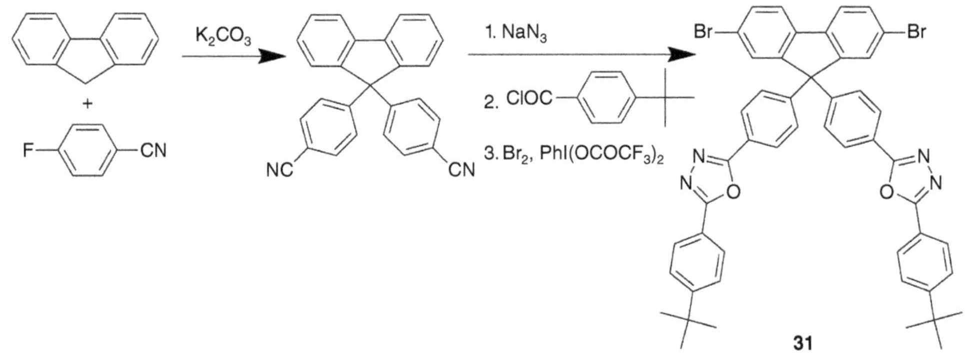

31

Figure 6.10 Synthesis of a fluorene with oxadiazole substituents

only a proportion of charge transporting units so as to optimally combine the good charge transporting qualities of the arylfluorene and the high PL efficiency of the alkylfluorene units.

The aryl substituents in the polymers shown in Figure 6.8 inhibit chain packing, as illustrated by the absence of any observable signs of organisation in their films when studied by Atomic Force Microscopy (AFM) [435]. Unlike PDAFs, they do not possess liquid crystalline phases, so that polarised emission cannot be obtained from them.

There exists one example of an aryl group being directly substituted onto the fluorene C9 position. As shown in Figure 6.10, this involves nucleophilic substitution of 4-fluorobenzonitrile with the fluorenyl anion, followed by conversion of the nitriles to oxadiazoles via tetrazole intermediates to make the monomer **31** with two electron-transporting aryloxadiazole groups at the bridgehead [436].

This monomer has been used to make the alternating copolymer **32** (Figure 6.11) whose EL efficiency in a single-layer device is considerably higher than for the PDAF homopolymer **9** (0.52% vs 0.2%) [436]. Even higher EL efficiency (1.21%)

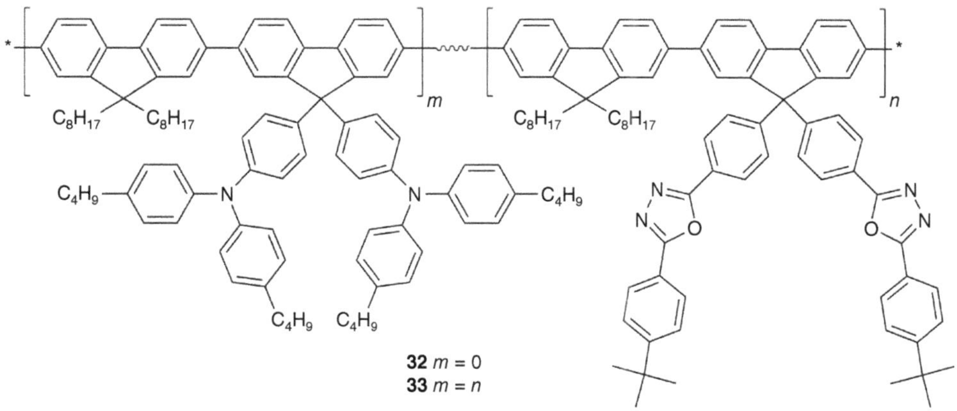

Figure 6.11 Fluorene copolymers with electron-accepting units at C9

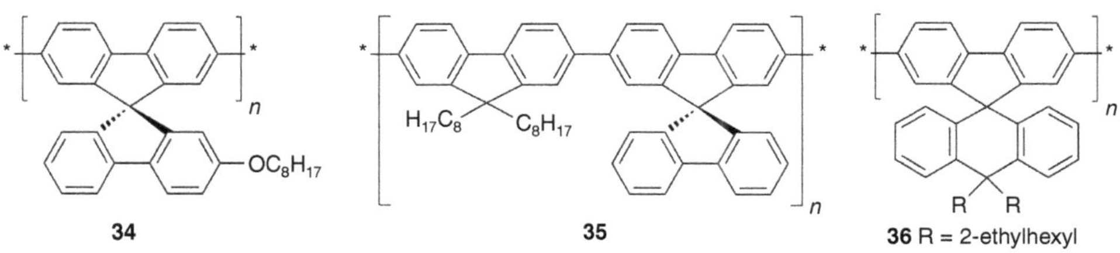

Figure 6.12 Polymers containing spirobifluorene and related units

has been obtained from the ternary copolymer **33** bearing both hole- and electron-accepting groups, due to it having better charge injection and transport of both types of charges [437]. As has been mentioned before, a balanced injection and mobility of charges is necessary for obtaining optimal device efficiencies and it is probable that, by varying the relative number and nature of the charge transporting units in such copolymers, even better efficiencies might be attainable. The long-term colour stability of these materials has not been reported, but they are likely to be more stable than the PDAF homopolymers and it should be possible to further enhance the stability without losing efficiency by replacing the dialkylfluorenes with diarylfluorene units.

Spirobifluorenes have been found to be promising materials for use in blue-emitting OLEDs [438]. They have an obvious structural affinity to diarylfluorenes, but are more crystalline and less soluble than diarylfluorenes, which may affect the solubility of polymers containing them. Soluble spirobifluorene-containing polymers (Figure 6.12) thus either contain solubilising side groups, as in **34** [439], or are copolymers such as **35** [440]. While the emission from **34** is reported to be stable, that from **35**, while more stable than that from the fluorene homopolymer, still becomes green upon heating in air at 150 °C. Polymers with

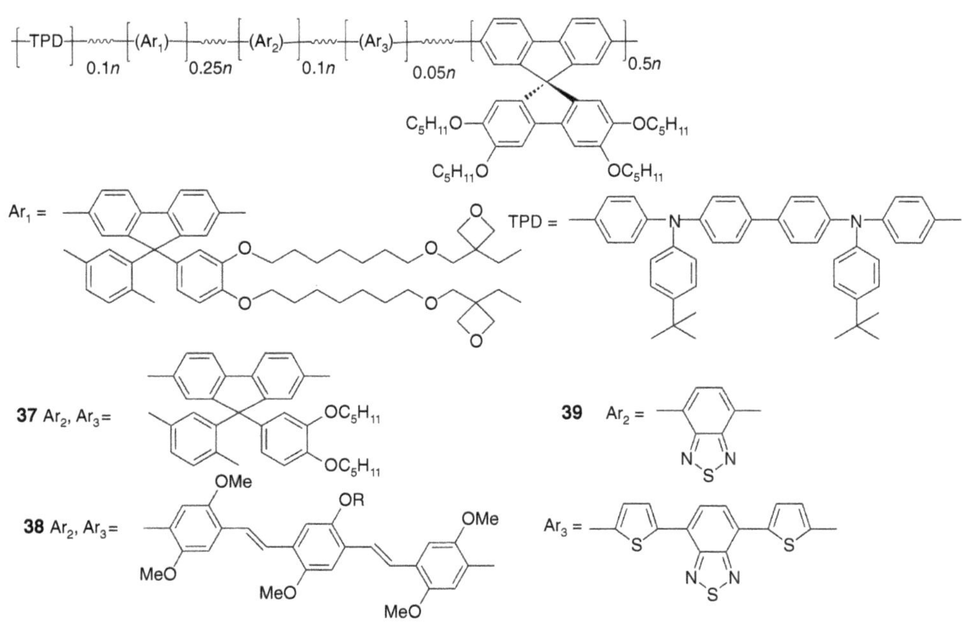

Figure 6.13 Cross-linkable blue-emitting copolymer containing spirobifluorenes

spirocycloalkyl groups also produce green emission upon annealing [441], whereas the polymer **36** with aryl groups at C9 displays stable blue PL and EL [442], again showing the greater stability of 9-aryl as compared to 9-alkyl fluorenes towards oxidation.

Suzuki cross-coupling of a spirobifluorene bisboronate with various dibromo-comonomers has been used to prepare cross-linkable copolymers **37–39** (Figure 6.13) which, respectively, produce efficient and stable blue (λ_{max} = 457 nm, 3 cd A^{-1}), green (λ_{max} = 507 nm, 7 cd A^{-1}), and red EL (λ_{max} = 650 nm, 1.1 cd A^{-1}) [443]. These are cross-linked by illumination with ultraviolet light in the presence of a photoacid to induce cationic polymerisation of the oxetane rings. Comparison of the EL properties of the cross-linked polymers with similar non-cross-linkable polymers showed that the cross-linking had no significant effect upon the luminescence of the materials but did enhance the device efficiency and device stability.

Polyfluorenes have not been used for other electronic applications as their large bandgaps (like those of PPPs) make them unsuitable for OPVs, while the charge carrier mobility of **9** has been variously measured as 4×10^{-4} cm^2 Vs^{-1} by Time-of-Flight technique [444] and as 2×10^{-4} cm^2 Vs^{-1} in an FET [445], which is too low for any plausible application. This is much lower than is seen for polymers such as polythiophenes (see Chapter 7), but the low HOMO level of PFs also makes them in principle less susceptible to aerial oxidation than polythiophenes.

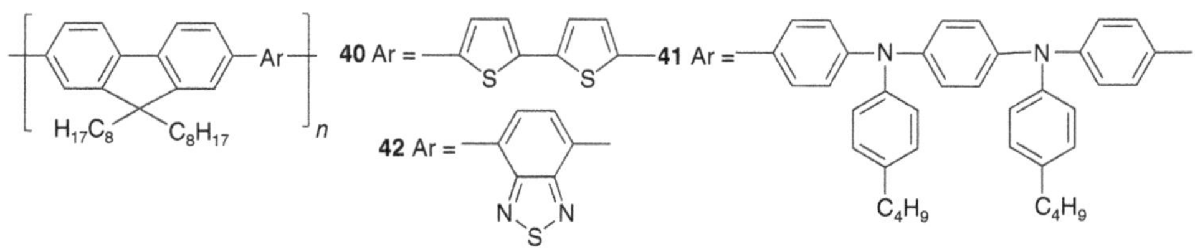

Figure 6.14 Fluorene copolymers for FETs and OPVs

One naively obvious way to combine together the best features of the two polymer classes is to make copolymers, and indeed the fluorene-bithiophene copolymer **40** (Figure 6.14) has proven a useful FET material. The incorporation of the thiophene units reduces the bandgap so that this material is a yellow-green emitter ($\lambda_{max} = 567$ nm) and it shows good packing and liquid crystalline behaviour which enables the chains in the polymer film to be aligned by rapid cooling from the mesophase. This alignment assists efficient charge carrier movement so mobilities of up to 0.02 $cm^2 Vs^{-1}$ have been obtained in FETs using aligned **40** [446]. It has been used as the active unit in a solution processed all organic FET in which the electrodes made of the conductive polythiophene PEDOT:PSS (see Chapter 7) were ink-jet printed while the other layers were deposited by spin coating [447]. Theoretical modelling has shown that the chain packing occurs with the bithiophene units stacking on top of the fluorene units, which is perhaps not intuitively the most obvious arrangement [448]. As the two units in this case are the same size, inter-chain packing occurs efficiently in **40**, but other copolymers with different sized units should not be able to pack so well. This result has obvious implications for the design of copolymers for FET or OPV applications where efficient packing may be an important criterion for good device performance.

Other fluorene copolymers have been studied as materials for OPVs. Here the fluorene usually acts as a donor-unit in a donor–acceptor copolymer. Blends of the copolymers **41** and **42** in which the comonomers are, respectively, hole-accepting and electron-accepting units have been the subject of particularly intense study. This research has not been driven by their being notably efficient producers of photocurrent but because fundamental insights into the nature of the photocharge generation processes in bulk heterojunction systems have been obtained from such studies.

Despite its relatively low power conversion efficiency, the poly(9,9′-dioctyl-fluorene-co-bis-N,N′-(4,butylphenyl)-bis-N,N′-phenyl-1,4-phenylene-diamine) (PFB) and poly(9,9′-dioctylfluorene-co-benzothiadiazole) (F8BT, **42**) polymer blend system was used for many years as a benchmark OPV system. Indeed, as the archetypal polymer:polymer blend system, the structure function relationship in PFB:F8BT OPV devices has attracted enduring interest over the years, with many researchers trying to determine the origin of the performance limitations. Early results found that spin-coating from chloroform produced more uniform films and more efficient devices than films spin-coated from xylene [449]. In particular, the films spin-coated from chloroform were found to exhibit phase-segregation features on a scale of tens of nanometres, whereas the films spin-coated from xylene were found to exhibit phase-segregation features of the order of tens of micrometres. It would be expected, given an exciton diffusion length of a few nanometres in polyfluorenes [450], that, for films where the length scale of the phase-separation appears to vary over orders of magnitude, device efficiency should vary by a similar magnitude [451]. However, despite the length scale of phase-separation varying over orders of magnitude, the device efficiency varied between highest and lowest by only a factor of two to three.

In order to probe the structure–function relationship more deeply, direct correlations between photocurrent generation and active layer morphology were obtained using near-field scanning photocurrent microscopy (NSPM). This technique is an adaptation of the near-field scanning optical microscope that allows both photocurrent and surface morphology to be mapped simultaneously with nanometre resolution [452].

NSPM measurements revealed that current is generated within the bulk and not (as might have been expected) at the boundaries of the micron-sized phase-segregated features, with the PFB-rich regions producing four times as much current as the F8BT-rich regions for illumination through a semi-transparent cathode. Scanning transmission X-ray microscopy (STXM) studies have been used to map the chemical structure of conjugated polymer blend films with a spatial resolution better than 50 nm. Using blends of poly(9,9-dioctylfluorene-co-N-(4-butylphenyl) diphenylamine) (TFB; an analogue of PFB) and F8BT, it was shown that, while there is little compositional variation in TFB domains, significant F8BT enrichment near the domain interface is observed in the F8BT domains [453]. STXM was also used to map the composition of different blend ratios; revealing that, for 1:1 blends, the PFB-rich domain is more inter-mixed than the F8BT domain. However, the minority phase of the 1:5 and 5:1 PFB:F8BT blend ratios is not pure, with up to 50% of each polymer present in the phase [454].

Subsequent STXM studies of this system focused on probing the nanoscale segregation in PFB:F8BT blends by using chloroform instead of xylene as the solvent. For the as-spun films, the STXM images reveal domain structures of dimension 85 nm, with photoluminescence quenching experiments indicating that these structures are intimately mixed. By contrast, upon annealing phase segregation occurs and pure domains appear within the structures [455]. It is also possible to control the phase segregation in PFB:F8BT films by varying the time in a solvent saturated atmosphere. In particular, the evolution of a secondary domain structure within both the PFB and F8BT-rich domains was observed using STXM mapping. Increased drying time resulted in lower OPV efficiencies; consistent with the increased domain purity observed in the STXM images [456]. The vertical phase structure has also been the subject of a number of studies. X-ray spectromicroscopy has been used to image the surface and bulk phase segregation simultaneously; revealing the presence of F8BT-rich phases penetrating through the film [457], with further confirmation provided by recent Raman microscopy results [458]. These phases provide a potential percolation pathway for charges. These results not only illustrate how processing conditions can profoundly influence the performance of polymers in OPVs, but also that the composition of phases inside the blends are more varied and the origins of the photocurrent are more complicated than might have been expected from the simple picture usually painted of such blends. Indeed, as we will see in Chapters 7 and 8, the issue of the structure–function relationships in organic electronic devices remains, in many cases, an open issue.

If one were to replace the carbon bridge in fluorene with a heteroatom which could not be oxidised to form a ketone, but whose incorporation did not change the electronic properties so as to affect the emission colour, then logically one would expect the resulting materials to display stable blue emission. This concept has stimulated research into polymers based on heterofluorenes such as carbazole or silafluorene.

6.2 Polycarbazoles and Other Heterofluorene-Based Polymers

Carbazole is isoelectronic with fluorene, but the nitrogen bridge induces two important differences in its properties. One is that carbazole is a good hole acceptor as the nitrogen pushes up the HOMO energy compared to fluorene, so that carbazole-based molecules have been widely investigated as hole-accepting or transporting materials. The other is to change the favoured sites for electrophilic substitution from the 2- and 7-positions in fluorene to the 3- and 6-positions, followed by the 1- and 5-positions if the 3,6-positions are blocked, in carbazole. As a result, whereas polymers prepared from fluorene are 2,7-linked, only poly(3,6-carbazole)s (Figure 6.15, **43**) or poly(1,8-carbazole)s (**44**) [459] are available by routes starting from carbazole, while the monomers for poly(2,7-carbazole)s (**45**) have to be synthesised in other ways [460–462].

Short, efficient routes to 2,7-dihalocarbazoles **46** have been developed (Figure 6.16) in which the key step is the reduction of a nitrobiphenyl to form the carbazole [463, 464]. This reductive ring-closure was originally achieved by a Cadogan reduction using triethyl phosphite [465], but more recently the less toxic triphenylphosphine has been found to be a better reagent [466]. The corresponding

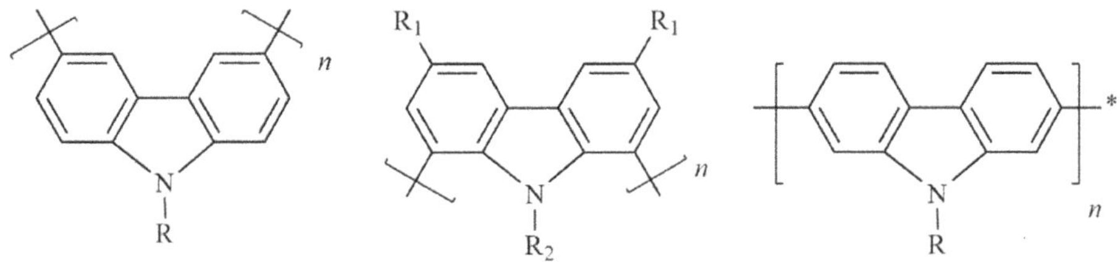

43 **44** **45**

Figure 6.15 Carbazole homopolymers and copolymers

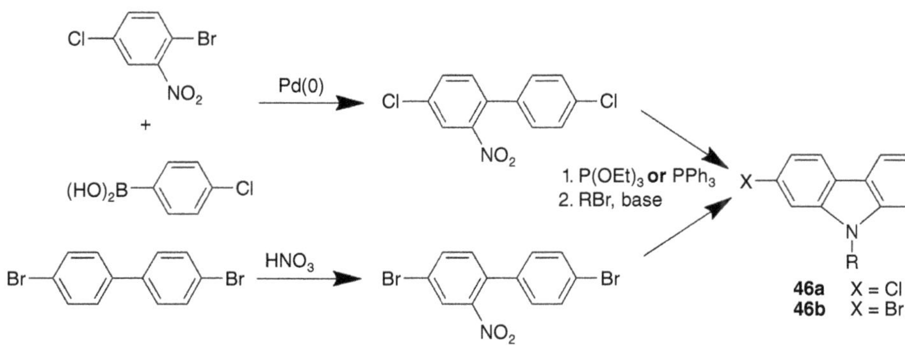

Figure 6.16 Routes to 2,7-dihalocarbazoles

homopolymers **45** are readily obtained by Yamamoto polycondensations of **46**, while the dibromide **46b** can be used in Suzuki polycouplings to make copolymers [463, 467–470]. The homopolymers **45** have only one solubilising alkyl chain per repeat units, so they are less soluble than PDAFs which possess two such groups per monomer unit; long branched alkyl chains are thus needed to obtain high molecular mass polymers [468, 470].

Polymers of 3,6-carbazoles (**43**) and 1,8-carbazoles (**44**) display blue or blue-green emission [459, 471], suggesting effective conjugation occurs through the nitrogens. The 2,7-carbazole polymers **45** produce blue emission similar to that of PDAFs in the solid state but, unlike PDAFs, no green emission appears after annealing or upon operating an LED [463, 467]. Stable blue emission is also seen from alternating copolymers **47** of 2,7-carbazole with fluorene [467]. However, the EL efficiencies for devices using these polymers are modest [472]. The emission colour can be varied by copolymerising with other units, for example, copolymers **48** and **49**, respectively, produce green ($\lambda_{max} = 500$ nm)[463] and red ($\lambda_{max} = 671$ nm) emission [467].

One potential problem with 2,7-carbazole-based materials is that the high electron densities at the 3- and 6-positions promote electrophilic reactions there. Electrochemical measurements show that this makes these materials unstable under electrolytic conditions [468]. The lifetime of devices using them is therefore likely to

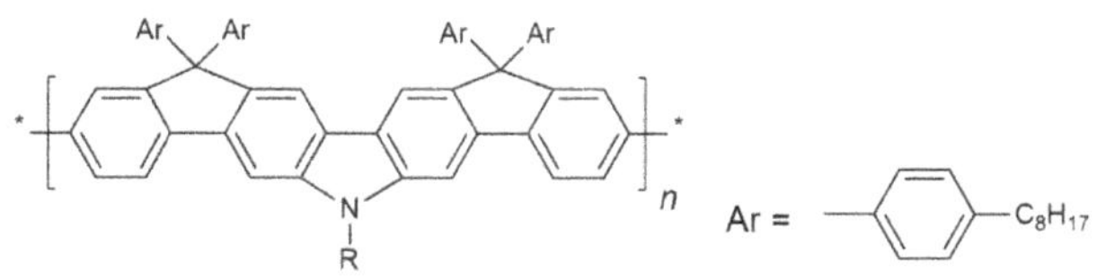

50 R = 2-ethylhexyl

Figure 6.17 Poly(diindenocarbazole)

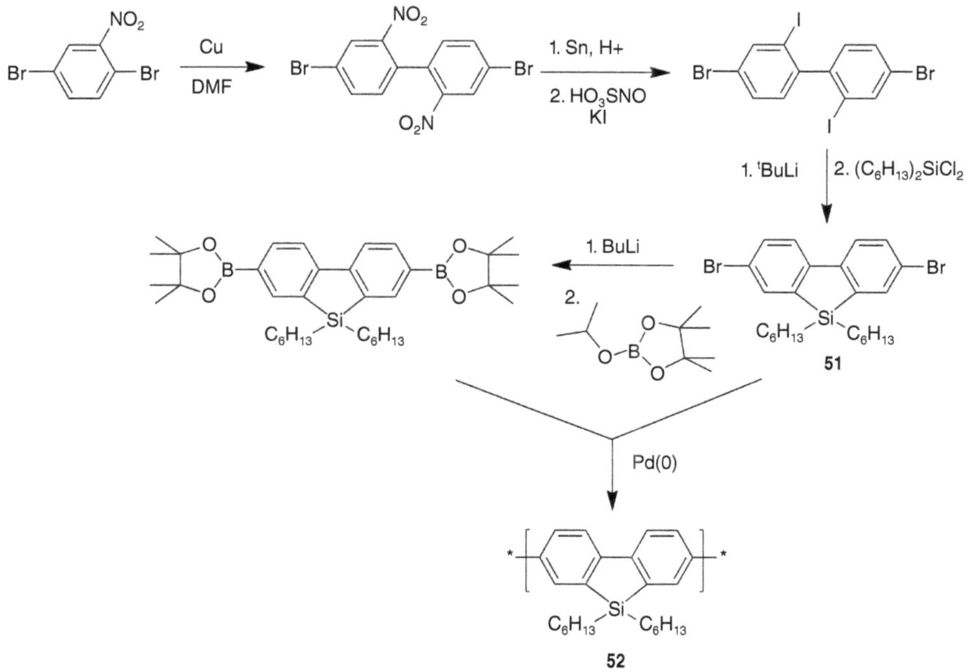

Figure 6.18 Synthesis of a poly(2,7-silafluorene)

be short. One way around this is to incorporate carbazole units into a ladder or stepladder structure in which the 3- and 6-positions are substituted by carbon or nitrogen bridging atoms. Ladder-type polymers containing carbazole units have been prepared. The fully ladderised materials display blue-green emission [473, 474], while blue emission has been obtained from a polydiindenocarbazole **50** (Figure 6.17), as will be discussed in more detail in Section 6.3 [475].

Silicon has similar electronic properties to carbon but does not undergo oxidation to form a silicone. 2,7-Dihalosilafluorenes **51** have been made by the route shown in Figure 6.18 and used to prepare a poly(2,7-silafluorene) **52** which displays similar EL emission ($\lambda_{max} = 425$ nm) to the analogous polyfluorene **8**, but with no green emission band. No degradation was detected upon heating **51** in air at 200°C for 16 h, whereas strong green emission is seen from **8** after only 4 h under the same conditions [476].

Carbazole-based polymers have become the subject of some interest recently as materials for OPVs [462]. A poly(2,7-carbazole) **45** (R = 2-decyltetradecyl) when blended with the fullerene PCBM showed only 0.07% efficiency, but the efficiency increased significantly to 0.6% when it was combined with a perylene dye acceptor (0.6%), with much superior absorption in the visible than PCBM [470]. The efficiency was later further increased to 0.75% by use of a vinazene as acceptor [477]. While these values were not terribly high, they were at the time of writing the best values yet reported for a polymer blended with a non-fullerene molecular acceptor. Because of their large bandgaps carbazole homopolymers are unlikely to produce efficiencies much above 1% in BHJ devices, but carbazole's good electron donating properties have made it an attractive unit for making low bandgap donor–acceptor copolymers, which we shall discuss in Chapter 8.

6.3 Other Stepladder-Type Polyphenylenes

Phenanthrene is a biphenyl with an ethene bridge. Both poly(2,7-phenanthrylene)s **52** and the 3,6-phenanthrene analogues **53** have been made from the diacylbiphenyl **54** (Figure 6.19) [478]. The 2,7-polymers are obtained by polymer-analogous McMurry coupling of the precursor poly(2-acyl-*p*-phenylene) s **55**. This process appears to proceed quantitatively as no residual ketone stretch can be seen in the IR spectrum. If one performs the McMurry coupling on **54** one obtains the phenanthrene **56** which undergoes bromination at the 3,6-positions. Thus, one can selectively obtain either isomer by changing the order of the reaction steps.

Whereas the 2,7-polymers **52** are analogues of PPP, the 3,6-polymers **53** are analogues of PPV, but, despite their major differences in conjugation pathways and chain conformations, these materials show similar optical properties, with blue PL and EL (λ_{max} = 400–425 nm) being obtained from both of them. The stability of the emission has not been reported but, in view of the known ease of oxidation of the phenanthrene 9- and 10-positions, their long-term stability must be open to some doubt.

In order to obtain pure blue emission from bridged phenylene structures one has to use a structure intermediate between the deep-blue emitting PDAFs and the blue-green emitting LPPPs. The first such structures to be investigated were the poly(tetraalkylindenofluorene)s (PIFs) which have monomers containing three bridged phenylenes [479]. Their PL emission, while still not a true blue, is slightly red-shifted compared to PDAFs with maxima around 430 nm. The phase behaviour of PIFs **57** and **58** (Figure 6.20) resembles that of the corresponding

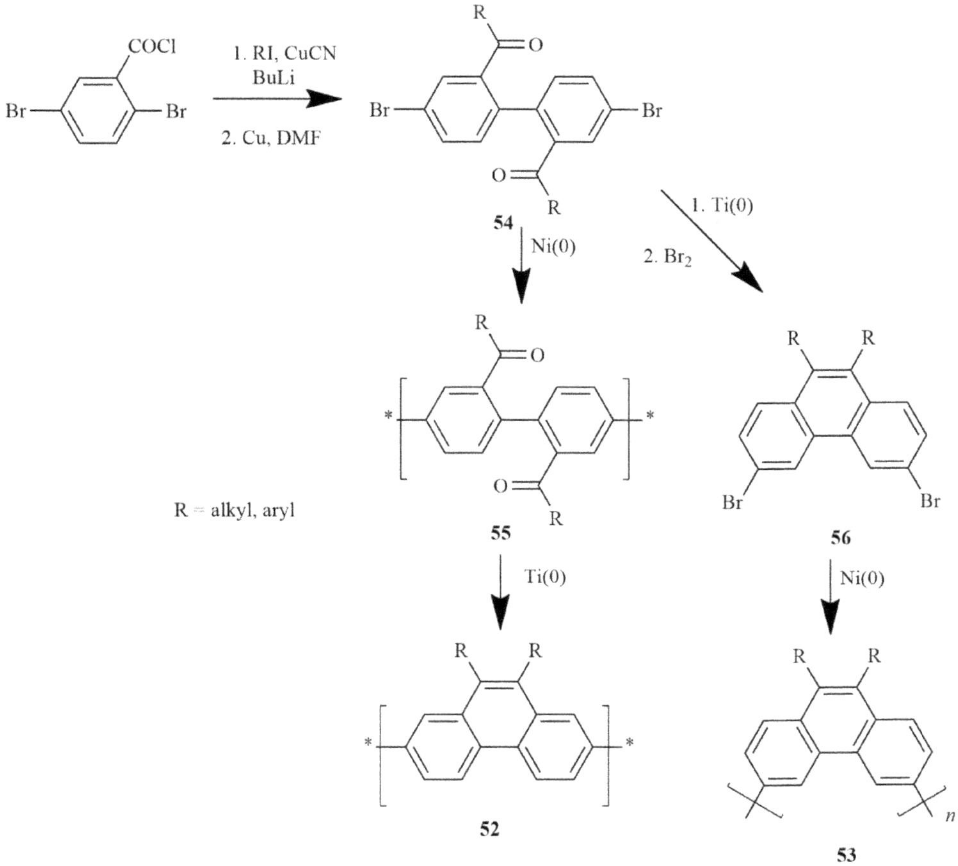

Figure 6.19 Synthesis of poly(2,7-phenanthrylene)s and poly(3,6-phenanthrylene)s

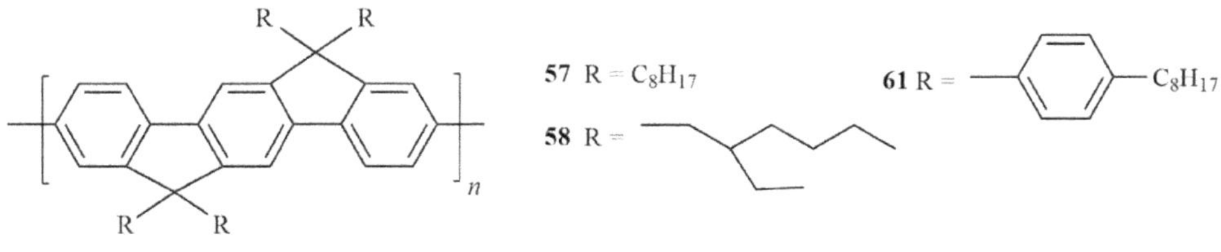

Figure 6.20 Polyindenofluorenes

PDAFs **9** and **10**, but with higher phase transition temperatures, so obtaining polarised emission is also feasible.

These polymers display similar problems with colour stability to those observed with PDAFs, with a strong green emission band around 560 nm being produced upon operation of an LED [480]. As with PDAFs the origin of this emission was originally suspected to be excimers, but was later shown to arise from ketone

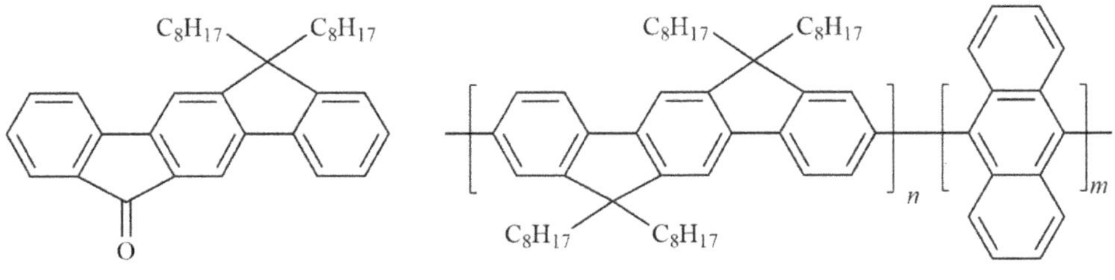

59 **60**

Figure 6.21 Model compound for defect and a copolymer which produces stable blue emission

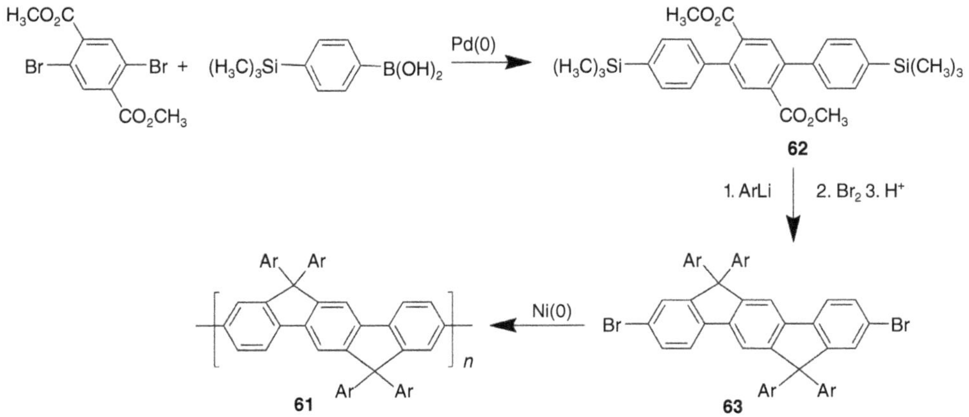

Figure 6.22 Route to poly(tetraarylindenofluorene)s

defects, as confirmed by synthesis of a model ketone **59** whose emission properties closely matched those of the green emission band [481]. Some of the approaches used to obtain stable blue emission from PDAFs have been successfully applied to PIFs. For example, a copolymer of PIF with anthracene **60** (Figure 6.21, n:m = 85:15) shows stable blue EL (λ_{max} = 445 nm), probably due to exciton confinement [482].

As aryl-substituted polyfluorenes have been found to be more stable than the alkyl-substituted polymers, it is not unexpected that the poly (tetraarylindenofluorene) **61** (Figure 6.20) displays stable blue EL [427, 483]. As shown in Figure 6.22, the key step in the synthesis is a Suzuki coupling of a dibromoterephthalate with 4-trimethylsilylbenzeneboronic acid to give a terphenyl **62**. Treatment of this with four equivalents of an aryllithium followed by electrophilic displacement of the silyl group with bromine and then ring-closure produced the tetraarylindenofluorene monomers **63**, which were then polymerised with nickel(0).

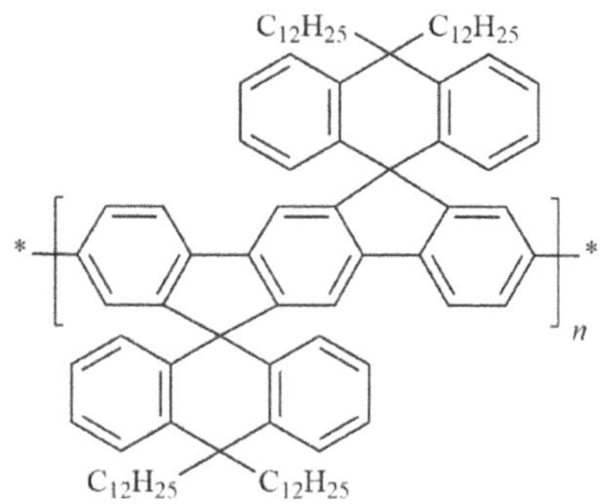

64

Figure 6.23 Polyindenofluorene with double spirolinkages

The octylphenyl polymer **61** produces blue emission ($\lambda_{max}=437$ nm) with no sign of the green emission band. Some unoptimised LEDs have been constructed using **61**, which produce stable blue emission, though with only moderate device efficiency [483]. Further device optimisation, for example, by using charge transporting layers, is still required to get high efficiency. Blending 0.3 wt% of the ketone **60** into **61** has been found to totally suppress the blue emission, demonstrating how even very low levels of defects can dominate the emission from conjugated polymers [481]. Stable blue emission ($\lambda_{max}=445$ nm) has also been obtained from polymer **64** (Figure 6.23) with spirolinkages at both bridgeheads [484].

The emission from PIFs is still violet-blue, rather than a pure blue and so, to obtain a purer blue emission colour, a poly(ladder-type tetraphenylene) **65** has been made by a route involving Suzuki coupling of the fluorenebisboronate **66** to the bromobenzoate **67** followed by addition of aryllithium and ring closure (Figure 6.24) [475]. This polymer produces pure blue PL ($\lambda_{max}=442$ nm), but the EL spectrum exhibits an extra band in the green at 510 nm. This does not appear to be due to defects from oxidation, as heating **65** in air produces a band at 575 nm but is thought to arise from an interaction between the polymer and the metal cathode, which might be controlled by introduction of a protective inter-layer. This is a useful reminder that excimers and oxidation are not the only source of extra emission bands and that chemical interactions between electrodes and active materials in organic electronic devices have significant effects upon their optical and electrical properties. It is likely that the band at 575 nm would also appear upon prolonged running of an LED. A fully arylated polymer **68** has been prepared by the same route [485]. EL measurements have yet to be reported but it is likely to be more resistant towards oxidation – whether it would also produce the band at 510 nm has yet to be seen.

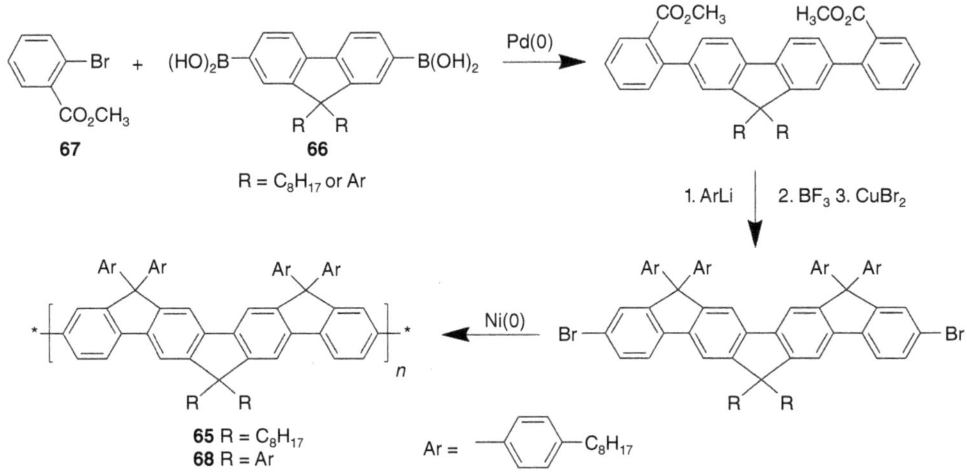

Figure 6.24 Synthesis of poly(ladder-type tetraphenylene)s

Substitution of a 2,7-carbazole bisboronate for a fluorene bisboronate in this synthesis produces the poly(diindenocarbazole) **50** (Figure 6.17) [475]. The PL emission from this polymer is slightly red-shifted compared to **65** ($\lambda_{max} = 445$ nm vs 442 nm) due to the higher HOMO (and thus lower bandgap) arising from the nitrogen bridge. Polymer **50** does not display an extra band at 510 nm in the EL spectrum, so that its EL is blue, nor does it produce a ketone emission band at 575 nm on heating in air. This material appears to be a very promising candidate for obtaining stable pure blue emission, though the efficiency of the unoptimised devices so far reported is only 0.1 cd A^{-1}.

The final gap in emission colours between PF and LPPP has been bridged by the preparation of ladder-type pentaphenylenes **69–70** using a route based on that for arylindenofluorenes but utilising a fluorene boronate instead of a phenylboronate in the Suzuki coupling (Figure 6.25) [486].

The emission from **69** is a pure blue with an emission maximum at 445 nm and a very small Stokes shift and well-resolved spectra very similar to those for LPPPs. Polymer **69** has a very high persistence length of 25 nm [342], which is actually greater than that measured for LPPPs, indicating the significant amount of defects that must be present in the ladder polymers. The blue EL emission from **69** is much more stable than from PDAFs such as **9**, but a long wavelength emission band has been found to appear rapidly upon heating it in air. By contrast the all-aryl polymer **70** is remarkably stable, with no green emission being observed even after prolonged heating in air [487]. This suggests the green emission band from **69** arises from oxidation of the alkylated bridgeheads, which might arise from the presence of some bridgehead hydrogens, due to incomplete alkylation.

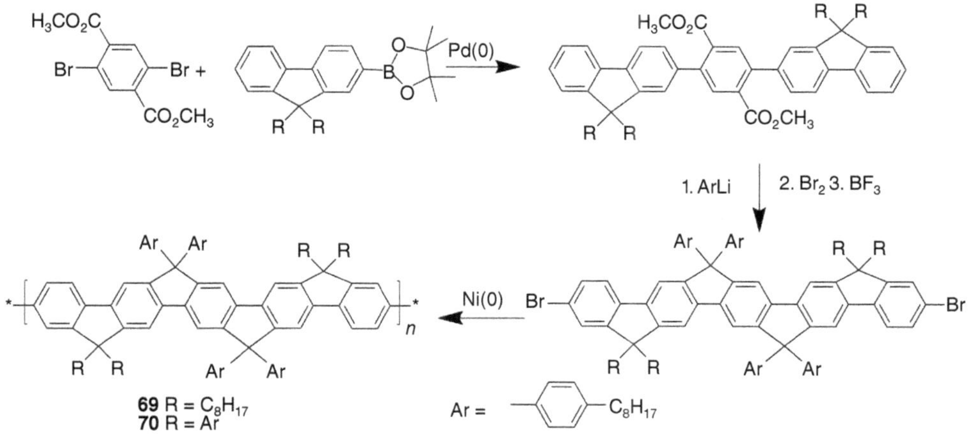

Figure 6.25 Synthesis of poly(ladder-type pentaphenylene)s

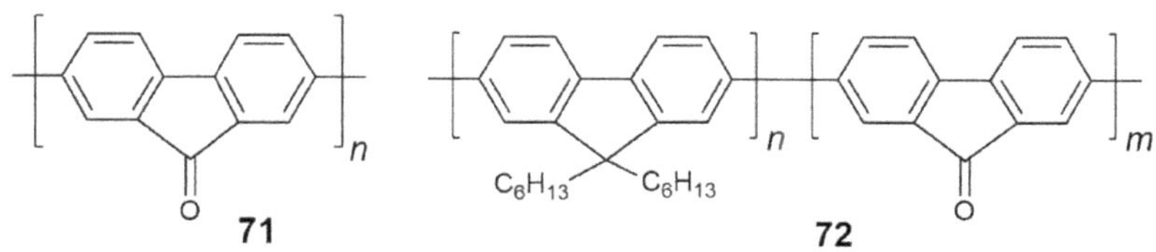

Figure 6.26 Fluorenone containing polymers with good electron injection

To date only unoptimised devices with moderate efficiencies have been reported for these polymers, but suitable incorporation of charge transporting units onto the monomer units should produce high efficiencies. It has been shown that triarylamine and oxadiazole charge-transporting substituents can be introduced as substituents onto the bridgeheads of ladder-type pentaphenylenes, but, when both groups were present on the same molecule, intra-molecular charge transfer occurred which produced undesirable red-shifts in and/or quenching of the emission [488]. A more promising approach would be to have electron and hole-transporting units on separate monomer units as in the polyfluorene **33** (Figure 6.11).

While the bridgehead ketones mentioned in Section 6.1 act as defects from the point of view of obtaining blue emission, they are conversely desirable structures to incorporate if one wants to improve electron accepting properties, that is, to make an n-type material. Polyfluorenone (Figure 6.26, **71**) is insoluble but films of it can be made via a soluble precursor [489] and have been found to enhance electron injection into PLEDs [490]. Unfortunately, the precursor conversion process produces acid residues which can dope the materials and/or corrode the electrodes, which make it of dubious utility for real applications due to the need to remove the residues. Copolymers **72** containing 3 mol% of fluorenone have been

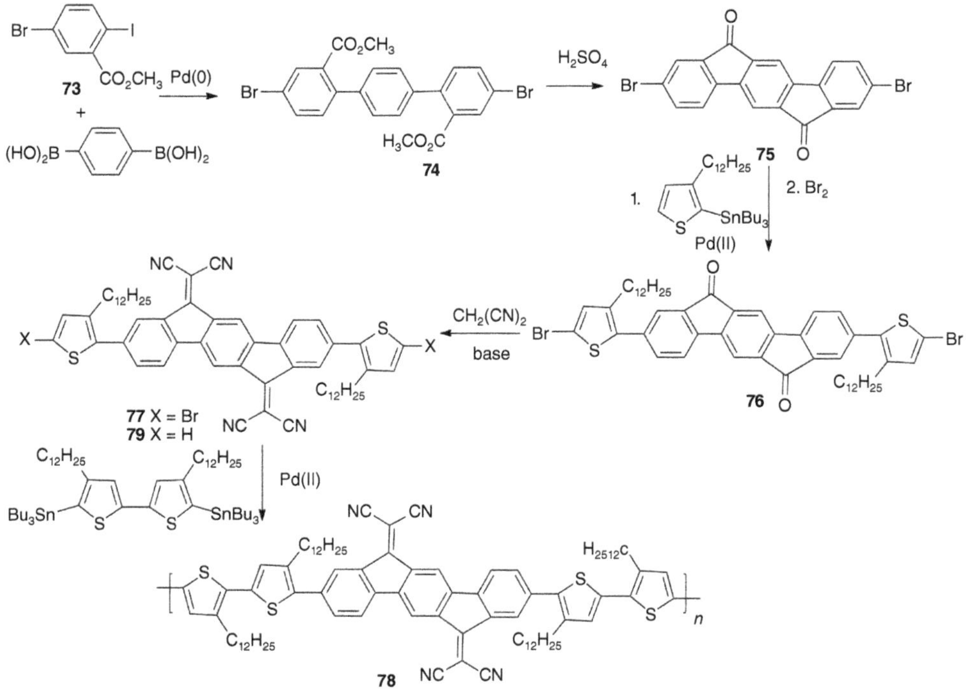

Figure 6.27 Synthesis of n-type indenofluorenedione-based materials

used in moderately efficient green-emitting PLEDs, in which the fluorenone units are thought to enhance the electron injection efficiency [491].

Polymers containing electron accepting indenofluorenedione units have been made as shown in Figure 6.27 [492]. Suzuki coupling of benzene-1,4-bisboronate with ester **73** occurs preferentially with the more reactive iodide to give the diester **74**. Acid-catalysed acylation then produces the diketone **75**. This is virtually insoluble but can undergo Stille coupling to produce the soluble dithienyl adduct **76**, which undergoes Knoevenagel condensation with malononitrile to produce the nitrile **77**. Further Stille coupling gave the low bandgap polymer **78** which displays ambipolar behaviour in air-stable transistors with equally high hole and electron mobilities ($\mu = 2 \times 10^{-4}$ cm^2 Vs^{-1}). The dehalogenated compound **79** displays high electron mobility ($\mu = 0.1$–0.16 cm^2 Vs^{-1}) in air-stable n-type transistors, suggesting that high mobilities should be obtainable from polymers containing it if suitable comonomers are chosen. The marked difference in mobilities between **78** and **79** probably reflects differences in packing affecting the charge transport. The air stability of **78** is somewhat unusual for a low bandgap polymer and is due to its relatively low HOMO energy of –5.51 eV.

A major disadvantage of the diketone **75** as an electron accepting unit for polymers is its very low solubility. To produce a more soluble electron accepting

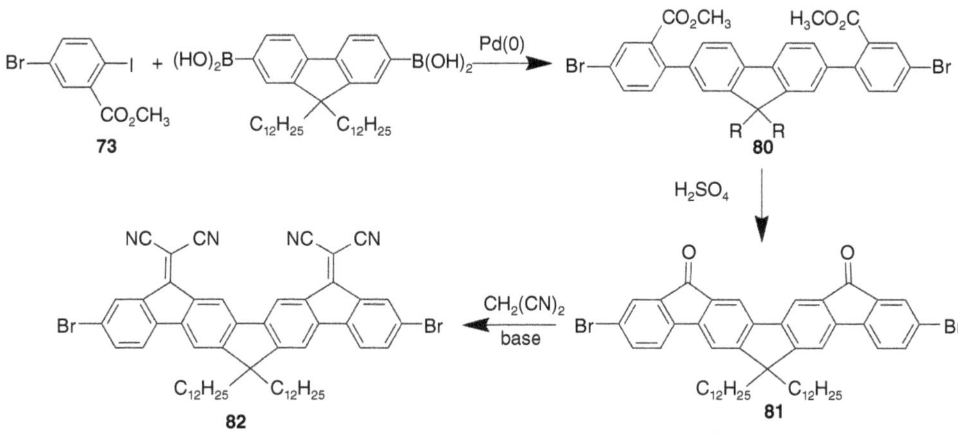

Figure 6.28 Synthesis of electron-accepting ladder-type tetraphenylenes

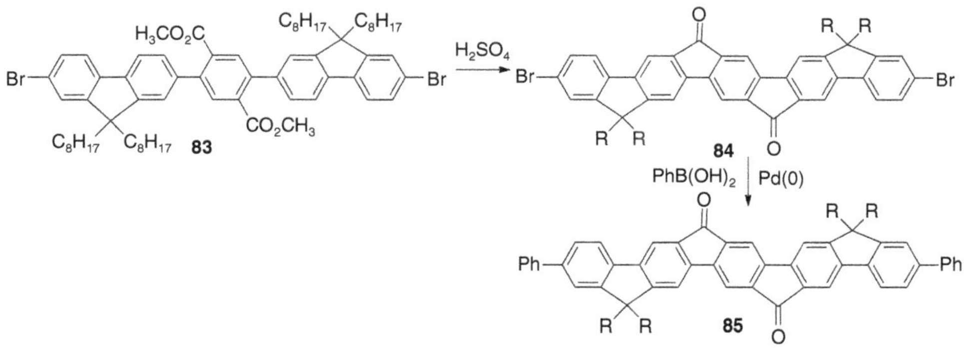

Figure 6.29 Synthesis of n-type ladder-type pentaphenylenes

unit, a variation of the synthesis of ladder-type tetraphenylenes (Figure 6.24) has been used as shown in Figure 6.28. Suzuki coupling of a fluorene-bisboronate with ester **73** occurs preferentially with the more reactive iodide to give the key intermediate **80**. This is then converted by the same methods as in Figure 6.27 into the diketone **81** and the nitrile **82** which both display good solubility in organic solvents [493].

The synthesis of ladder-type pentaphenylenes (Figure 6.25) can also be varied in a similar way to produce n-type materials. Thus, as shown in Figure 6.29, acid-mediated ring closure of the pentaphenylene **83** produces the diketone **84** in good yield [486]. In this case it was specifically noted that the reaction did not work in the absence of halogen substituents. This need for halogen substituents has not been reported for the similar ring-closures shown in Figure 6.25 and may be a specific feature of this system. The diphenyl compound **85** was estimated by cyclic voltammetry (CV) to have a LUMO energy of –3.53 eV, suggesting materials

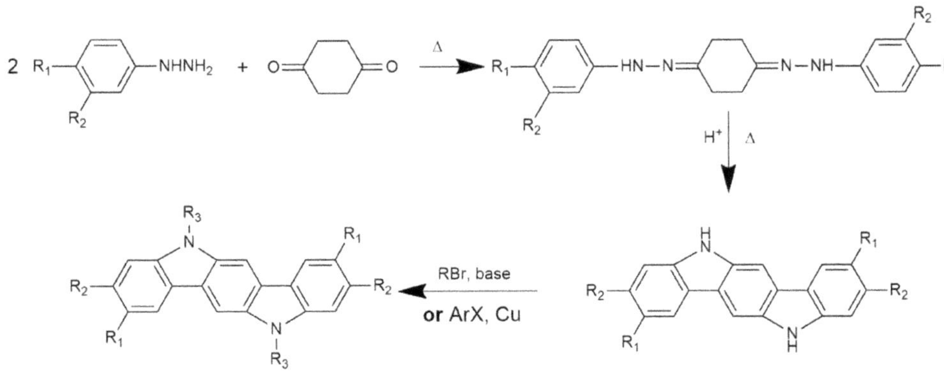

86a R_1, R_2 = H, R_3 = alkyl
86b R_1 = Cl, R_2 = H, R_3 = alkyl
86c R_1 = H, R_2 = Cl, R_3 = alkyl
86d R_1, R_2 = H, R_3 = 4-octylphenyl

Figure 6.30 Fischer indolisation route to indolo[3,2-b]carbazoles

containing this unit might be good electron acceptors in solar cells. The low HOMO energy of –5.71 eV for **85** suggest such materials are likely to be air-stable.

While incorporating ketone bridgeheads into bridged phenylenes serves to enhance their electron accepting and transporting properties, nitrogen bridges enhance the injection of holes. As we have seen for carbazole versus fluorene this is made apparent by the rise in the HOMO energy level. Ladder-type oligophenylenes with all-nitrogen bridges can be thought of as ladder-type oligo-anilines which might well have high charge carrier mobilities (polyaniline is a well-known conducting polymer) and so be suitable materials for field-effect transistors. The synthesis of such materials with up to four nitrogen bridges has been reported [494]. A major problem in the synthesis of such materials is the formation of the nitrogen bridges, since Cadogan reductive ring closure of nitrophenylenes proceeds with poor yields and/or a lack of regioselectivity when applied to the synthesis of materials with two or more nitrogen bridges, and intramolecular Ullmann coupling or Buchwald-Hartwig amination also presents synthetic problems. This problem can be overcome for the smaller oligomers by using other routes. Thus, the N,N'-dialkylindolo[3,2-b]carbazoles **86** have been prepared by a double Fischer indolisation reaction (Figure 6.30) [495–497]. The *meta-* and *para-*linked polymers **87** and **88** (Figure 6.31) were then obtained by Yamamoto polycondensation of **86b** and **86c**, respectively [497]. Polymers **87** can also be made by oxidative polymerisation of **86a**.

These compounds have been studied as active materials in organic FETs which has revealed some interesting structure–property relationships. The dialkyl oligomer **86a** (R_3 = dodecyl) has a charge carrier mobility of only 1.2 × 10^{-4} $cm^2\ Vs^{-1}$, which is increased by substitution with chlorine. Substitution *para* to the

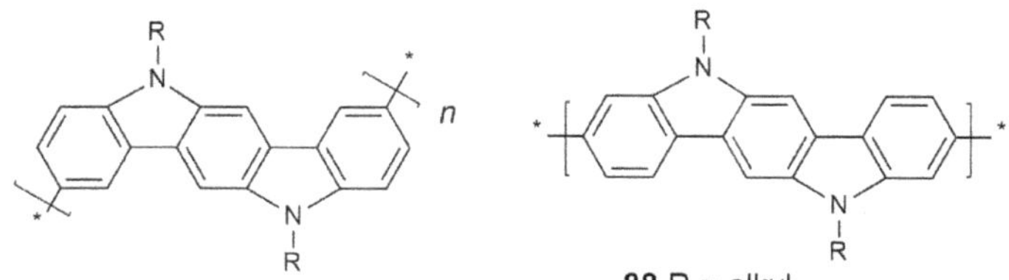

Figure 6.31 Poly(indolo[3,2-b]carbazole)s

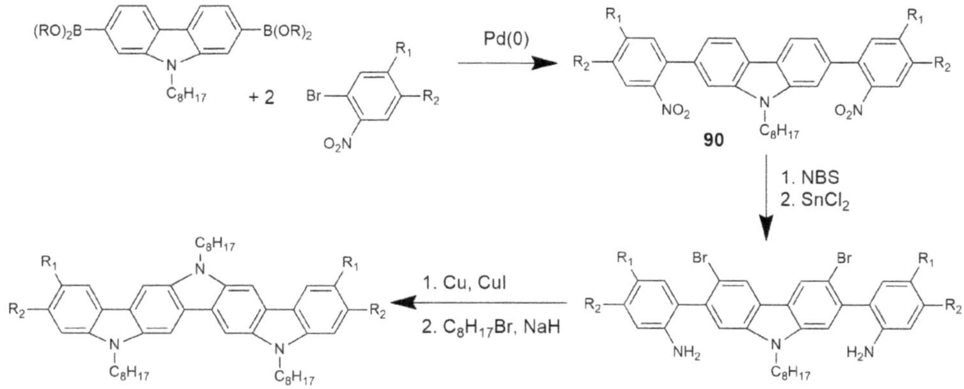

Figure 6.32 Synthesis of diindolocarbazoles

nitrogen as in **86b** leads to higher mobilities than *meta*-substitution as in **86c** (0.14 vs 0.01 $cm^2\ Vs^{-1}$) [496]. The N,N-diaryl oligomer **86d** also has an excellent mobility value of 0.12 $cm^2\ Vs^{-1}$ [495]. The mobilities obtained for the polymers depended not only upon the position of the linkages but also upon the method of polymerisation, with the highest mobility obtained being 0.02 $cm^2\ Vs^{-1}$ for **87** (R_3 = octyl) prepared by oxidative polymerisation [497]. The values for the polymers **87** made by the Yamamoto method were 2–3 orders of magnitude lower, while no FET activity could be detected from polymers **88**. This difference in mobility with polymerisation method may reflect possible doping during oxidative polymerisation. The conductivity of the *meta*-coupled polymer **87** (R_3 = 2-ethylhexyl) is also significantly higher than the *para*-coupled **88** (0.04 vs 0.0002 S cm^{-1} for oxidised films) [498].

As an example of the sorts of synthetic problems encountered for larger oligomers with multiple nitrogen bonds, the synthesis of diindolocarbazoles **89** had to be performed using an internal Ullmann coupling (Figure 6.32), since Cadogan reduction of the dinitro-compounds **90** gave a mixture of isomers [498].

Yamamoto polycondensation of **89a** and **89b** gives the polymers **91** and **92**, respectively (Figure 6.33) [498]. The conductivity values for these show the same

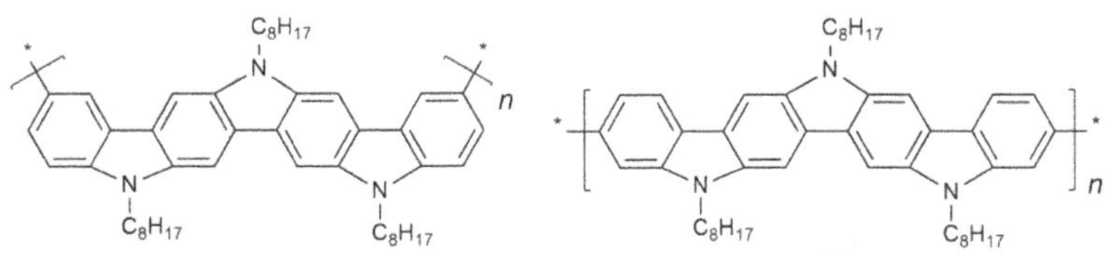

91 **92**

Figure 6.33 Poly(diindolocarbazole)s

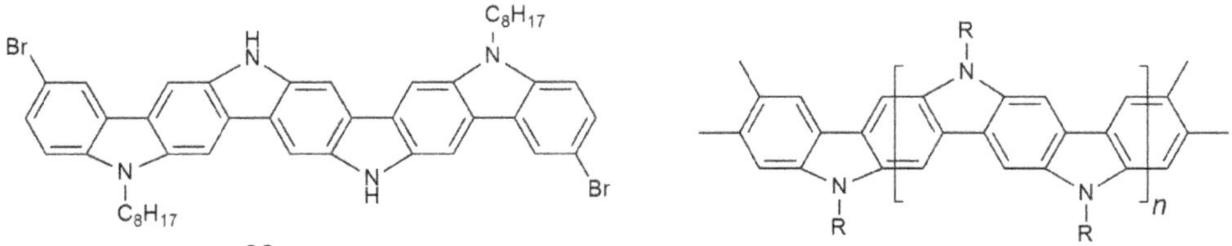

93

94 R = alkyl, aryl

Figure 6.34 A ladder-type (tetraaniline) and poly(aniline)

trends as for **87** and **88**, with the values for *meta*-linked polymer **91** being higher than for the *para*-linked **92**, in both the singly (0.04 vs 0.01 S cm^{-1}) and doubly oxidised states (0.5 vs 0.04 S cm^{-1}). These cases suggest that conjugation through the nitrogen bridges is desirable for obtaining high conductivity and high charge carrier mobility, presumably because more of the charges in the oxidised forms are to be found there. The difference in packing arising from conformational differences may also play a role here but theoretical calculations will be needed to determine the relative contributions of charge distribution and packing to the overall electrical properties.

The largest ladder-type oligo(p-aniline) yet made is the oligomer **93** (Figure 6.34) with four nitrogen bridges, which was made using an intra-molecular Ullmann coupling [494]. The optical bandgap of **93** is 2.42 eV, with a small Stokes shift consistent with the rigid structure, while the oxidation potential is 0.58 V, corresponding to a HOMO energy of –5.38 eV, so it should be stable towards aerial oxidation. Further materials will need to be made before reliable extrapolation to the properties of a fully nitrogen bridged polyphenylene **94** can be performed, but these numbers already suggest that such a material might have interesting optoelectronic properties. An alternative way to determine the properties of such a polymer is to extrapolate from ladder-type polyphenylenes containing a mixture of carbon and nitrogen bridges and so it is to these we now turn.

Poly(ladder-type hexaphenylene)s **95** and **96** in which the monomers contain, respectively, one and three nitrogen bridges have been prepared by Suzuki

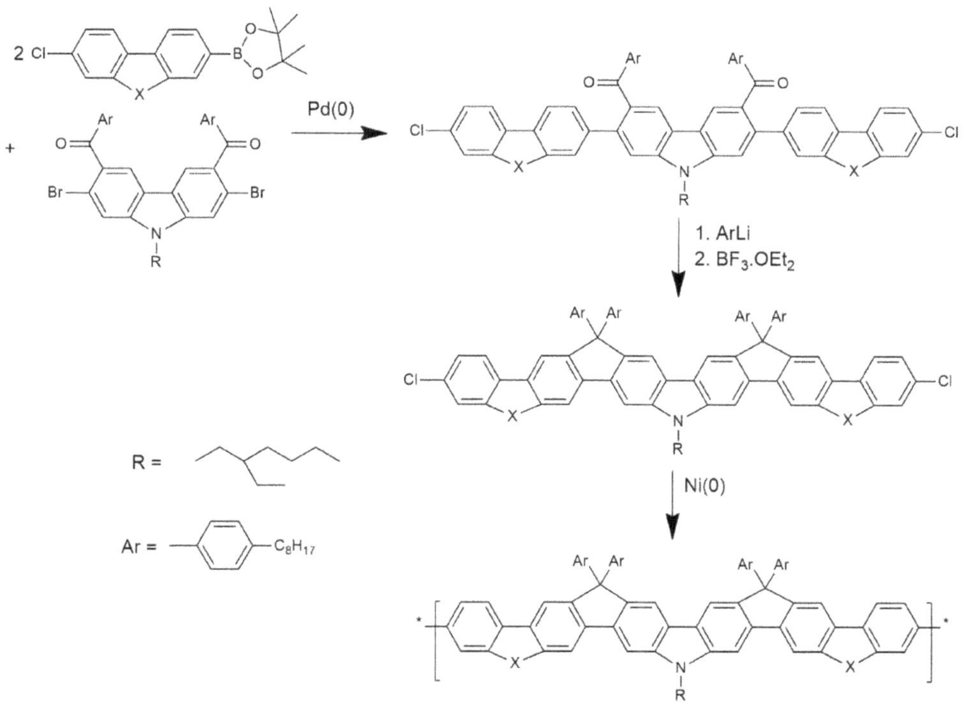

95 X = NR **96** X = $C(C_8H_{17})_2$

Figure 6.35 Synthesis of poly(ladder-type hexaphenylene)s with 1 and 3 nitrogen-bridges

coupling of a diacylcarbazole with fluorene and carbazole boronates, respectively, followed by ring closure, and Yamamoto polycondensation (Figure 6.35) [373, 374, 499]. As would be expected due to the greater number of electron-donating nitrogens, the emission from **95** is red-shifted compared to that from **96** (λ_{max} = 470 vs 454 nm), with the reduced bandgap arising from a higher HOMO energy level for **95** than **96** (–5.40 vs –5.46 eV). Due to their absorbing across more of the visible spectrum than the polycarbazole **45** mentioned, solar cells constructed using blends of these polymers with a fullerene display efficiencies of 0.74% and 0.56%, as compared to 0.07% for **45**. Even higher efficiencies have been achieved using a perylene dye as an acceptor due its better absorption in the visible spectrum compared to fullerene [462].

There has been some recent interest in making phenylenes bridged with other heteroelements (e.g. phosphorus), but this research is still in its infancy and to date no significant results have flowed from it. It is an area that may show much promise in the future. An area that has produced significant results has been heterocycle-based polymers of which thiophene-based polymers have been the most important and it is to these we will now turn.

7
Polythiophenes

Polythiophenes are by far the most widely studied class of heteroaromatic-based conjugated polymers. This is because polymers based on thiophene (and fused thiophenes) have been found to have exceptionally high charge carrier mobilities and to produce excellent performance as the active components in organic FETs and as electron donors in OPVs. As we shall see, they are also the source of some major advances in the development of synthetic methods for preparing regular conjugated polymers.

7.1 Polythiophene

Electropolymerisation of thiophene (**1**) produces insoluble but robust films of polythiophene (**2**) which show conductivity of up to 100 S cm^{-1} after doping [500]. However, this material is rather ill-defined and contains a mixture of 2,5- and 2,4-linkages (Figure 7.1). Polythiophene can be made containing only 2,5-linkages by oxidative or reductive chemical polymerisation of suitable 2,5-disubstituted thiophenes but these methods produce intractable powders from which films can be produced only by vacuum deposition of low molar mass materials – a method which is impractical for making films for potential commercial applications [501]. To date, there is no reported soluble precursor route for making polythiophene films, which remains a useful goal for future synthetic efforts in view of the high thermal stability and good electrical properties of this polymer. Attention has, therefore, concentrated on processable derivatives bearing solubilising groups – in particular poly(3-alkylthiophenes) (P3ATs).

7.2 Poly(3-alkylthiophene)s

P3ATs can be readily prepared by oxidative polymerisation of the monomers. The most commonly used oxidant is iron(III)chloride, as shown in Figure 7.2 [501, 502].

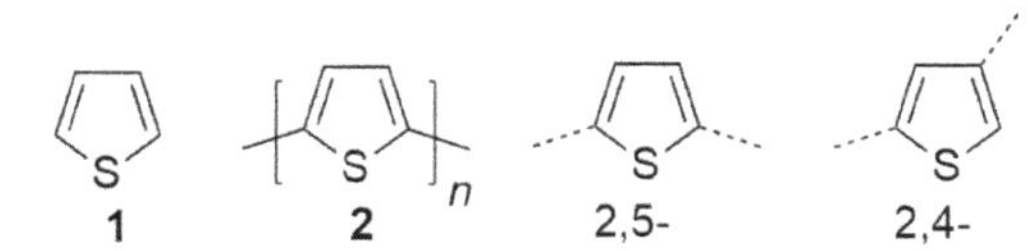

Figure 7.1 Regioisomers in electropolymerised thiophene

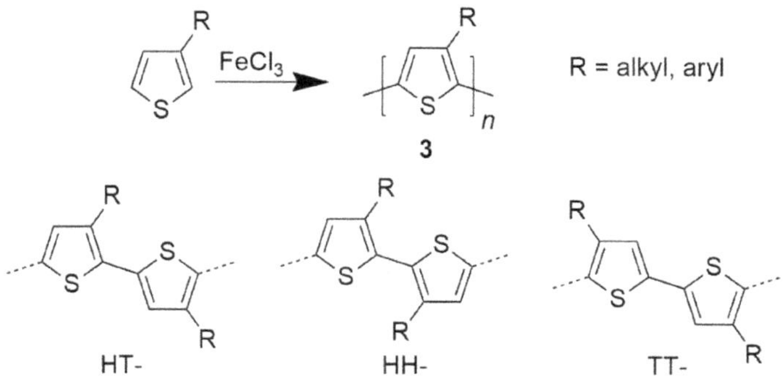

Figure 7.2 Regiorandom polyalkylthiophenes by oxidative polymerisation

This method seems to produce only 2,5-linked polymers **3**, which show good solubility with alkyl substituents that are C_5 or longer. However, a new question of stereochemistry arises in these polymers as three regioisomeric couplings can occur between adjacent thiophenes – 2,5- (head-to-tail, HT), 2,2- (head-to-head, HH) or 5,5- (tail-to-tail, TT). The 2,5- or HT-coupling is the thermodynamically and kinetically preferred arrangement but oxidative polymerisation tends to produce regiorandom materials with a high degree of HH- and TT-linkages within the polymers. The degree of regularity in P3ATs is defined as the molar percentage of HT-linkages, with an all-HT polymer being regarded as totally regioregular. For P3ATs made using iron(III) chloride the regioregularity is dependent upon the reaction conditions but typically ranges from 70–80%. Much higher regularity (94%) is obtained from polymerisation of 3-arylthiophenes due to a combination of thermodynamic (greater stabilisation of a cation at the 2-position by the aryl group) and kinetic (steric repulsion between aryl groups during 2,2 coupling) factors which hinder HH-coupling [503].

The effects of HH-linkages are to reduce conjugation as the steric repulsion between the alkyl chains on adjacent thiophenes results in pronounced torsion of the polymer backbone. This results in an increase in bandgap and is a useful way to tune the emission properties of P3ATs, as demonstrated by the shift in emission for the polymers **4–7** (Figure 7.3) made by oxidative polymerisation of the oligothiophene monomers. As the distance between the H–H linkages increases, the emission colour red-shifts from blue-green ($\lambda_{max} = 460$ nm) for the

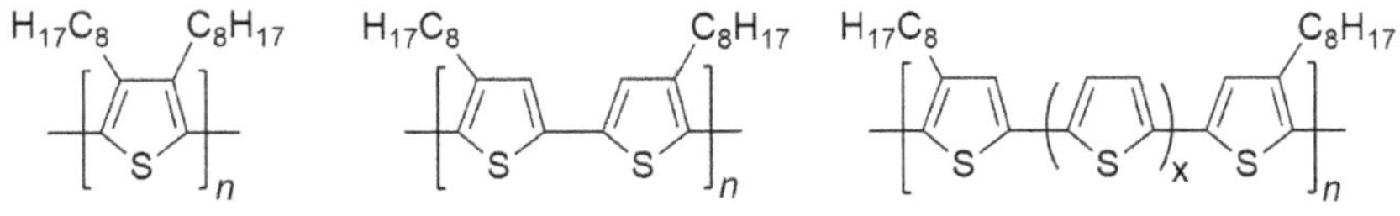

4 5 6 x = 1 7 x - 2

Figure 7.3 Polyalkylthiophenes with controlled stereochemistry

polydialkylthiophene **4** to yellow-orange (λ_{max} = 550 nm) for the poly(dialkylquaterthiophene) **7** [504].

Another effect of such linkages is to hinder efficient chain packing, which makes charge transport less efficient and so results in reduced charge carrier mobility and conductivity. The regioregular poly(arylthiophene)s mentioned at the beginning of this section will also not show good chain packing, as the bulky aryl groups will hinder inter-chain interactions. By contrast, a regioregular P3AT in which all the units are linked in head-to-tail (HT) fashion will show good chain packing and higher charge carrier mobility and conductivity. As a result, methods have been developed to make regioregular P3ATs.

7.3 Routes to Regioregular P3ATs

Some very innovative synthetic methods have been developed to make regioregular polythiophenes, in particular poly(3-hexylthiophene) (P3HT, **7**), which has been the most intensively studied of all P3ATs (P3HT) [501]. As shown in Figure 7.4, routes to regioregular P3HT using metal–halide cross-coupling reactions were developed independently by the groups of McCullough [505, 506] and Rieke [507, 508], who prepared 2-halo-5-metallo-3-hexylthiophenes **8** and **9** which then underwent nickel-catalysed coupling reactions to produce P3HT with over 95% HT linkages. Of these, the McCullough route using a nickel-mediated Kumada coupling has been the more widely used, as the Rieke method relies upon the use of specially activated zinc to produce the monomer for the Negishi polycoupling, and this reagent seems to be difficult to prepare and handle. Similar routes using Stille [509] or Suzuki polycoupling [510] of 2-halo-5-metallothiophenes have been used to make regioregular P3ATs, but seem to offer no advantages over the McCullough method and suffer, respectively, from problems of toxicity of tin reagents and experimental difficulty in purification of the boronate.

A disadvantage of these methods is that very pure monomers **8** or **9** are required to obtain high molar masses and the conditions for their synthesis and handling are rather demanding. More recently, the McCullough group developed a much simpler Grignard metathesis (GRIM) method for preparing regioregular P3HT [511]. This method (Figure 7.5) involves reacting the readily available dibromide

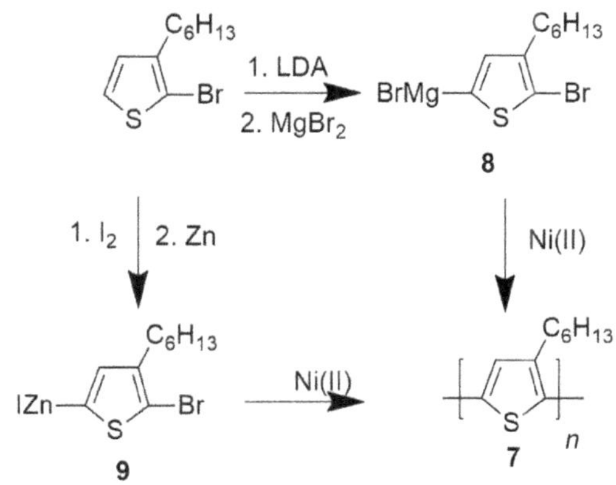

Figure 7.4 Rieke and McCullough routes to regioregular P3HT

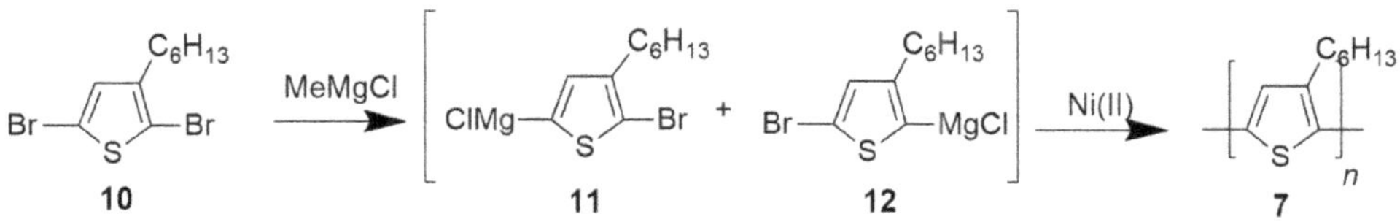

Figure 7.5 GRIM route to regioregular P3HT

10 with a commercially available methylmagnesiumhalide. Nickel-catalysed cross-coupling of the resulting mixture of thiophene Grignard intermediates **11** and **12** then produces regioregular P3HT. That the mixture of metallothiophenes produces only HT coupling has been explained as arising from a combination of thermodynamic and kinetic factors [512]. This method is experimentally very straightforward and can be carried out on a large scale, making it suitable for the industrial preparation of P3HT.

A particularly interesting feature of the GRIM synthesis is that it is quasi-living in nature, which permits good control of the molar mass of the polymer produced [513] and selective end-capping with functional groups is possible by introducing suitably functionalised Grignard reagents, at the end of the synthesis which can react selectively with one or both end groups [514, 515]. Such end-capping can be useful for making block copolymers by attaching an initiator for the growth of the second block or a functional group to which a suitably functionalised polymer can be bound. End-capping may also assist in binding the polymers to a substrate surface or to an inorganic semiconductor nanoparticle. This feature of the process is explained by the mechanism shown in Figure 7.6, which shows that the growing polymer has two distinct termini at which further reaction is possible – generally the reaction can occur selectively at the living end bearing the nickel or at both, depending upon the choice of end-capping reagent. One interesting point to note is

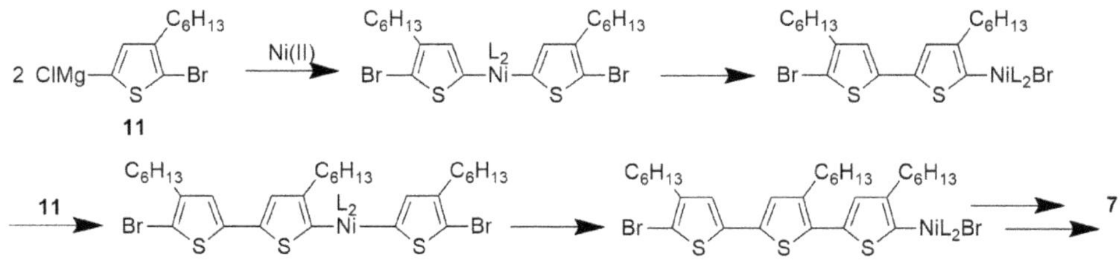

Figure 7.6 Mechanism of the GRIM polymerisation

that the very first thiophene–thiophene coupling occurs tail-to-tail, so every P3HT chain made by this route thus contains a defect – a situation reminiscent of what has been found in the Gilch synthesis of poly(arylene vinylene)s. Very recently, however, it has been reported that, by using new nickel initiators, totally defect-free polymers are attainable, and the first studies on them suggest that the removal of the defect improves conjugation length and increases crystallinity of the polymer, but as yet the effects of this upon device performance are undetermined, though it may reasonably be surmised it will enhance charge carrier mobility and, thus, transistor performance [516].

The GRIM route is an example of a chain-growth condensation polymerisation which offers the possibility of obtaining well-defined polymers with reproducible and controllable limited molecular weight range (polydispersity) [517]. It has been successfully applied to the synthesis of other classes of polymers, including polyphenylenes and polyfluorenes, and though the results are not yet comparable in molar mass to the best materials obtained by the currently standard procedures for making such polymers, it is probable that, with further optimisation of reaction conditions, it will be possible to use it to make very well-defined, high molar mass conjugated polymers. This illustrates how developments in the synthesis of one class of polymers can be used to advance the synthesis of other types.

7.4 Effects of Regioregularity upon Electronic Properties of P3ATs

One obvious effect of regioregularity on the properties of P3ATs is that it greatly enhances their conductivities from a few S cm^{-1} up to 1,000–2,000 S cm^{-1} [501]. Similarly, the charge carrier mobility increases by several orders of magnitude, with values of 0.05–0.2 cm^2 Vs^{-1} being recorded for regioregular (rr-)P3HT (**7**) in transistors [518, 519]. These effects can readily be attributed to improvements in chain packing and alignment due to the enhanced ability of the polymer chains to form regular two-dimensionally ordered arrays with layers of π-stacked polymer chains separated by inter-digitated alkyl chains. Increasing the chain length diminishes the mobility, probably because the longer chains increase the distance

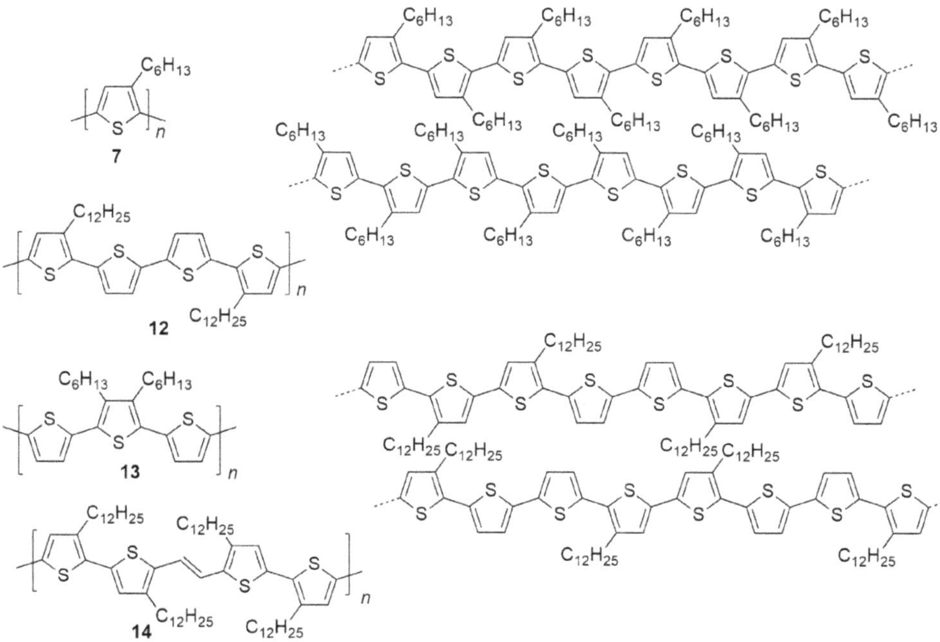

Figure 7.7 Structures of P3HT, PQT-12, PHTT, and PTVT and inter-layer packing of P3HT and PQT-12

between the layers of polymer chains and so make charge hopping between them less efficient. Increased molar mass assists in improving charge carrier mobility in P3HT, due to there being fewer grain boundaries which disrupt charge movement within the polymer films of the higher mass polymer [520].

An excellent example of how a combination of understanding of solid-state packing can be used to design a material with optimised charge carrier mobility is presented by the polyquaterthiophene PQT-12 (Figure 7.7, **12**) which can easily be made by oxidative polymerisation of the monomer [521]. This material has a very high charge carrier mobility of 0.12 $cm^2\ Vs^{-1}$ and is also reported to show better stability in device performance upon storage or operation in air than P3HT. As can be seen from Figure 7.7, PQT-12 like rr-P3HT can form lamellae with inter-digitated alkyl chains between the layers. An important feature of the conformation of **12** is that the alkylthiophenes are linked T–T so the backbone remains planar. Both these features form a part of the structural design. The longer alkyl chains in **12** are required to obtain high solubility as there are fewer alkyl chains than in P3HT. The correlation between efficient packing and high charge carrier mobility has recently been thrown into doubt by the discovery that polymers such as **13** containing 3,4-dialkylthiophene units display charge carrier mobilities comparable (0.10–0.12 $cm^2\ Vs^{-1}$) to those of **7** and **12** but without the same degree of long range order [522]. This suggests that strong π−π stacking between polymer chains which, from XRD analysis, is lacking in **13** and is not always necessary for high charge carrier mobility, though it is advantageous when it does occur. Like PQT-12, polymer **13** is reported to show greater stability towards air then P3HT, with devices showing less of a decline in performance (especially less drop in on/off ratio) upon prolonged operation in air. This may reflect a lower tendency towards aerial oxidation, which causes doping with a concomitant increase in the off current. The highest charge carrier mobility yet seen in a thiophene-based polymer is 1.05 $cm^2\ Vs^{-1}$, recently reported for the polymer **14** containing a vinylene linkage between H–T bithiophene units [523]. The presence of the vinylene unit prevents steric interactions between the substituents on the middle thiophene units from causing backbone torsion so the polymer remains fully conjugated and packs well, as shown by the ordered morphology visible in AFM images.

7.5 PEDOT as an Electrically Conducting Polymer and Its Application in Transistor-Based Sensors

While alkylthiophene polymers such as P3HT and PQT-12 have proved extremely successful as TFT materials, they have not been much used as conducting polymers as they suffer from problems of instability when highly doped. The polythiophene most widely used for that purpose has been poly(3,4-ethylenedioxythiophene-2,5-diyl) (PEDOT, Figure 7.8, **15**) [524]. PEDOT is easily made by oxidative polymerisation of the electron-rich monomer. This is usually used as an aqueous suspension of PEDOT doped with poly(sodium styrene sulphonate) (PSS) (**16**). The anions in the PSS form a complex with the cations on the oxidatively doped PEDOT chains which serve to stabilise them. This mixture of PEDOT:PSS is now commercially available in a variety of formulations, formerly under the name Baytron® produced by H. C. Stark but now under the trade name CLEVIOS® produced by Heraeus. PEDOT:PSS is widely used as a hole-injecting layer in devices such as OLEDs and OPVs, usually on top of an ITO electrode, where it serves to smooth out the electrode surface and to improve hole injection into the active layers. The conductivity of these formulations varies dramatically, ranging

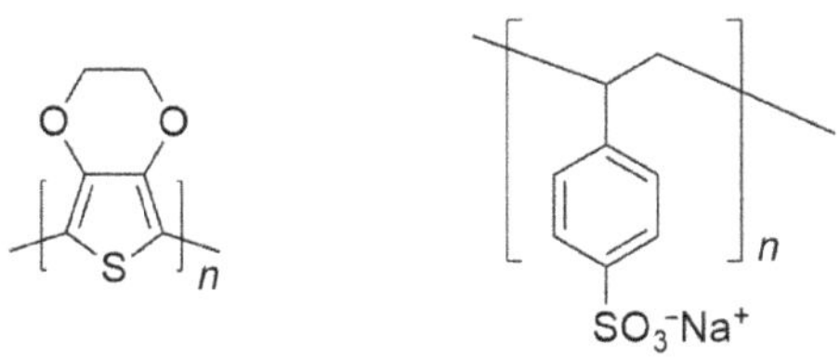

15 **16**

Figure 7.8 Structures of PEDOT and PSS

from 0.1–1 S cm^{-1} (for selective charge injection/transport layers in OLEDs/OPVs) to over 800 S cm^{-1} following the addition of a few percent of dimethyl sulphoxide (for charge transport/electrode layers in organic electronic devices).

Structurally, PEDOT:PSS forms spherular grains (of average diameter 50–80 nm) consisting of tangles of PEDOT segments adhering to single PSS chains [525]. The structures consist of a hydrophobic PEDOT-rich core surrounded by a PSS-rich shell with no overall alignment of the PEDOT segments within grains. As well as being used as an electrode buffer layer in multi-layered organic electronic devices, PEDOT:PSS is also widely used in electronic devices in its own right with a number of important current and emerging technological applications [526], including thermoelectric devices [527], wearable devices [528] and sensors [529].

PEDOT:PSS has been widely used in sensor applications; commonly as part of an organic electrochemical transistor structure. Early device architectures employed PEDOT:PSS as the semiconducting channel with a PDMS well structure used to hold both the analyte solution and a platinum gate electrode [530] with glucose sensing subsequently achieved by adding enzyme (glucose oxidase) to the well [531]. Integrated sensors were fabricated with the enzyme mixed into a phosphate buffer solution and spin-coated to form a film on top of the PEDOT:PSS semiconductor layer joining the OECT source and drain electrodes [532]. Triglyceride, a biomarker for some serious diseases, sensing has been achieved by immobilising three different enzymes (lipase, glycerol kinase and glycerol-3-phosphate oxidase) onto a working electrode fabricated using a PEDOT:PSS/gold nanocomposite material [533].

7.6 Optical Properties of P3ATs and Their Applications in OLEDs and Sensors

As mentioned in Section 7.1, the emission colour of polythiophenes can be controlled by the choice of substituents and in particular by using large substituents or H–H linkages to induce twisting of the backbone to induce blue-shifts in the emission. In this way emission colours ranging from blue through to red have been obtained from polythiophenes [150, 534]. A major problem with polythiophenes as emissive materials, however, is that their solid-state PL efficiencies are generally very low, which causes their EL efficiencies to also be low. For example, P3OT (Figure 7.9, **17**) has a solid-state PL efficiency of only 0.04 compared to 0.27 in solution. Poly(aryl-thiophene)s suffer this problem to a much lesser degree; thus the aryl polymer PDOPT (**18**) (obtained with high regioregularity from oxidative polymerisation, as explained in Section 7.2) displays only a relatively small drop in PL efficiency from 0.37 to 0.24 in going from solution to thin film. This seems to

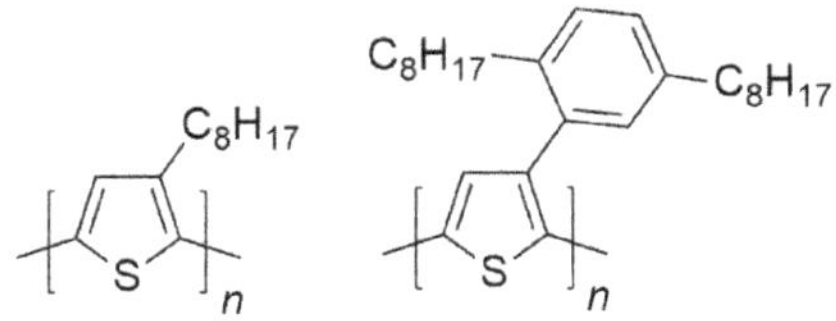

17 18

Figure 7.9 P3OT and PDOPT

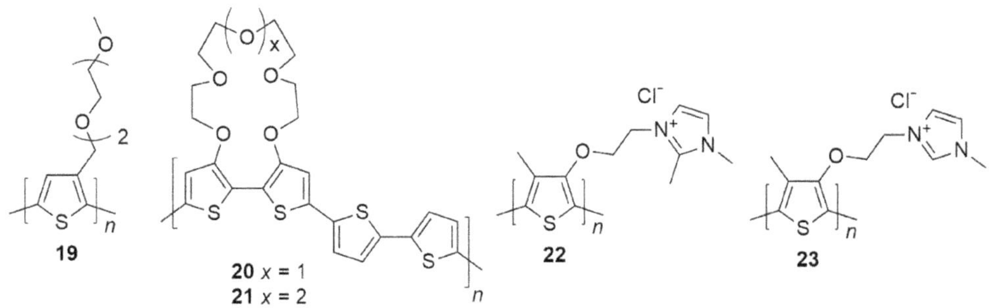

Figure 7.10 Polythiophenes for sensors

be due to efficient non-radiative decay pathways existing in polythiophenes, accentuated in regioregular P3ATs by good packing assisting exciton migration. There is no evidence that regioregularity improves the EL efficiency of P3ATs as the improvement in charge mobility seems to be cancelled out by the greater efficiency of non-radiative decay pathways [150]. Thus, the good packing, which makes the regioregular polymers so good for FET applications, makes them unsuitable for LEDs.

That the emission colour from polythiophenes is strongly affected by backbone torsion provides a mechanism for them to be used in sensors. Interactions between substituents on side-chains of alkylated polymers and the desired analytes can induce twisting of the backbone with a detectable change in absorption (i.e. colour) or emission spectra and PL efficiency. Thus, polythiophenes can be used for either colorimetric or fluorimetric sensors.

Polymers such as **19** (Figure 7.10) with ethylene oxide sidechains act as sensors for metal ions as the sidechains wrap themselves around the metal ions, inducing a twist in the backbone and a change in absorption maximum (and hence colour). The effect is much greater for large ions such as Hg^{2+} or Pb^{2+} ($\Delta\lambda_{max} = 176$ nm) than for alkali metals ($\Delta\lambda_{max} = 11$ nm), thus displaying a useful chemoselectivity [535]. The polymers **20–21**, in which the ethylene oxide units link a bithiophene unit to form a crown ether, show selectivity between different alkali metals as the crown ether–metal binding efficiency depends upon the size both of the ether ring and of the metal ion. As a result, **20** shows the largest shift in absorption maximum with sodium, while for **21** this occurs for potassium [536]. The polymer **22** bearing imidazolium units on the sidechains acts as a selective sensor for iodide ions, in which the presence of the anion induces not only a change in colour but also a quenching in the fluorescence, since the halide ion induces a planarisation and aggregation of the polymer chains [537]. The related polymer **23** has been used for DNA detection [537–539]. The mechanism behind this is that DNA strands are polyanions which can bind electrostatically to polycations like **23**. If a probe DNA single strand is complexed in this way to **23**, then in the presence of the complementary DNA strand a ternary complex is formed between the polycation **23** and the resulting double-stranded DNA. This induces a change in absorption spectrum and also an enhancement in fluorescence by the same planarisation and aggregation effects just mentioned for halide detection using polymer **22**. This effect is not seen for non-complementary (mismatched) DNA as the ternary complex cannot form. The sensitivity of the test can be enhanced by attaching a chromophore as a fluorescent label to the probe DNA strand. In this case the complexation of the probe and target DNA strands leads to an enhancement in the fluorescence from the attached chromophores as efficient fluorescent resonant energy transfer (Förster transfer) between the thiophene backbone and the label can occur within the triple complex but not within the original polymer–probe complex [540]. A similar mechanism has been used to produce an efficient sensor for DNA using a fluorene-based polycation [541, 542].

7.7 Polythiophenes as Donor Materials for OPVs

While P3HT has proven of little use as an LED material, due to its low solid-state PL efficiency [534], its relatively modest bandgap (*ca.* 2.1 eV) has made it of interest as a solar cell material. The combination of P3HT as an electron donor with PCBM as an acceptor is one of the more intensively studied BHJ combinations in polymer solar cell research [244, 543], with over 1,000 publications on the use of this combination appearing between 2002 and 2010 [544].

It has been reported that the solar cell performance of P3HT is enhanced by factors that are known to enhance its charge carrier mobility, such as increased molecular weight [545] and/or increased regioregularity [546]. More recently, defect-free 100% regioregular P3HT has been shown to provide further OPV performance improvements of around 10% [547].

On the other hand it has also been reported that a small decrease in the effective regioregularity of P3HT appears to confer greater thermal stability upon the bulk heterojunction solar cell with PCBM as an acceptor [548]. Efficiencies of up to 5% have been obtained from devices using a regioregular P3HT:PCBM blend, and there has been considerable investigation into methods for enhancing the photocurrent by methods such as thermal or solvent annealing, which aim to increase the order of the polymer chains and promote charge transport within the blend [543, 544]. However, the theoretical model developed by workers at Konarka suggests that the maximum possible efficiency from these blends is only a little above 5%, so any further improvement in efficiency from them is likely to be very limited [543]. Nonetheless, the insights gained into the behaviour of polymer–acceptor blends, on the relationships between the blend morphology, the degree of order in the blend and the device efficiencies and, most importantly, on the methods by which these can be enhanced, have been of great value in the ongoing efforts to improve the efficiency of organic solar cells.

For the highly studied P3HT:PCBM system [544], the diffusion of PCBM in the P3HT matrix has long been recognised as a key parameter, since it governs the formation of an intricate film structure that incorporates both crystalline and amorphous P3HT regions [549]. In particular, it is the phase segregation of PCBM that influences ordering of the polymer regions and hence device function [550]. However, only a few experimental techniques are capable of following the diffusion of fullerene in the P3HT matrix. Scanning transmission X-ray microscopy (STXM) was used to provide an early measure of the PCBM diffusion constant [551]. More recently, it has been shown that the diffusion of PCBM in the P3HT matrix is actually more complicated than originally thought; with the formation of PCBM-rich phase segregated features in these systems occurring via the diffusion of PCBM crystallites rather than molecular PCBM [552], resulting in a difference in the calculated diffusion constant [553].

Efficiencies of over 7% have recently been reported using rr-P3HT with another fullerene acceptor, with the best results coming from devices containing vertically aligned cross-linked nanorods of a fullerene molecule attached to one electrode [554].

7.8 Polythiophenes as Active Layers in Sensors

In Chapter 3, we discussed how PAV polymer materials had been incorporated into devices as biosensors. P3HT has also been extensively used as the semi-conducting layer in sensors based on organic thin film transistors, with a number of excellent reviews on the subject [555, 556]. Here we discuss some interesting case studies across the range of sensing applications that P3HT has been used for. In the case of lab-on-a-chip applications, compact, low cost, and practical fluorescence detection platforms incorporating an organic photodetector for fluorescence immunoassays for cardiac markers myoglobin and CK-MB have been developed based on P3HT:PCBM blends as the organic detection system [557].

The ability to biologically tailor the sensing interface in organic thin film transistors was highlighted by Magliulo et al., who showed biotin detection at concentrations as low as 15 pM by using a functionalised inter-layer between the source and drain [558]. The inter-layer consisted of streptavidin (SA), a protein which has an extremely high affinity to biotin, and the organic semiconductor P3HT. The sensitivity arises from changes in the P3HT layer's electronic properties upon binding of biotin by the SA. Successful sensing trials were also conducted in which the SA layer was replaced firstly with a biotin antibody and secondly with the enzyme horseradish peroxidase, indicating a promising platform for a variety of biological recognition elements.

Pressure sensors employing electrically conductive rubber have also been fabricated based upon space-charge limited transistors (SCLT) with P3HT as the active layer [559]. Impressively, the response time of the sensor to the pressure is less than 22 ms. Li and Lambeth showed that when using a nanostructured rr-P3HT as the active layer, OTFTs show sensitivity to 10 analytes, although the device responses are different for polar and non-polar gases [560]. They attributed the sensing behaviour to grain boundary effects, which influences charge transport in the OTFT.

Enzymes which catalyse charge carrier-liberating reactions are well-suited recognition elements for OTFT-based biosensors due to their inherent selectivity. This principle has been applied to develop a range of OTFT sensors. Glucose sensors based on OTFTs containing glucose oxidase and P3HT as the semiconducting layer have been demonstrated by Elkington et al., with sensitivities in the μmol concentration range [82] and structures that can be entirely fabricated using ink-jet printing [561].

7.9 Fused Thiophene-Based Materials

Fused thiophenes have been used as building blocks in polymers for use in high mobility FETs [562]. The most widely studied class of fused thiophenes for such purposes have been polymers containing thienothiophenes [563]. Polymer **24** (Figure 7.11), which has an obvious structural affinity to PQT-12 (**12**), displays the highest mobility for such polymers of 0.63 $cm^2 Vs^{-1}$ [564]. The alkyl chains can also be attached to the thienothiophene as in **25**, but the mobility obtained is a little lower (0.3 $cm^2 Vs^{-1}$) [565]. Similar mobilities are reported for the dithienothiophene copolymer **26** [566] and the fused quaterthiophene copolymer **27** [567]. Fused systems containing both thiophene and benzene rings have also been investigated as units for polymers with high charge carrier mobilities. The best results have been obtained from polymers **28** (0.25 $cm^2 Vs^{-1}$) [568] and **29** (0.13 $cm^2 Vs^{-1}$) [569]. Attaching alkyl groups onto the benzene ring in structure

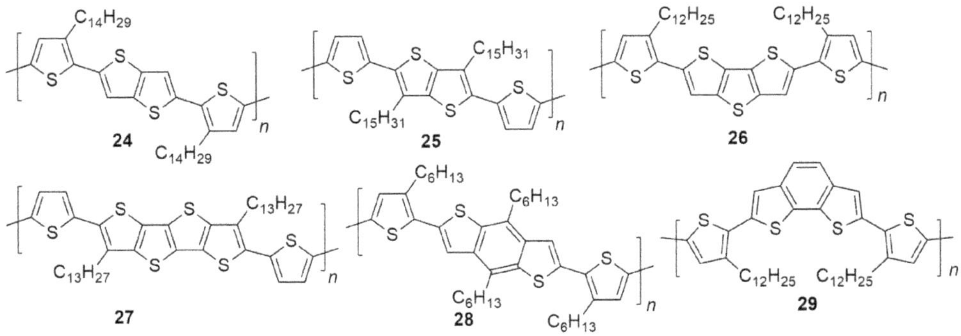

Figure 7.11 High charge-carrier mobility polymers containing fused thiophene units

29 has been found to sharply reduce the mobility, probably because it adversely affects the packing.

Given their electron-rich nature, thiophenes make good electron donors and so they are widely used as donor units in copolymers with other units, especially in donor–acceptor copolymers to which we will turn our attention in Chapter 8.

8

Other Arylene-Based Polymers

In Chapters 2–7 we have discussed mainly homopolymers of phenylenes or heteroarylenes, but an obvious way to fine-tune the properties of conjugated polymers is to combine the two types of units together in copolymers, especially in alternating copolymers. The theoretical basis for this approach is that when one conjugatively links two units together orbital mixing occurs so that the new orbital energies reflect the higher of the two HOMO and the lower of the two LUMO energy levels for the individual units. This is the basis of the well-known donor–acceptor alternating copolymer design for low bandgap materials. If one combines an electron rich unit with a high HOMO which will be a good electron donor (D), with an electron poor unit with a low lying LUMO which is naturally a good electron acceptor (A), then the resulting donor–acceptor (D–A) copolymer will have a smaller bandgap (Figure 8.1).

This copolymerisation approach enables one to design materials with appropriate bandgaps for LEDs – it is particularly useful for making near-infrared (NIR) emitters – and solar cells, where low bandgap materials are needed in order to absorb as much as possible of the solar spectrum in order to get efficient harvesting of the incident solar energy. Also, it enables one to produce materials with appropriate frontier orbital energies for obtaining efficient charge injection from electrodes, and in the case of solar cells for enabling efficient electron transfer between donor and acceptor. With respect to transistors this approach also has the potential to allow ambipolar charge transport, i.e. to produce materials which are equally good at transport of holes (usually through having a high HOMO) and electrons (which usually requires a low LUMO).

There have been a very large number of copolymers made containing different arylene or heteroarylene units for use in applications such as LEDs [150], solar cells [244, 570, 571] and transistors [571, 572]. Space does not permit an exhaustive discussion of all the various materials here, to which the readers are referred to the extensive reviews in the literature cited here, but we will give some illustrative examples of how the properties of materials can be optimised for

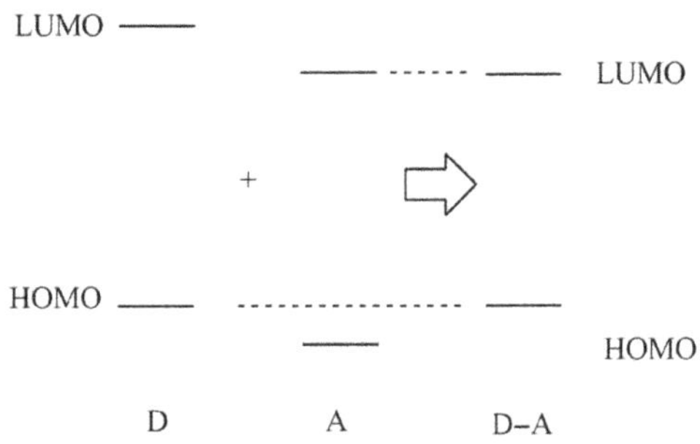

Figure 8.1 Molecular orbital rationale for the donor–acceptor low bandgap copolymer design

various applications by synthetic design. Since new materials are constantly being developed and optimised, the examples we give, while in some cases being the best performing materials at the time of writing, will almost certainly not be the record holders when this book is read or even when it is published. But we trust that the principles we will be discussing will still be valid and will enable our readers to elucidate how the then best performing materials will have been designed and how they may start to design even better materials themselves.

8.1 Block Copolymers

There are three types of copolymers that can be made as illustrated for fluorene-thiophene copolymers in Figure 8.2: block copolymers like **1** in which there are distinct sections each consisting of just one monomer unit, random (or statistical) copolymers **2** in which the two monomers are found randomly distributed along the polymer chain, and alternating copolymers such as **3** in which the two units occur in strictly alternating fashion (note that the units here mentioned do not have to be monomeric, e.g. the fluorene-bithiophene alternating copolymer **3** has alternately one fluorene and two thiophenes).

Of these, the block copolymers are both the most difficult to make and the least studied. With recent advances in chain-growth polymerisation methods [517], well-defined blocks of polythiophenes and other conjugated polymers are now accessible in which the chain end is still 'living', enabling block copolymers to be made [573]. However, so far there has been little attention paid to such materials as these structures offer no advantages for transistor or LED applications. They do have potential for use in solar cells as they will tend to undergo nanophase separation to form *in situ* bulk heterojunctions between the blocks. The first example reported of a donor–acceptor block copolymer being tested in OPVs was a copolymer **4** (R = 2-octyldodecyl) containing a rr-P3HT donor block and a

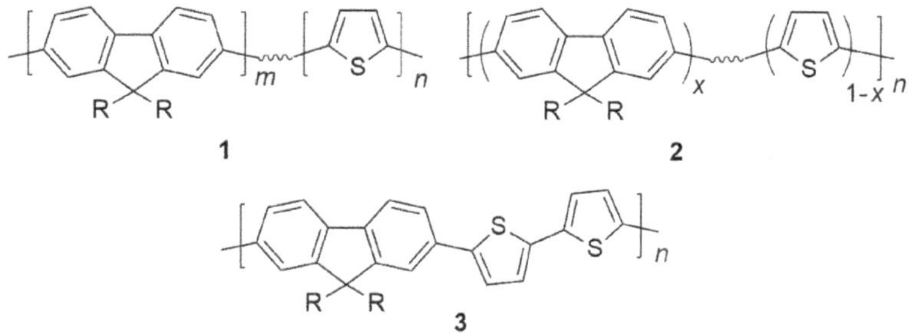

Figure 8.2 Examples of block, random and alternating fluorene-thiophene copolymers

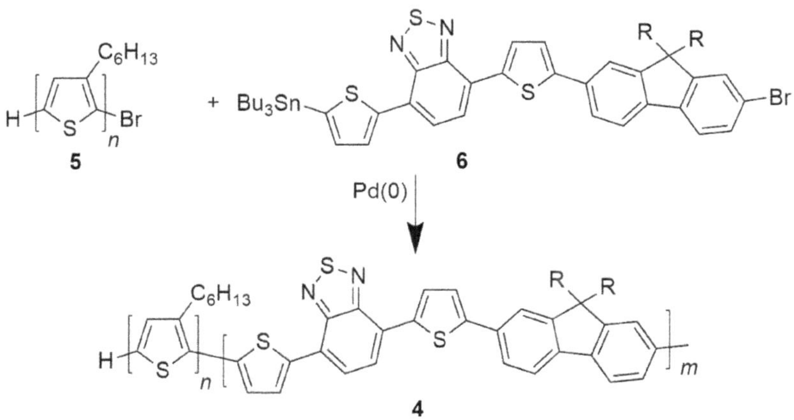

Figure 8.3 Synthesis of a P3HT–PFTBTT block copolymer

fluorene-thiophene-benzothiadiazole (PFTBTT) copolymer as the acceptor block [574]. The synthesis of this copolymer (shown in Figure 8.3) illustrates the synthetic problems inherent in making such materials. Here the bromo-end capped P3HT **5** made by the GRIM method (see Chapter 7) is used as a macro-co-monomer with the mono-tin reagent **6** in an AB-type Stille polycoupling. There are major problems associated with this method. Firstly, monomers like **6** are usually hard to make – in this case the marked difference in reactivity between the thiophene-α-position and the fluorene-2-position enables selective stannylation at the former. Secondly, there is no certainty that a macromonomer like **5** will react as the active site is only a small portion of the polymer surface and may even be hidden if the polymer is capable of coiling up upon itself. As a result, the synthesis in this case produces an inseparable mixture of the copolymer **4** (53 mol% by NMR analysis) and the homopolymers P3HT (5 mol%) and PFTBTT (42 mol%). Atomic force microscopy (AFM) studies showed that the mixture underwent phase separation on annealing to produce domains of, respectively, P3HT rich and P3HT poor content. As a 1:1 weight ratio of the homopolymers had been found to give the best performance in OPVs [575], P3HT was added to the mixture to produce blends for device studies containing approximately that weight ratio of the two components. Though the device efficiencies of the resulting ternary blends were lower (<1%) than for a 1:1 blend of the homopolymers, the ternary blends showed much more stability in performance after annealing. This is attributed to the block copolymer acting as a compatibiliser hindering phase-separation of the two homopolymers.

A similar polymer with straight chains on the fluorene (**4**, R = octyl) has been made by adding **5** to an AA–BB Suzuki polycondensation of a fluorene bisboronate and a bis(5-bromothienyl)-benzothiadiazole [576]. This contained about 14 wt% of P3HT as an impurity. When used as an active layer in an OPV an efficiency of 3.1% was obtained without any acceptor being added as the copolymer showed microphase separation to form nanodomains of P3HT and PFTBTT. While this does show that the block copolymer concept has some validity, the results are much lower than from the best alternating copolymers discussed in Section 8.4, whose synthesis is much simpler.

While block copolymers offer a fascinating challenge for synthetic chemists to prepare, the device results obtained to date do not seem to justify the effort required for their synthesis, and so the main emphasis for practical applications has been on the much more accessible statistical and alternating copolymers. While statistical copolymers offer a chance to fine-tune the properties of the polymer by varying the ratio of the comonomers, the difficulty in controlling and reproducing the distribution of the units within the polymer chains means that the alternating copolymers have been the dominant type of copolymer made and evaluated. This is because they are readily made by transition-metal mediated cross-coupling reactions such as Suzuki, Stille, Negishi and Kumada couplings for which as we have seen in previous chapters optimised synthetic protocols have been developed to produce reproducible high molar mass polymers with low polydispersities. Generally AA–BB type coupling methods are preferred to AB couplings as not only are the monomers usually significantly easier to prepare and purify, but also for these couplings it is often important that the halide be on the electron-rich and the metal be on the electron-poor unit (or vice versa) in order to maximise the efficiency of the coupling reaction. For example, thiophene-boronates are notoriously prone to loss of the boron so it is usually better to have the halide on the thiophene when making a copolymer containing thiophenes by Suzuki coupling.

8.2 Copolymers for LEDs

The utility of copolymers for tuning bandgaps is best illustrated by copolymers of wide bandgap phenylene-based units with low bandgap co-monomers such as

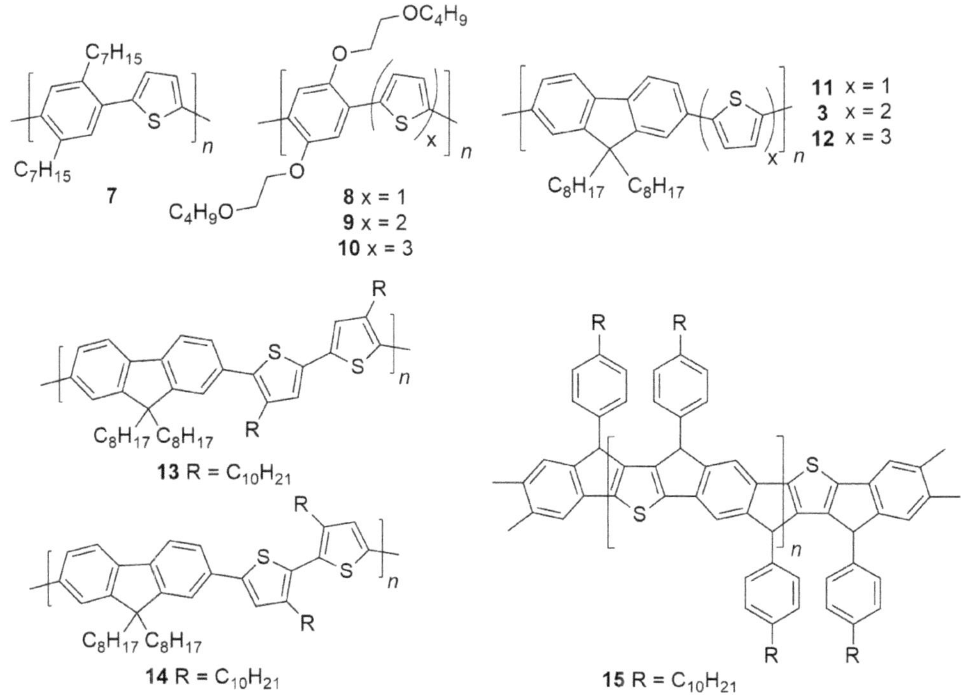

Figure 8.4 Some phenylene-thiophene copolymers

thiophenes. For example whereas polyphenylenes are violet-blue emitters, the alternating copolymer **7** (Figure 8.4) produces bright turquoise emission (λ_{max} = 500 nm) [577]. Increasing the amount of thienyl units red-shifts the emission as seen for the phenylene copolymers **8–10** where the emission maximum goes from 520 nm for **8** through 543 nm for **9** to 547 nm for **10** [578]. The solution PL efficiency also drops with increasing thiophene content from 82% for **8** to 42% for **9–10**. Similar effects are seen for the fluorene copolymers **3**, **11**, and **12** with emission colour changing from blue-green for **11** (λ_{max} = 480 nm) to yellow green for **3** and **12** (λ_{max} = 543 and 567 nm, respectively) [579]. Alkyl substituents on the thiophenes can induce blue-shifting due to torsion of the backbone so that the copolymers **13** and **14** are both blue-green emitters (λ_{max} = 490 and 493 nm) [580]. The ladder-type copolymer **15** is an orange emitter (λ_{max} = 536 nm) [581].

The thiophenes in these copolymers lower the bandgap, primarily by raising the HOMO energy levels as thiophene rings are electron rich and so thienyl units are good electron donors, a point we shall come back to when discussing copolymers for OPVs. Another way to reduce the bandgap of phenylene-based copolymers is to use comonomers which have low LUMO energies, that is, are good electron acceptors. Thus, the fluorene-benzodithiadiazole copolymer (F8BT, **16**, Figure 8.5) is a green emitter. The good electron accepting properties of the BT unit make this

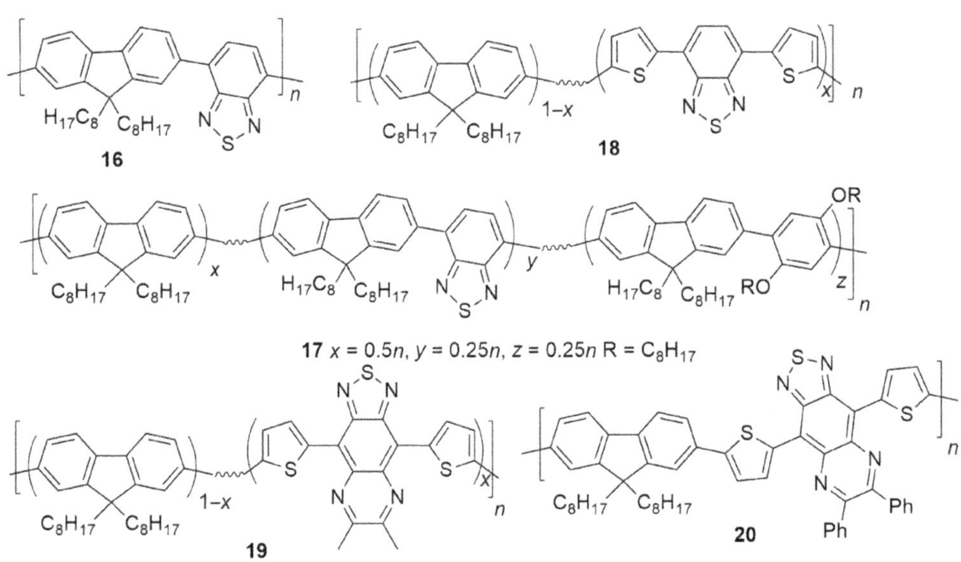

Figure 8.5 Low bandgap copolymers of fluorene with electron accepting units

an efficient emitter in LED with reported device efficiencies of up to 3.8% [582–585]. Even higher efficiencies of up to 6% (18.5 cd A^{-1}) have been reported for a random copolymer **17** [586]. When thiophenes are attached to an electron acceptor the result unit is usually a low bandgap chromophore and copolymers containing such chromophores are also low bandgap materials. For example, the fluorene-dithienylbenzodithiazole copolymers **18** are red emitters with emission maxima varying between 613 and 672 nm depending upon the composition [587–590]. If stronger electron donors than BT are used the resulting materials can be so low bandgap as to become infrared emitters, as exemplified by the statistical copolymers **19** ($\lambda_{max} = 734–780$ nm) [591] and the alternating copolymer **20** ($\lambda_{max} = 970$ nm) [592, 593].

In order to control the emission colour, it is not necessary to make an alternating copolymer. Statistical copolymers of fluorenes (and other large bandgap moieties) with other lower bandgap emissive chromophores can show emission from the co-monomers even when they are present in low mol%.

This is well illustrated by the behaviour of copolymers of fluorene containing perylene-based dye units (Figure 8.6) in the backbone (**21**), as end groups (**22**) or as sidechains (**23**) [409]. As was mentioned in Chapter 6, the EL from random copolymers of fluorene such as **21** with only 1–5 mol% of perylene dyes incorporated into the backbone occurs entirely from the perylene chromophores [409]. By changing the dye unit, the emission colour of the polymers was varied from green through to red, and this approach has since been used to obtain emission colours across the visible range and into the near-infrared from other

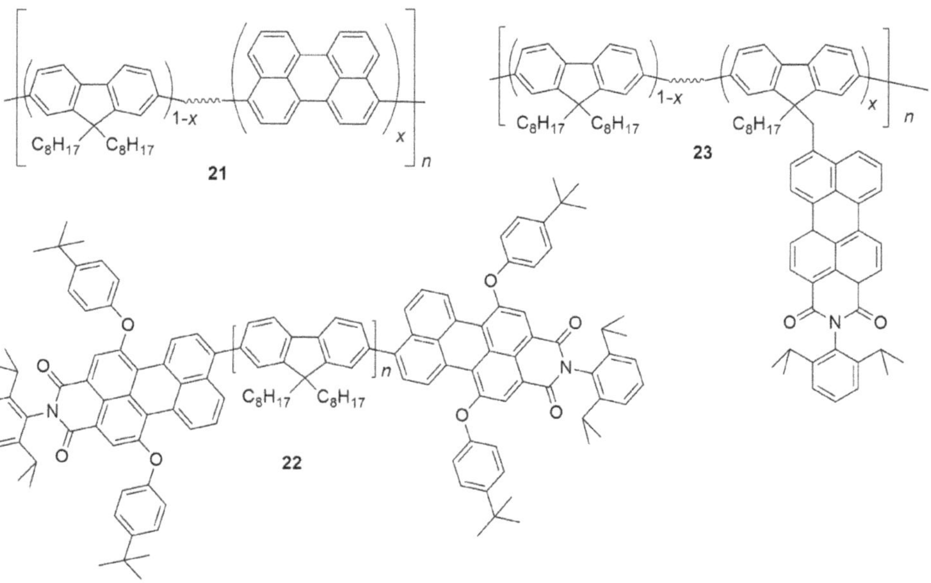

Figure 8.6 Examples of polyfluorenes with chromophores attached

blue-emitting polymers based upon phenylenes by using a wide variety of emissive comonomers [150]. The use of phosphorescent units to obtain electrophosphorescence (EL from triplets) will be discussed in Chapter 11. If significantly lower amounts of comonomers are used, emission from both the fluorene-backbone and the dye units occurs, which, as we will see in Chapter 12, is a promising method for obtaining white emission. Similar behaviour is observed from polymers such as **22** and **23** with the dyes, respectively, as end groups or as sidechains. This complete suppression of the fluorene emissions appears to be due to the dyes acting as charge trapping agents, as the solution and solid-state PL spectra for these polymers show some emission from both fluorene and dye units. In solution, the dye emission dominates from the copolymers **23** with higher molar amounts of dyes (up to 33 mol%) incorporated, while the PL spectra from the polymers like **21** and **22** with low molar fractions of dyes is dominated by the blue emission from the fluorene units. The emission from the dye dominates in the solid state for all these polymers due to efficient energy transfer, which implies that inter-chain transfer is dominant in these polymers. This has been confirmed by ultrafast spectroscopy measurements, which indicate that inter-molecular energy transfer is faster than intra-molecular transfer, in which exciton hopping along the backbone is the rate-limiting step [594]. Interestingly, it is possible to get energy transfer from the low bandgap substituents to the high bandgap polymer backbone as part of the process of up-conversion in which the absorption (sequential or simultaneous) of two or more low energy photons by the substituent leads to the emission of a single high energy photon from the polymer backbone. Up-conversion has been observed from a poly(ladder-type pentaphenylene) end capped with palladium porphyrin units [595]. This end capping is reported to enhance the efficiency of the process compared to that seen from blends of the uncapped polymer with the porphyrin [596].

8.3 Copolymers for FETs

The highest reported mobilities from conjugated polymers have increased dramatically in recent years, with values over 10 $cm^2\ Vs^{-1}$ now being reported, whereas values an order of magnitude smaller were state-of-the-art when we started writing this. This is a result not just of design of new structures but also of advances in control of molecular weights and dispersities, better purification of materials, and especially of new processing methods being developed which enhance structural order and thus mobility in the solid state. While it is not yet possible to say that definite structural design rules can be obtained, some indications have emerged of possible factors that can be designed into a material to obtain good mobility.

As of writing, the highest mobility values that have been reported have been from ordered films of the dithienocyclopentadiene copolymers **24** and **25** (Figure 8.7) [597, 598]. Values of over 20 $cm^2\ Vs^{-1}$ have been reported in FETs, and the intrinsic hole mobility for these polymers has been estimated to be >40 $cm^2\ Vs^{-1}$. Such values are comparable to those reported for single crystals of acenes such as rubrene or pentacene [599], which, unlike the solution processed films of the polymers, have to be grown by difficult and/or expensive thermal evaporation techniques. These results suggest that conjugated polymers may thus be superior to the small molecule materials for potential applications, due to their lower processing and thus lower device fabrication costs.

The very high mobilities reported for these polymers have arisen in part from the alignment of the polymer chains to form aligned fibres within the films and in fractionation of the polymer so as to obtain samples with narrow molar mass

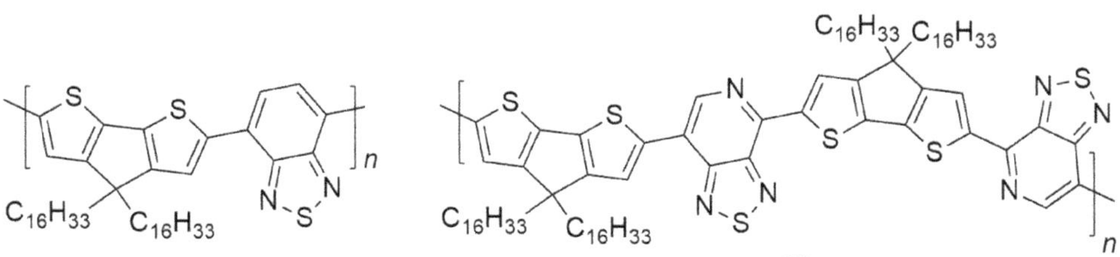

24 **25**

Figure 8.7 Copolymers with exceptionally high mobility values

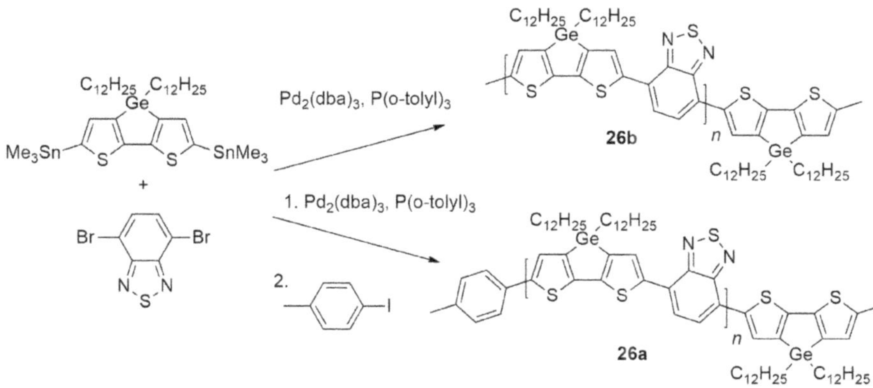

Figure 8.8 End capping a copolymer to increase order

ranges (low polydispersity). Both of these approaches have been shown with other polymers such as P3HT (see Chapter 7) to enhance mobilities in devices.

For polymer **25** it was reported that, without alignment, the average field effect mobility increased from 0.7 to 1.9 cm^2 Vs^{-1} when average molar mass (and thus chain length) increased from 30 to 300 kg mol^{-1} (kDa) but, for aligned films, while there was a general trend of increasing mobility from 9.1 to 14.6 cm^2 Vs^{-1} with increasing molar mass, an exceptional average value of 16.3 cm^2 Vs^{-1}, with a highest value of 23.7 cm^2 Vs^{-1}, was obtained for the fraction with an average molar mass of 50 kDa [597]. The model that has been advanced to explain the higher charge carrier mobilities seen with higher molar masses in P3HT and other polymers is that the polymers contain regions of high crystallinity where the polymer chains are closely packed and connected by amorphous regions [600, 601]. The need for high molar masses thus arises from the need to have efficient connections between the crystalline domains. One plausible reason for the exceptionally large effect of alignment for the 50 kDa sample of **25** is that, as one increases, the molar mass aligning the polymer chains becomes slower so that achieving ordered films is more difficult. This suggests that there may need to be a compromise between the improved charge transport seen with longer polymer chains (high molar mass) and the superior processability of smaller chains. Narrow polydispersity is likely to be a necessary condition for achieving high mobility as a narrower range of polymer chain lengths should enhance the order within the film. It was recently reported that end capping polymer chains with 4-methylbenzene rings increased the order and charge transport in polymer **26a** made by Stille coupling (Figure 8.8) compared to the uncapped polymer **26b**, which from NMR and mass spectral analysis appears to have hydrogens at chain termini, with hole mobilities increasing from 0.077 to 0.60 cm^2 Vs^{-1} [602]. Whether this effect is general and whether other end capping groups will also assist chain ordering is as

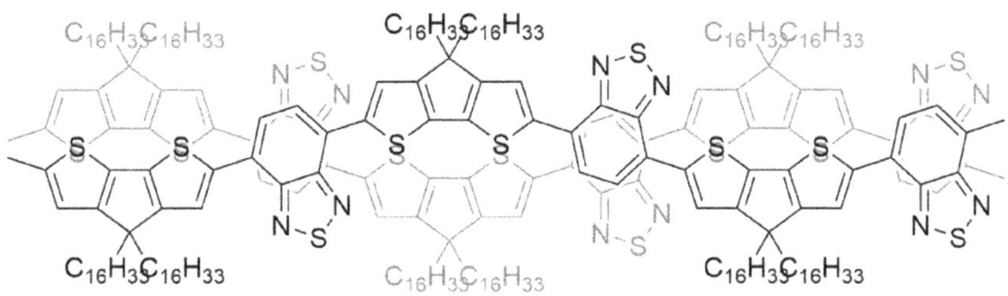

Figure 8.9 Packing of polymer chains in **24** – top layer in grey, bottom layer in black

yet undetermined, but certainly attention should be paid to the nature of any end groups in a conjugated polymer, and end capping to remove possibly reactive end groups is generally advisable.

Achieving high degrees of order within the film appears thus to be the obvious way to obtain high mobility, but studies on the packing of various polymers and attempts to correlate this with the mobilities suggest the picture is actually more complex, and that there may in fact be more than one way to build high mobility into a polymer film.

The full details of the inter-chain packing have been reported for **24** [603], which has been found by a combination of solid-state NMR and X-ray diffraction (XRD) analysis to pack, as shown in Figure 8.9, with the donor and acceptor units stacking atop each other with their orientations in alternating layers reversed, but without the extensive face-to-face interaction between the units seen in P3HT. This stacking allows the cancelling of the internal dipoles of the heterocyclic units. This dipole cancelling has been deliberately designed into polymer **25** where the N atoms in the pyridinothiadiazoles have been inserted in a regioregular fashion.

As mentioned, this type of packing, however, is not what is seen in P3HT (Chapter 7) nor in other copolymers. As has already been mentioned in Chapter 6, a fluorene-bithiophene copolymer **3** displays a good charge carrier mobility in FETs [446] and has been used to make solution-processed all-organic devices [447]. One proposed reason for the good mobility of this polymer is that computer simulations indicate that the polymer chains can stack closely with the fluorene units stacking on top of the similarly sized bithiophene units, which serves to minimise alkyl chain interactions, producing rather well-ordered films with a marked fibrillar morphology visible in AFM images [448]. Also, the films of **3** can be aligned by heating it into a nematic liquid crystalline phase and then cooling rapidly to trap the order. By contrast, the chains of an indenofluorene-bithiophene copolymer (Figure 8.10, **27**) cannot efficiently stack in this fashion as the two monomer units are of different sizes and shapes and so this polymer produces amorphous films with mobilities around 10^{-5} cm^2 Vs^{-1} [604]. The similarly sized units in the terthiophene copolymer

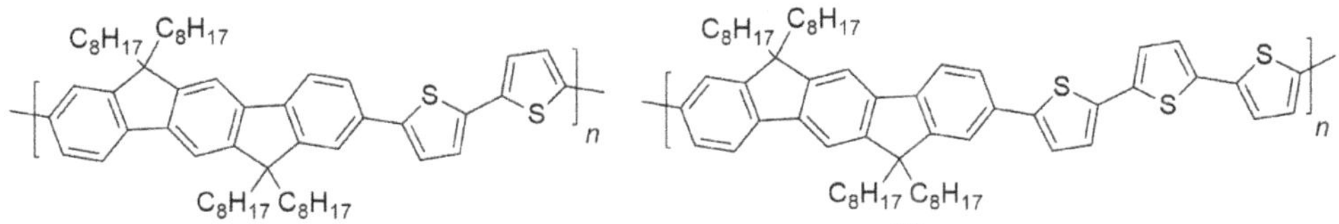

Figure 8.10 Alternating indenofluorene-oligothiophene copolymers

28 can stack efficiently so this material displays a fibrillar morphology similar to that seen for **3** [448], but its mobility as measured in FETs is only 10^{-4} cm^2 Vs^{-1} [604], demonstrating that a fibrillar morphology may be a *necessary*, but is not a *sufficient* condition for high mobility. The modest molar mass (<20 kDa) of the polymer may be a factor in its having only such a modest mobility, and the melting into a nematic phase technique used to align **3** is not feasible for **28** as the phase transition temperature is too high. However, it seems likely that even high molar mass **28** will not match the mobility seen for **3**, let alone **24** or **25**, suggesting that other factors besides packing may be involved.

Further complicating the picture are the results obtained from copolymer **29** [605]. This polymer has a nearly amorphous microstructure but displays field-effect mobilities of up 2.5 cm^2 Vs^{-1}. From molecular dynamics simulations the explanation for this was found to be that the polymer backbone has a highly planar conformation allowing extremely efficient, almost defect-free, intra-chain charge transport, which is not affected by the disordered nature of the alkyl side-chains. Occasional close encounters between different chains due to slight bending of the chains allow for inter-chain charge movement through $\pi - \pi$ interactions [606].

High hole mobilities (>1 cm^2 Vs^{-1}) have also been obtained from other donor–acceptor type copolymers, including **30** [607, 608], **31a** [609], **32** [610, 611], and **33** (Figure 8.11) [612, 613]. The highest mobility yet reported from these is 12 cm^2 Vs^{-1} for the copolymers **33a** [613] and **34d** [614]. A number of common factors can be found in all these high mobility materials – they contain strong electron accepting units such as benzodithiazole, diketopyrrolopyrrole or isoindigo units attached to electron donating thiophene-based units, and they form ordered films with closely packed polymer chains.

As the intrinsic mobilities of these materials remain unreported, it cannot yet be assessed if there is any advantage to incorporating particular donor or acceptor units into polymers to maximise mobilities, but studies on changing sidechains in some of these materials have produced some interesting results suggesting possible ways to fine-tune the properties.

It is interesting to note that the siloxane-terminated sidechains in polymer **32** reduce the inter-chain stacking distance from 3.76 Å for the alkyl-substituted

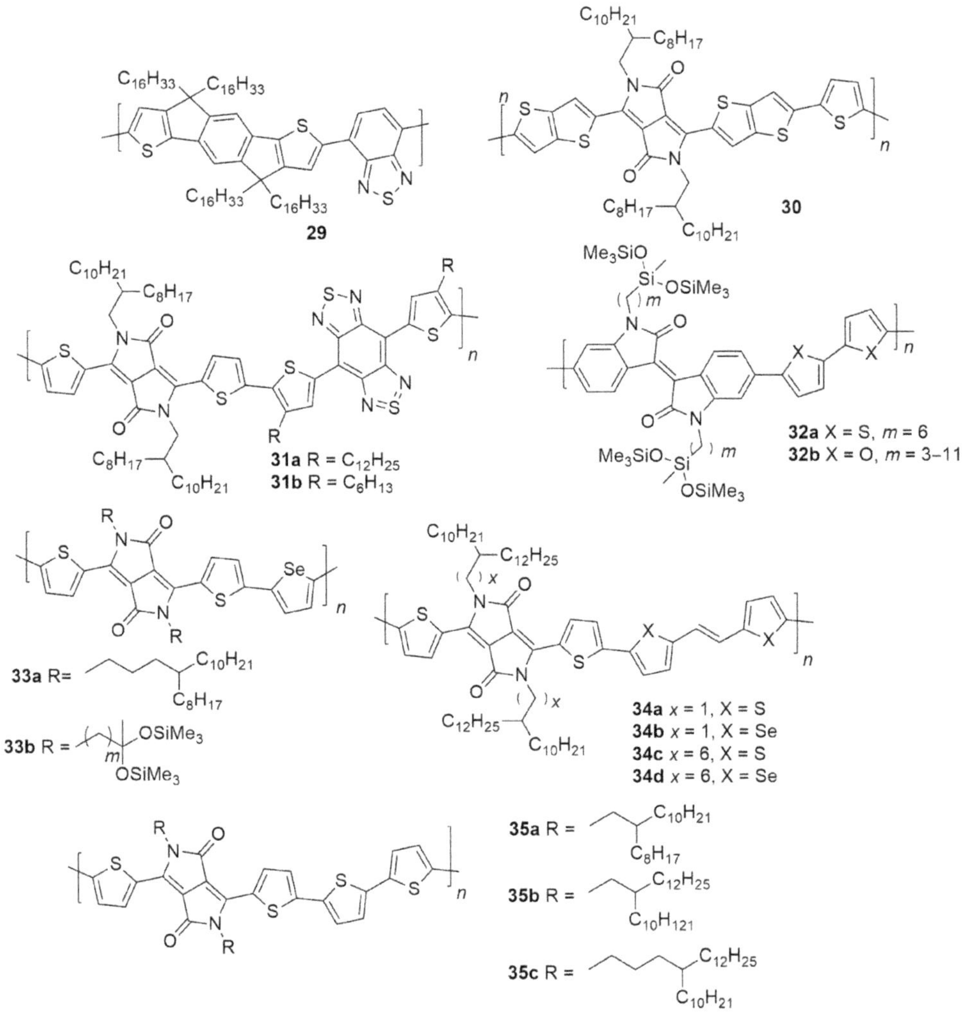

Figure 8.11 Copolymers showing exceptionally high hole mobility in FETs

analogue to 3.58 Å, and so enhanced the mobility from 0.5 up to 2.0 $cm^2\ Vs^{-1}$ [610]. A similar decrease in packing distance with increase in charge carrier mobility has been reported by replacing alkyl chains with oligo(ethylene glycol) chains in a fluorene dithienobenzothiadiazole copolymer [615]. In this case the mobility increase was from 1.5×10^{-6} up to 1.1×10^{-5} $cm^2\ Vs^{-1}$, but as yet there is no report of this approach being adopted for polymers with high mobilities. These results do strongly suggest that sidechain modification is another possible tool for controlling the properties of conjugated polymers for FETs. In this regard it is noteworthy that for polymers **32b** the highest mobility results were obtained for the polymer with $m=9$ (4.5 $cm^2\ Vs^{-1}$) [611], while for **33b** the best results

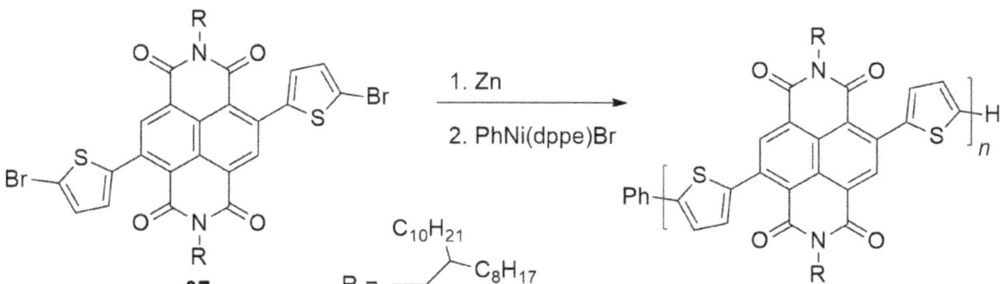

Figure 8.12 Synthesis of a high electron mobility polymer by chain growth polymerisation

(8.84 $cm^2\ Vs^{-1}$) were obtained for the polymers with m = 4 [612]. In both cases polymers with shorter or longer sidechains also showed high but slightly lower mobilities. By contrast, polymers **34c**, **d** showed higher mobilities (10–12 $cm^2\ Vs^{-1}$) than the corresponding polymers **34a**, **b** with slightly shorter sidechains (2.8–4.4 $cm^2\ Vs^{-1}$) [614], indicating that for conjugated polymers there is a definite optimum side-chain length range which may be fairly narrow. Recently it has been reported for copolymers **35a–c** that the mobility increased from 2.1 to 6.9 $cm^2\ Vs^{-1}$ as the sidechain size and the distance to the bifurcation point in the branched sidechain increased, which was associated with closer packing of the conjugated chains [616]. Whether this effect is general, or peculiar to these polymers, has yet to be established.

High electron mobilities have also been obtained from donor–acceptor polymers containing good n-type units [617]. Until recently, the best purely n-type material reported was the naphthalenetetracarboxybisimide-dithiophene copolymer **36**, which has been used to make printed n-channel transistors with electron mobilities of up to 0.85 $cm^2\ Vs^{-1}$ [618]. This polymer has recently been made by a novel chain-growth polymerisation method (Figure 8.12) enabling excellent control of the molar mass, polydispersity and the chain endgroups [619]. The unusual feature of this reaction is that the monomer **37** is first reduced with zinc, and the resulting radical anion then undergoes polymerisation with nickel(II) to form the polymer by a chain growth mechanism similar to that of the GRIM method discussed in Chapter 7.

Currently the best electron mobility reported is for the diketopyrrolopyrrole copolymer **38** (Figure 8.13) made by Stille polycoupling, for which a highest electron mobility of 6.3 $cm^2\ Vs^{-1}$ (average value was 4.54 $cm^2\ Vs^{-1}$) has been reported [620]. The hole mobility was also high, with maximum and average values of 2.78 and 2.20 $cm^2\ Vs^{-1}$, respectively, so this material should be classified as ambipolar rather than n-type (*vide infra*).

In both these materials, the electron affinity of the polymers is attributable to the presence of carbonyl groups containing highly electronegative oxygen atoms.

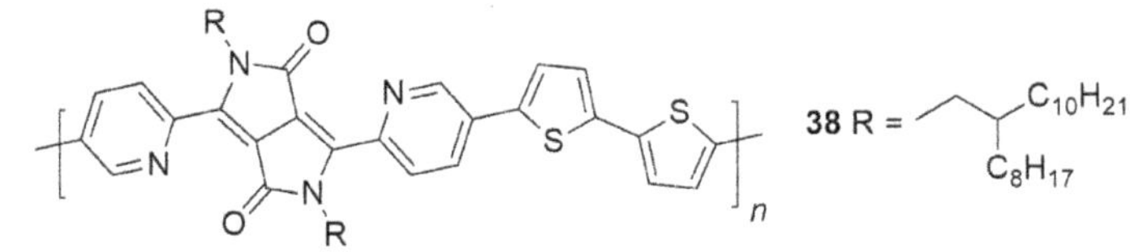

Figure 8.13 Diketopyrrolopyrrole copolymer with very high electron mobility

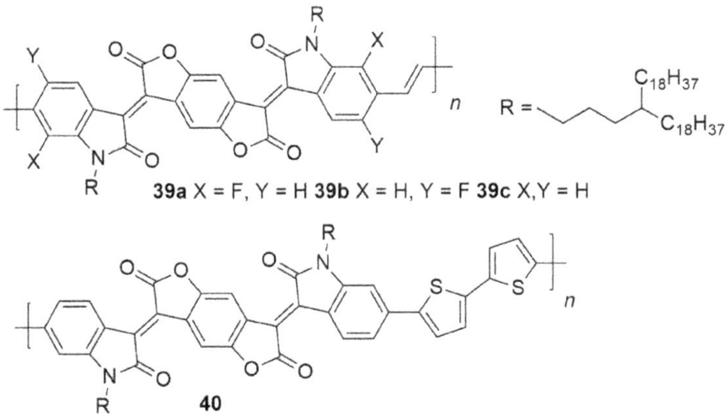

Figure 8.14 Benzodifuranone polymers with very high electron mobilities

Another obvious approach to increase the electron affinity of polymers is to attach fluorines which are even more electronegative. The most successful example of this for transistor applications is the fluorinated polymer **39a** (Figure 8.14) made by Stille coupling of a dibromoarene with distannylethene, for which an electron mobility of 1.7 $cm^2\ Vs^{-1}$ in ambient conditions has been reported [621]. The polymer is reported to be stable in air during operation of the transistors, which is attributed to the electronic effects of the fluorines. The position of the fluorine atoms was important as the isomeric polymers **39b** showed much lower electron mobility (0.62–0.81 $cm^2\ Vs^{-1}$), which was even lower than for the unfluorinated polymer **39c** (0.84–1.10 $cm^2\ Vs^{-1}$). Similar electron mobility values have been reported for the dithiophene copolymer **40**, for devices prepared in a glove box [622]. Devices prepared and operated in air showed slightly lower electron mobilities of only up to 1.4 $cm^2\ Vs^{-1}$. By contrast, preparing devices in an ambient environment was found to significantly enhance the hole mobilities (from 10^{-2} up to 0.47 $cm^2\ Vs^{-1}$), which was attributed to oxygen doping. Thus, depending upon fabrication conditions, one obtains either n-type or ambipolar behaviour.

The first electron mobilities over 1 $cm^2\ Vs^{-1}$ reported in FETs were from polymers **31a** and **31b** (Figure 8.11) which both display values of up to 1.3 $cm^2\ Vs^{-1}$, though only **31b** had an average mobility in devices above 1 $cm^2\ Vs^{-1}$ [609]. An even higher value of 1.56 $cm^2\ Vs^{-1}$ was later reported for **30** [608], while the

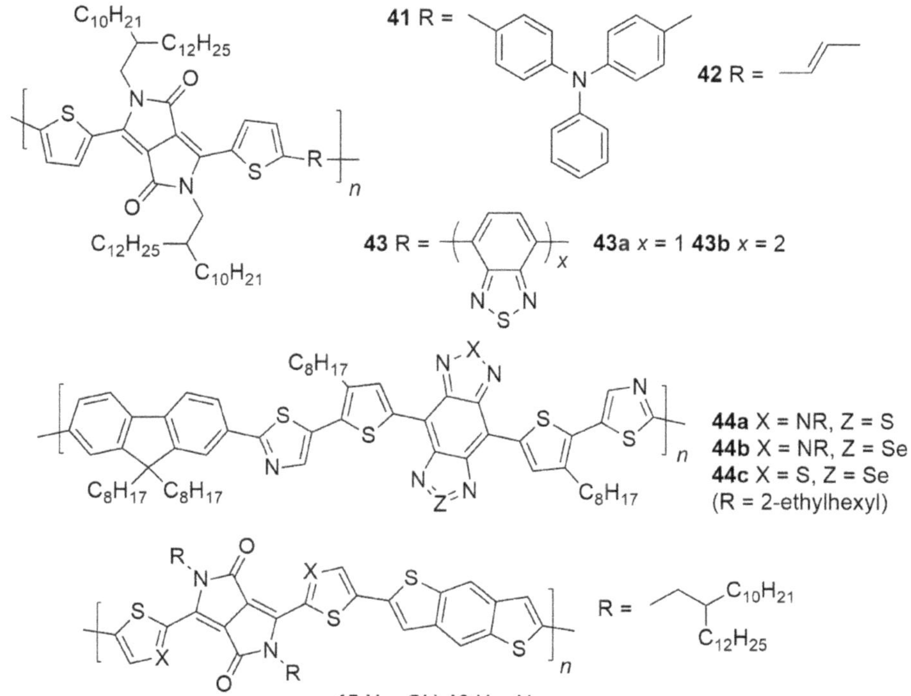

Figure 8.15 Copolymers with variously p-type, n-type and ambipolar transport characteristics

best value yet reported from the polymers shown in Figure 8.11 is 4.34 $cm^2\ Vs^{-1}$ for **32a** [612]. It is interesting to note that the greater alkyl chain length in **31a** increases its hole mobility slightly compared to **31b** but reduces its electron mobility. This possession of both high electron and high hole mobility means **30**, **31a**, **b**, **33**, and **38** are ambipolar rather than n-type in behaviour, as is polymer **40** for devices prepared and operated in air. Depending upon the exact application one may require a material to be either p-type (hole transport dominant), n-type (electron transport dominant) or ambipolar (hole and electron transport roughly equal). An example of how the transport properties of copolymers may be controlled by changing the co-monomer is provided by the diketopyrrolopyrrole (DPP) copolymers **41–43** (Figure 8.15). Copolymer **41** contains a hole-transporting triphenylamine comonomer and so not surprisingly shows p-type behaviour ($\mu_h = 3.3 \times 10^{-3}\ cm^2\ Vs^{-1}$) [623]. The vinylene-copolymer **42** is also p-type with a much higher hole than electron mobility ($\mu_h = 0.17\ cm^2\ Vs^{-1}$, $\mu_e = 0.017\ cm^2\ Vs^{-1}$) [624]. Copolymer **43a** with an electron accepting benzothiadiazole commoner is ambipolar with almost identical hole and electron mobilities ($\mu_h = 0.1\ cm^2\ Vs^{-1}$, $\mu_e = 0.09\ cm^2\ Vs^{-1}$) [625], while the copolymer **43b** with two benzodithiazole units is n-type as it shows much higher electron than hole mobility ($\mu_h = 6.5 \times 10^{-4}$ cm^2 Vs^{-1}, $\mu_e = 1.3 \times 10^{-3}$ cm^2 Vs^{-1}) [626]. The notably higher mobility values for **43a** compared to the other copolymers are a function of its better packing, which is reflected in it showing a more ordered morphology in AFM and in XRD studies than **41** and **43b** (no data is as yet available for **42**). To date there have been no reports on modelling or other studies to determine how these polymer structures correlate to the packing and thus mobility.

Another way to tune the polarity of the charge transport is seen in copolymers **44** containing benzobithiadiazole and analogous benzobisheterocycles (Figure 8.15) [627]. Polymer **44a** with a thiadiazolobenzotriazole unit is a purely p-type material ($\mu_h = 0.7$ cm^2 Vs^{-1}), **44b** with a selenadiazolobenzotriazole unit is ambipolar (μ_h and μ_e both ca. 10^{-3} cm^2 Vs^{-1}) and **44c** with a selenadiazolobenzothiadiazole unit shows dominant n-type charge transport ($\mu_h = 7.3 \times 10^{-3}$ cm^2 Vs^{-1}, $\mu_e = 0.09$ cm^2 Vs^{-1}). A similar switch from p-type to ambipolar transport is seen for the benzodithiophene-diketopyrrolopyrrole copolymers **45** and **46** [628]. Copolymer **45** containing thiophene units is a p-type material ($\mu_h = 0.72$ cm^2 Vs^{-1}, $\mu_e = 3.4 \times 10^{-3}$ cm^2 Vs^{-1} in thin films), whereas **46** with thiazole units is ambipolar ($\mu_h = 0.14$ cm^2 Vs^{-1}, $\mu_e = 0.11$ cm^2 Vs^{-1} in thin films). When the polymers are assembled into nanowires, both hole and electron mobilities increase by an order of magnitude for both **45** ($\mu_h = 7.42$ cm^2 Vs^{-1}, $\mu_e = 0.04$ cm^2 Vs^{-1}) and **46** ($\mu_h = 5.47$ cm^2 Vs^{-1}, $\mu_e = 5.33$ cm^2 Vs^{-1}). These results suggest that electron mobilities matching the hole mobilities reported above for polymers **24** and **25** might be attainable for suitable fractionated and oriented samples of n-type or ambipolar polymers, and that if measured their intrinsic electron mobilities may turn out to match those for holes.

With ambipolar materials it should also be noted that, for some applications, the absolute magnitude of the mobilities may not be as important as that the two mobilities closely match each other – that is, that the active material is as good an n-type as p-type material. That said, however, it is definitely true that possessing high hole and electron mobilities is still desirable for ambipolar materials so based on mobility criteria polymers such as **29**, **30a**, **32** and **46** with mobilities >1 cm^2 Vs^{-1} for both carrier types, are currently the best such polymers. A feature which has yet to be reported for these polymers is their stability in each operating mode, as one major problem with organic n-type materials is device stability and again it may be more important that the stability in both modes be similar than that the mobilities be high. It will be interesting to see if the mobility values recorded for the best p-type materials can be matched by n-type and ambipolar polymers. Device mobilities have been improving for the past two decades and now reach ~ 5 cm^2 Vs^{-1} [629]. However, although hero devices with mobilities >1 cm^2 Vs^{-1} have been reported, these have been for devices that have either been for 'one-off' transistors or, at best, for a few carefully engineered devices. In general, the vast majority of materials have mobilities that are below 1 $cm^2\ Vs^{-1}$ and thus the applications that are available to printed electronics are still limited. Indeed, it has been argued that, with the exception of OLED displays (where mobilities of ~1 $cm^2\ Vs^{-1}$ are sufficient for video addressing rates), the seminal application for organic electronics has yet to be identified [630, 631]. While the progress in improving electron mobilities still lags a little behind that in hole mobilities in absolute terms, the relative differences between the best hole and electron mobilities have narrowed considerably from orders of magnitude to factors of less than 5 [632]. Despite these improvements, the vast majority of organic electronic materials are still p-type and, although improvements in n-type mobilities have occurred, these are still far behind their p-type counterparts. As such, most printed analogue circuits are built from unipolar p-type materials, which results in low gain due to imprecise output currents in the conventional current mirror circuits used in operational amplifiers.

Not only is the absolute value of mobility an issue in printed organic electronic FETs but so is the variability of the device mobility, which is typically around $\pm 30\%$. These variations can lead to large differences in threshold voltage, which in turn drive errors in the oscillation frequency of ring oscillator circuits and gain variations in differential amplifiers. However, recent work has shown that it is possible to reduce the variability of FET device performance with careful material design and processing [633].

8.4 Copolymers for OPVs

When selecting materials for OPVs, a number of criteria are important, as was discussed in Chapter 1. First is the bandgap with the ability to absorb strongly over a wide range of the solar spectrum, especially in the visible and near-infrared parts of it, being necessary for high efficiency. Too low a bandgap is also inefficient as then much of the energy from absorption of the shorter wavelength photons (i.e. UV through blue light) will be lost as heat or by other non-radiative pathways. Current models suggest a bandgap of 1.3–1.75 eV is optimal for a material for use in a single-layer device, while for tandem devices two materials with bandgaps of 1.1–1.5 eV and 1.4–1.75 eV for top and bottom layers, respectively, are predicted to be optimal [543]. The second criterion is that the LUMO energy of the material should not have too low or too high an offset from the LUMO energy of the other component in the bulk heterojunction blend. As usually the polymer is the electron donor, it has been argued that its LUMO should be at least 0.3 eV higher than the acceptor in order for electron transfer to be efficient as per Markus theory but not much higher than that in order to minimise the energy lost in the electron transfer.

However, the minimum LUMO offset for high efficiency OPVs has been the subject of debate within the physics/chemistry literature. Strictly speaking, it is $\Delta E_{CS} = \Delta E_{LUMO} - E_{EX}$ that needs to be sufficient to provide the driving force for charge separation, where ΔE_{CS} is the energy offset driving charge separation, ΔE_{LUMO} is the difference in LUMO energies and E_{EX} is the exciton binding energy [634]. Thus, given that E_{EX} is typically estimated to be around 0.3 eV for an organic exciton, the requirement that $\Delta E_{LUMO} > 0.3$ eV is equivalent to $\Delta E_{CS} > 0$. Interestingly, it has recently been shown that for certain non-fullerene acceptors (NFAs) it is possible to achieve effective charge separation with $\Delta E_{LUMO} \sim 50$ meV; substantially below the empirical value of 0.3 eV [635].

Thus, too strong an acceptor in a donor–acceptor copolymer may be undesirable as it lowers the LUMO and the overall bandgap too much. The open circuit voltage (V_{oc}) from a bulk heterojunction OPV is equal to the energy difference between the HOMO of the donor and the LUMO of the acceptor less the exciton binding energy. As a result, a lower HOMO energy level means a higher V_{oc} and thus a higher potential power output. Too high a HOMO is also undesirable because it makes the material susceptible to oxidation by air which results in unstable output and short device lifetimes. Thus, too strong a donor unit in a donor–acceptor copolymer is also generally undesirable as it raises the HOMO too much. If one is using a fullerene such as $PC_{61}BM$ or $PC_{71}BM$ as acceptor, an ideal polymer should thus have HOMO energy around –5.4 eV and LUMO around –3.9 eV, which means a V_{oc} of about 0.9–1.0 V. The best solution seems to be a relatively poor donor with a moderately strong acceptor – the so-called weak donor-strong acceptor strategy [636–639]. The strength of the acceptor is not just a function of the number and type of heteroatoms, but also of its tendency to form quinonoid resonance forms as these lower energy forms also help lower the polymer bandgap. An example of how this applies in practice is to compare the orbital energy levels of the copolymers **31** described in Figure 8.11 containing the very strong acceptor benzobisthiadiazole (BBT) with those of the analoguous copolymers **47** and **48** (Figure 8.16) containing, respectively, benzodithiazole (BT) and benzene rings [609]. Comparison of the quinonoid forms **49–51** for the BBT, BT and benzene copolymers shows that the BBT quinine form **49** is most stable as it contains two aromatic rings, while the benzene quinonoid form **51** is least stable as it contains no aromatic rings. As a result, the bandgaps of copolymers **31** (0.6 eV) are much lower than those of **47** (1.35 eV) and **48** (1.5 eV).

One interesting and important point about such donor–acceptor copolymers that has been determined from molecular orbital calculations is that, while the HOMO is spread evenly across both units, the LUMO electron density seems to be concentrated rather strongly on the acceptor units, so that the LUMO energy of the copolymer is determined largely by the strength of the electron acceptor [640, 641].

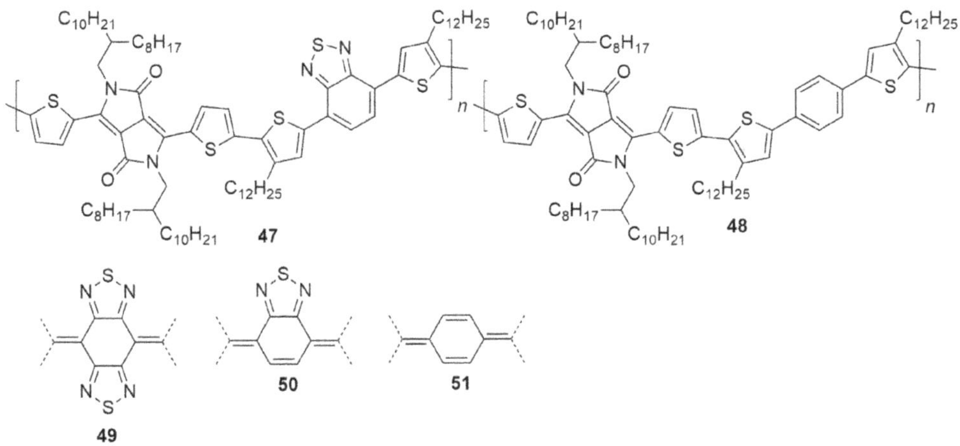

Figure 8.16 Copolymers bearing acceptor units with varying quinonoid character

This dramatic lowering of the bandgap with very strong acceptors such as BBT means that even donor–acceptor oligomers based on such systems can be quite low bandgap molecules producing near infrared emission, and combining BBT with a strong donor can produce polymers with bandgaps in the 0.5–0.6 eV range [641]. Such polymers are unsuitable for OPVs as their LUMO is too low to allow them to donate electrons efficiently to standard acceptor materials and their HOMO is too high for them to be good acceptors and, even if a suitable partner could be found the voltage obtainable would be too small. Materials with very low (<1 eV) bandgaps are, however, useful in photodetectors where high output currents are not needed. Indeed, the area of organic electronic materials applied to photodetectors has received increasing interest for industrial purposes. In particular, the use of low band-gap organic materials for near-IR sensing has attracted attention, driven by a variety of applications such as optical communications, surveillance and chemical sensing [642], with a number of excellent reviews available on the topic [643].

The electron accepting properties of BBT can be tuned by replacing one or both of the thiadiazole rings with other heterocycles to produce units such as thiadiazoloquinoxaline (Figure 8.17, **52**) [644], pyrazinoquinoxaline (**53**) [644], thia-diazolobenzotriazole (**54**) [645, 646] and triazoloquinoxaline (**55**) [645]. Comparing the properties of oligomers containing these units enables one to derive some structure–property relationships. For instance, replacing the sulphur in thiadiazole with a nitrogen to give a triazole seems to raise the bandgap mainly by raising the LUMO, that is, the triazole is a weaker electron acceptor. The plus side is that one gains higher solubility as one can attach an alkyl group on the middle nitrogen of the triazole. One of the advantages of the quinoxaline structure in some of these units is that not only do the substituents provide solubility but one can also

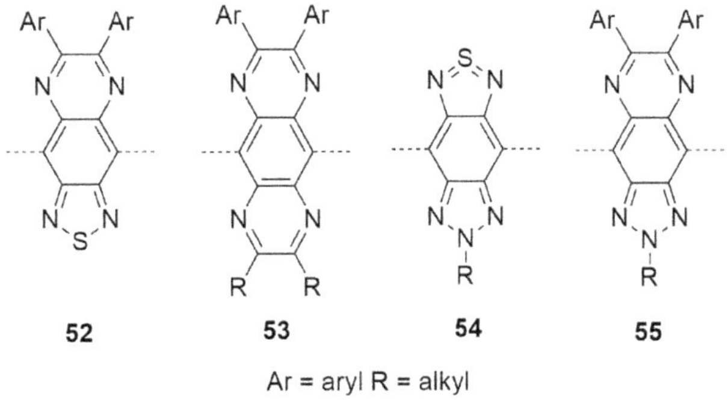

Figure 8.17 Heterocyclic units of variable electron accepting properties

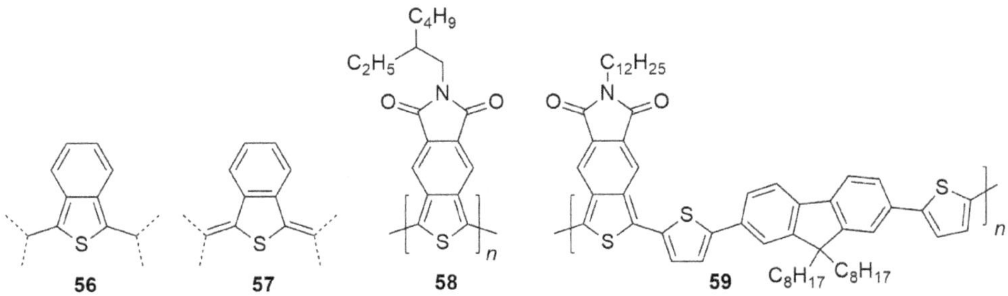

Figure 8.18 Isothianaphthene resonance forms and some isothianaphthene-based polymers

seek to tune the electronic properties by choosing electron donating or accepting groups, for example, the red-shifting effects of thienyl versus phenyl substituents on the optical spectra of derivatives of **52** have been studied and explained in terms of their relative effects upon the quinonid resonance structures [647].

One aspect of the use of strong acceptors in donor–acceptor molecules that must be considered is that they may show two well-separated strong absorption bands, of which the lower energy one is attributable to internal charge transfer but with limited absorption between them. This is a disadvantage when designing molecules for OPVs, as there one wants strong absorption across as much of the spectrum as possible. For example, comparison of the spectra of oligomers of **52** and **53** with the analogous BBT derivative showed that the former had larger bandgaps but displayed strong absorption over a greater section of the visible spectrum [644].

While the influence of the quinonoid resonance forms upon the bandgap must be taken into account in choosing components for a copolymer, polymers in which the ease of formation and stability of the quinonoid structure is the main cause of their having a low bandgap do not do well in solar cells as they usually have too high a HOMO. For example, polymers based on isothianaphthene **56** (Figure 8.18)

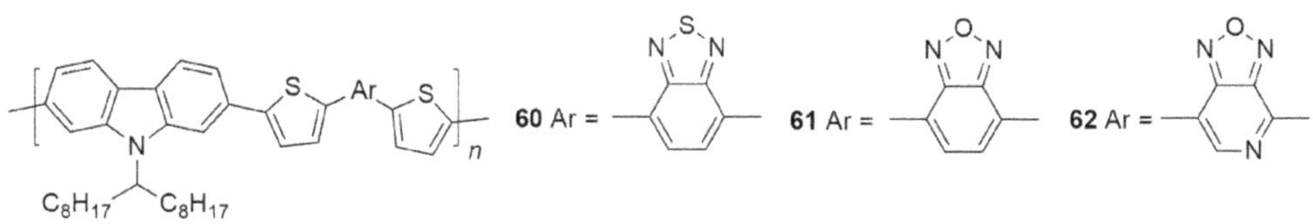

Figure 8.19 Carbazole-based polymers for solar cells

have been investigated as low bandgap materials as these units have stable quinonoid forms **57**. These forms are low energy due to the aromatisation of the benzene ring in them, with the result that some isothianaphthene-based polymers (e.g. **58**) display high conductivity as they have bandgaps smaller than 1 eV [648]. The performance from isothianaphthene-based copolymers as donors in solar cells, however, has been disappointing, with the highest efficiency so far reported being only 1.6% for a device using copolymer **59** [649]. The main cause of this low efficiency seems to be the mediocre absorptivity of the polymer, which is possibly due to excessive localisation of electron density on the imide units, that is, away from the conjugated main chain. The donor–acceptor concept thus remains the best way to tune the bandgap to obtain suitable materials.

As already mentioned, a very large number of donor–acceptor polymers have been made and tested in solar cells and some structure–property relationships which might serve as guides towards efficient materials have begun to emerge [244, 570, 637, 638, 650]. The main problem remains that, as yet, there is no reliable method for predicting the solid-state structure, let alone the behaviour in the solid state, of a given molecule. While some necessary parameters such as orbital energy levels and electron densities can now be fairly reliably predicted using established theoretical models, others such as charge carrier mobility cannot be so predicted, as they depend upon the solid-state packing. Thus, while we can usually eliminate candidate structures that probably cannot produce high device efficiencies, we are still at the stage where for plausible structures we must actually make the materials and test them, then seek to rationalise the results. Space does not permit describing all the various donors and acceptors that have been investigated so we will look at two classes of materials that have recently attracted attention as producing markedly high efficiencies in devices and at some of the attempts to optimise their performance by synthetic design.

First we will look at carbazole-based copolymers which came to prominence as they were the first class of polymers to produce efficiencies of above 6% in BHJ solar cells [462]. This was achieved using the dithienyl-benzothiadiazole copolymer PCDTBT (**60**) (Figure 8.19), which has a bandgap of 1.9 eV and HOMO and LUMO energies of –5.5 eV and –3.6 eV, respectively. Using the model developed at Konarka for predicting potential device efficiency suggests that this polymer when blended with PCBM has a potential maximum device efficiency of about 6.5%. The first reported devices produced only 3.6% efficiency [651], but subsequent device optimisation brought this up to 6.1% [652] and then to a certified value of 6.45% [653], which meant these devices have an internal quantum efficiency of close to 100%. Later using $PC_{71}BM$ efficiencies of up to 7.5% have been reported for single-layer devices and up to 8.9% in tandem cells, with promising lifetime data [654]. To attempt to improve on these results, new polymers **61** and **62** were designed whose orbital energies suggested efficiencies of 8–10% could be possible but the device results were disappointing with device efficiencies of only 2.4% and 0.8%, respectively, being obtained [655]. The main cause was the low currents in the devices which correlated with their having lower charge carrier mobilities than **56** and poorer polymer packing as shown by Wide-Angle X-ray Scattering (WAXS) experiments. This may partially be due to the low molar masses obtained for these materials, but probably is due to some other factor which is not yet fully understood, that inhibits the efficient packing of these materials. What this reiterates is the important point that the right orbital energies and band gap may be *necessary* but are not *sufficient* conditions for obtaining high efficiency in BHJ solar cells, and that other, often as yet poorly understood, factors may prove decisive in determining the overall effectiveness of materials in devices.

While carbazole copolymers were the first to improve upon the efficiency of P3HT:PCBM blends, the biggest advance in efficiencies has come from donor–acceptor copolymers with benzodithiophene (BDT) as a donor unit. The first of these to show high efficiency was copolymer PTB7 (**63**) (Figure 8.20) with a fluorinated thienothiophene as acceptor. PTB7 was initially reported to produce

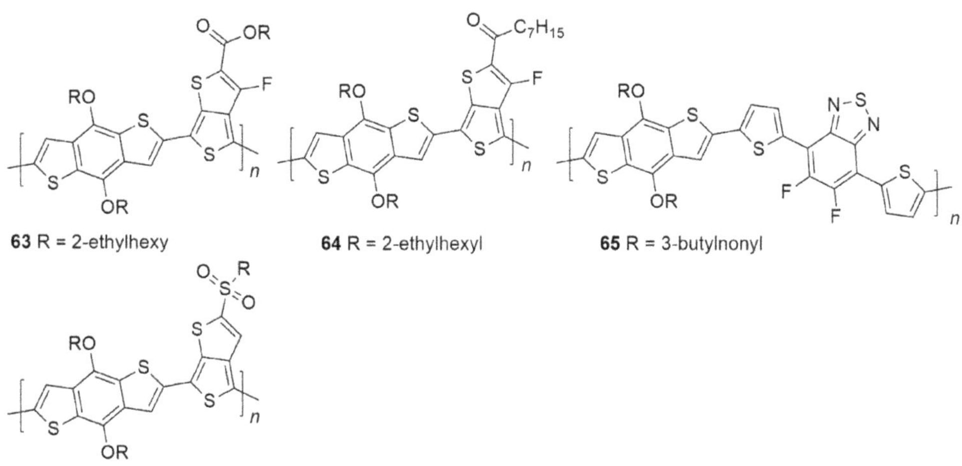

63 R = 2-ethylhexy

64 R = 2-ethylhexyl

65 R = 3-butylnonyl

66 R = 2-ethylhexyl

Figure 8.20 BDT-based copolymers with high OPV performance

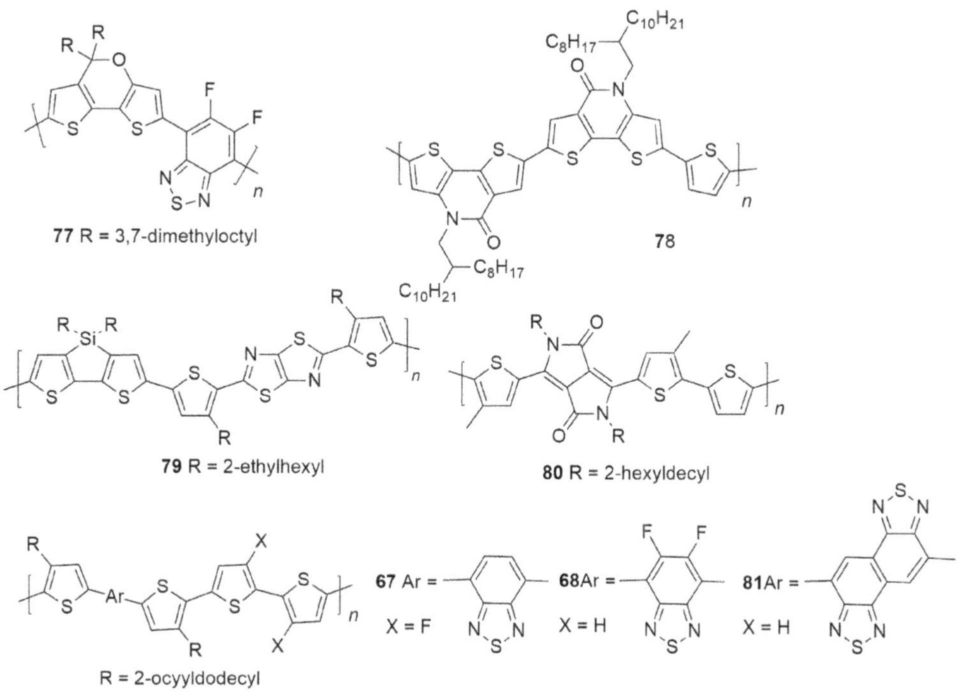

Figure 8.21 Other polymers used to make OPVs with >10% efficiency

efficiencies of up to 8.37% with a confirmed certified value of 7.65% [656]. Later device modifications involving using a metallopolymer cathode inter-layer [657] or a nano-patterned film of PTB7 and $PC_{71}BM$ [658] produced efficiencies of over 9%. The similar polymer **64** has been reported to display up to 7.7% efficiency but the maximum certified value is only 6.75% [67].

One notable feature of these BDT polymers is that a fluorine substituent markedly enhances the device performances as similar polymers without this unit display notably lower efficiencies [67, 637, 650]. This may be a more generally applicable design feature as the copolymer **65** with a fluorinated benzothiadiazole unit also displays efficiency above 7% [659]. This improvement seems to be due to the inductive effect of the fluorine atoms making the new units better electron acceptors. A similar effect may explain the high efficiency (6.2%) seen from the sulphonyl-substituted copolymer **66** [660]. A fluorine on the donor unit in a donor–acceptor copolymer may also be of assistance as exemplified by copolymers **67** and **68** (Figure 8.21), which both show higher device efficiencies (7.2% and 6.4%, respectively) than the corresponding unfluorinated copolymer (1.8%) [661]. Device optimisation has produced efficiencies of over 10% for both these polymers [662].

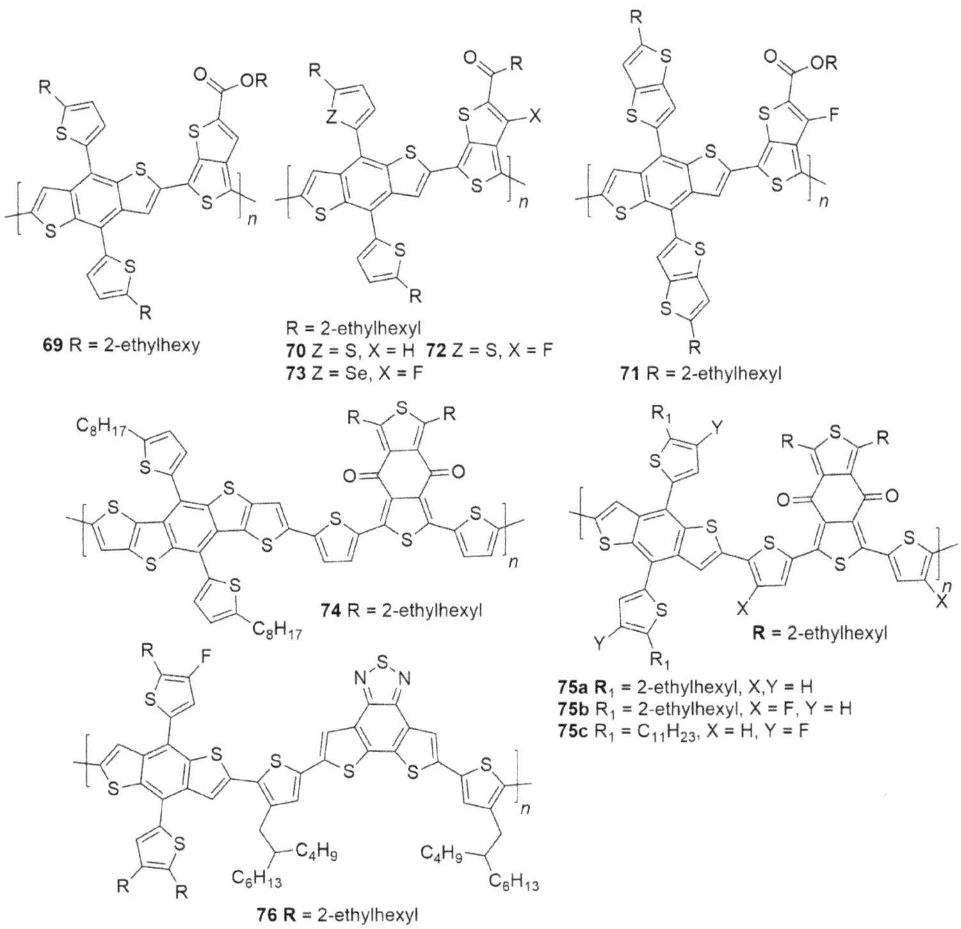

Figure 8.22 BDT-copolymers with 2-D conjugation

One approach that has been suggested to improve the efficiency of an OPV is to increase the optical absorptivity of the polymer donor. For BDT copolymers this has been achieved by attaching thienyl instead of alkoxy substituents to the benzene ring with the idea that the 2-D conjugation will increase the optical cross-section and may also improve packing as the bigger π-systems will enhance π–π interactions between the molecules. This approach has proven successful with the reported charge carrier mobilities, J_{sc} and PCE values for unoptimised devices using the copolymers **69–70** (Figure 8.22) all being higher than for the corresponding polymers with alkoxy substituents, with the best initially reported efficiency value being 7.59% for **67** which is comparable to the values first seen for the fluorinated polymers **63–64** [663]. The fluorinated polymer **71** produced efficiencies of 7.44% and 7.71% in standard and inverted configuration cells with $PC_{71}BM$ as an acceptor, and 8.66% in a tandem cell with P3HT:ICBA as the bottom cell [664]. Recently, extremely good results have been obtained from devices using PTB7-Th (**72**). Using a nano-patterned film of **72** and $PC_{71}BM$ [658] or a dual-doped zinc oxide nanofilm [665] produced efficiencies of 10.1% and 10.31%, respectively, which to the best of our knowledge are currently the highest efficiencies reported for single-layer OPVs using fullerene acceptors. An optimised inverted device using **72** with a photoconductive cathode inter-layer has been reported to show an efficiency of 10.5% [666], while a tandem cell using two layers of **72** of different thicknesses (a homo-tandem cell) has been reported with an efficiency of 11.3%, surpassing that seen from any device using **63** or **64** [667]. Efficiencies of up to 6.5% were obtained using **63** with the same acceptors. The selenophene copolymer **73** also shows excellent device performance, with 8.8% efficiency obtained from blends with $PC_{71}BM$ without use of solvent additives or special inter-facial layers [668]. A homo-tandem cell produced 9.9% efficiency. The π-system of BDT can also be extended by fusing more thienyl groups on to the ends as in copolymer **74**. This material has been used to make a device with 9.7% efficiency with $PC_{71}BM$ as an acceptor [669].

What has enabled recent further significant advances in the maximum efficiency of polymer-based organic solar cells has been the development of new non-fullerene acceptors. These lie outside the scope of this book but the increase in efficiencies seen with these appears to arise from their stronger absorption of light and/or their higher charge carrier mobilities than fullerenes. As the obtained efficiencies of over 13% exceed the maximum possible efficiencies predicted using the model developed at Konarka for OPVs using fullerenes [543], probably due to the assumptions built into that model, for example, about energy losses during electron transfer and absorption and charge carrier mobilities of the materials, being no longer valid, a new model will need to be developed to enable better prediction of maximum efficiencies. To the best of our knowledge, no attempts have been made to systematically test these new acceptors with the donors used in the previously reported high efficiency devices. The donors used in the new higher efficiency cells were polymers **75a–c** which combine the donor unit of **71** with the acceptor unit of **74**. PBDB **75a** produced 13% efficiency with a non-fullerene acceptor due its high V_{oc} (0.88 V), J_{sc} (20.5 mA cm^{-2}) and FF (0.72) [670], while a higher value of 13.2% was obtained using another acceptor with PFBDB-T (**75b**), due to a higher V_{oc} (0.94 V), J_{sc} (19.6 mA cm^{-2}) and FF (0.72) [671]. These were the first single OPV cells to exceed 13%. The highest value of 16.5% was achieved using **75c**, whose V_{oc} (0.87 V) was lower, but had a much better J_{sc} (25.4 mA cm^{-2}) and a slightly better FF (0.75) [672]. The best tandem cell reported uses two layers of **75b** each with a different non-fullerene acceptor [673]. By careful device optimisation a certified efficiency of 19.5% ($V_{oc} = 1.91$ V, $J_{sc} = 14.21$ mA cm^{-2}, FF = 0.72) was obtained – the best uncertified efficiency was 19.64%. At the time of writing, the highest efficiency reported for a polymer solar cell is from a device using polymer **76**, which incorporates a bisthienylbenzothiadiazole acceptor unit [674]. In conjunction with a non-fullerene acceptor, which coincidentally also contained a benzothiadiazole unit, a certified efficiency of 17.6% was achieved with a V_{oc} of 0.842 V, J_{sc} of 26.67.6 mA cm^{-2} and FF of 0.784. The best cell showed 18.22% efficiency due to a higher V_{oc} of 0.859 V and J_{sc} of 27.7 mA cm^{-2} despite a slightly lower FF (0.766). Further improvements in voltage and/or FF for such cells would make efficiencies of 20% or maybe even 25% is possible.

While BDT copolymers have produced many of the best device results, other donor units have also shown impressive results. For example, some of the most efficient tandem cells yet reported use the dithienopyran-based donor–acceptor polymer PDTP-DFBT **77** (Figure 8.21) which also contains a fluorinated-acceptor. A single-layer cell using **77** plus $PC_{71}BM$ produced 7.9% efficiency while a tandem cell with P3HT:ICBA produced 10.6% efficiency which made it the first polymer-based solar cell to exceed 10% efficiency [675]. Later a tandem cell with 10.2% efficiency was reported consisting of two layers of **77** plus $PC_{71}BM$ of different thickness, with a MoO_3-based inter-layer [63]. The corresponding inverted single-layer device with MoO_3 as a hole transport layer showed up to 8.1% efficiency. A copolymer **78** containing a structurally similar lactam unit has been used to make a device with 9.1% efficiency [676].

More recently, even higher efficiencies have been reported from triple-junction cells. Cells using layers of P3HT, PTB7 and **77** [677] and copolymer **79**, PTB7 and the copolymer **80** [678] have been reported to produce, respectively, efficiencies of 11.55% and 11.83% which were the highest efficiencies then reported for polymer-based solar cells. More recently it was reported that a tandem cell using cells based upon **70** and a polymer similar to **74** with non-fullerene acceptors and $PC_{71}BM$ had a certified efficiency of 17.3% for the best device (average 16.9%), which at the time of writing is the most efficient tandem OPV device yet reported [679].

A recent report that efficiencies of over 10% can be obtained from the relatively simple dialkylquaterthiophene copolymers **67**, **68** and **81** (Figure 8.21) shows that complicated co-monomers are not necessary for obtaining very high efficiencies [662]. The reported high efficiencies are claimed to be the result of optimisation of the film morphology, suggesting that, with appropriate processing, very high efficiencies may be obtainable from a range of other copolymers. This has commercial significance as one of the main potential obstacles to the commercial use of polymers such as **63** or **72** is their relatively lengthy, inefficient and costly synthesis.

Indeed, the cost of organic photovoltaics has been gaining increasing focus over the past few years with several publications (and the references cited therein) discussing the economics of printed solar cells [680–682]. Consequently, the importance of the synthetic cost of the organic electronic materials used in organic photovoltaic devices is attracting increasing attention. Recent work has highlighted the importance of economic cost and, in particular, the costs of materials synthesis, as a third key experimental parameter that needs to be optimised (alongside efficiency and lifetime) in the development of new organic photovoltaic materials [683]. Moreover, a recent study comparing efficiencies with the synthetic complexity of the polymers for use in solar cells concluded that none of the high performing materials reported up to 2013 was likely to prove commercially viable due to the complexity of their synthesis [684]. A similar calculation has not been performed on materials for FETs but it is probable that a similar conclusion might be reached there as well.

While the recent advances have pushed the efficiencies for organic PV devices well over the 10% mark currently widely considered necessary for commercial viability for most applications, some room for further improvement clearly remains. This will involve improving both the cost of the materials by improving the efficiency of the synthesis route and/or improving any or all of the following device parameters: V_{oc}, J_{sc} and/or FF. The first is easily attainable by lowering the HOMO of the donor and/or raising the LUMO of the acceptor. An example of the latter approach is shown by the recent development of new fullerene derivatives which possess higher LUMO energies than PCBM [685]. These have already enabled the efficiency of devices using P3HT as donor to rise from 5 to 7% [554], as mentioned in Chapter 7, and similar increases may well be possible for other polymers. Good efficiencies of up to 8.3% have been obtained using **72** with non-fullerene acceptors consisting of fused perylene-dimide units [686], suggesting that such acceptors might also be possible alternatives to fullerenes, though their synthesis may prove more expensive. The use of high mobility n-type polymers as acceptors has also been found to be promising with 4% efficiency being obtained from a blend of the n-type polymer **36** with the thiophene-quinoxaline copolymer **82** (Figure 8.23) [687] and 7.7% being obtained from a blend of the similar n-type polymer **83** with a BDT-copolymer **84** [688]. All-polymer solar cells with efficiencies >10% have been obtained using a mixture of **36** (R = 2-octyldodecyl) with **85** (V_{oc} = 0.87, J_{sc} = 15.57 mA cm^{-2}, FF = 0.73) [689]. Given the large LUMO difference between these (0.7 eV) it is likely that better matched polymers might give much higher efficiencies and there is obviously no reason why the efficiencies of all-polymer cells should not match those seen using non-polymeric acceptors. These advances are due to improvements in n-type polymers as acceptors, as has been reviewed by Andersson and colleagues [690].

The other approach is exemplified by, for example, replacing a 2,7-carbazole with a 2,7-fluorene unit as fluorenes have lower HOMOs than carbazoles. They also generally display better solubility as they possess two solubilising sidechains per

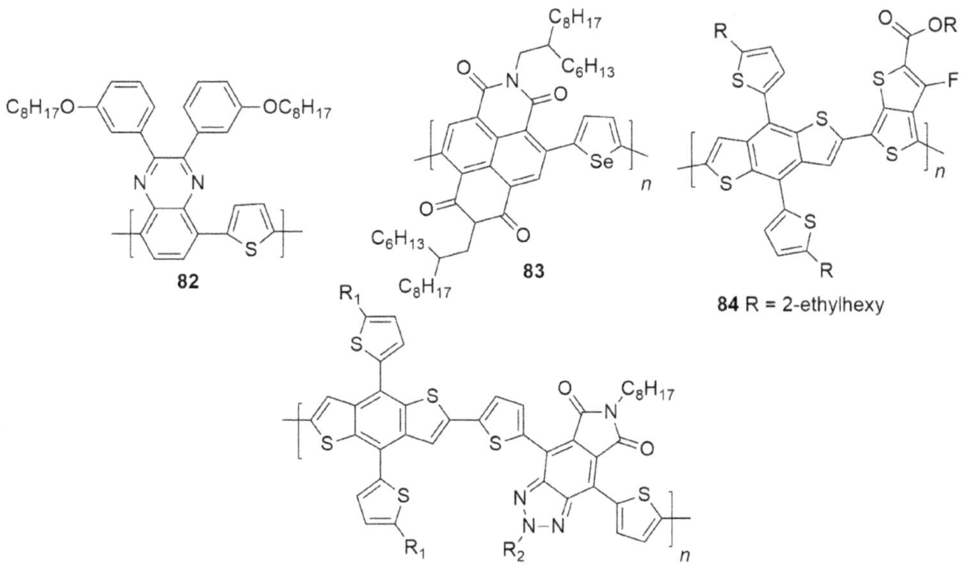

82 83 84 R = 2-ethylhexy

85 R_1 = 2-ethylhexyl, R_2 = $(CH_2)_6Si(CH_3)(OSiMe_3)_2$

Figure 8.23 Polymers used to make high efficiency all polymer solar cells

monomer unit instead of one. The best efficiency obtained for polymers **86** (Figure 8.24) which are the fluorene-based analogues of the best performing carbazole polymer **60**, is only 4.5% [570, 691]. While this lower value might reflect the fact that these polymers have yet to undergo the same degree of device optimisation as the carbazole material, comparison of the device data does suggest there is an inherent problem in the structures themselves. Their V_{oc} values are higher than for the carbazole copolymer **60** (0.97–1.03 V *cf.* 0.89 V) as expected, but the J_{sc} and FF values are notably lower, which has been ascribed to poorer packing as the two side-chains will tend to take up more space than the single substituent on the carbazole. That the efficiency varies notably (between 2.8 and 4.5%) with the side-chain also suggests that the chain packing is affecting the efficiency.

One recent suggestion to improve the packing in fluorene-based polymers is to make the monomer units more planar as in polymer **87**, for which an efficiency of 6.2% has been obtained, compared with 3.1% for the corresponding fluorene-copolymer **88a** [692]. Another is to replace the alkyl chains on the fluorene unit with oligo(ethylene glycol) chains, which has been found to produce an increase in device efficiency from 2% in **88b** to 4.04% in **88c**, which is associated with an increase in charge carrier mobility in the polymers due to closer $\pi-\pi$ stacking (4.1 vs 4.1 Å inter-chain distance) [615]. That fluorene-based copolymers can produce very high V_{oc} values is proven by the copolymer **89**. A cell using this with $PC_{61}BM$ produced an exceptionally high voltage of 1.11 V, though the efficiency

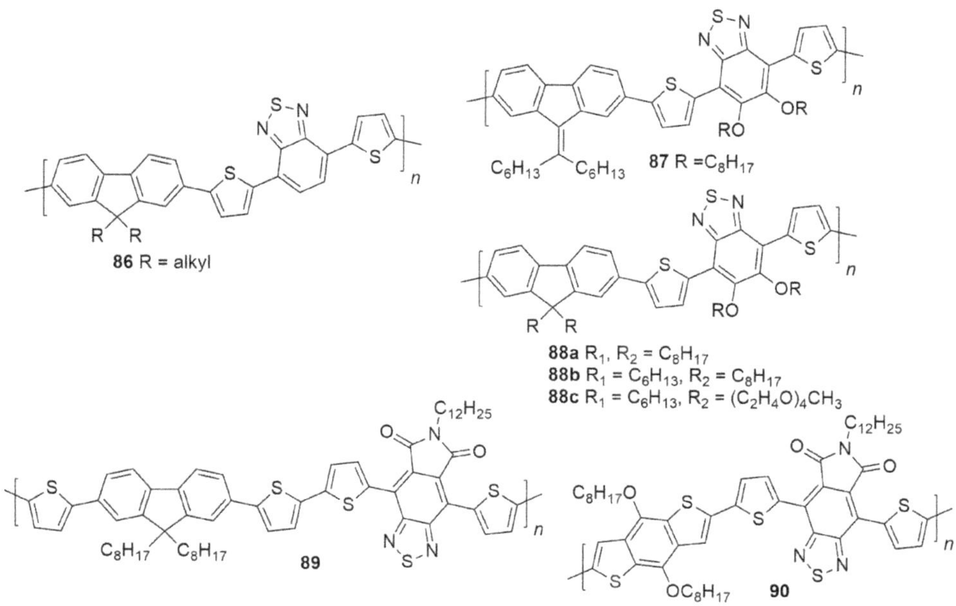

Figure 8.24 Fluorene-based copolymers for OPV

was only 1.61%, which was attributed to poor polymer packing. By contrast, the BDT copolymer **90** produced 3.42% efficiency in combination with $PC_{61}BM$, but with a voltage of only 0.75 V [693].

One area that needs to receive more systematic attention is the question of enhancing device lifetime by molecular/synthetic design. In the only systematic study yet reported on the relative photostability of conjugated copolymers with various donor and acceptor groups, it was found that the thienothiophene units used in the BDT polymers, for example, **63**, discussed earlier in this section confer lower long-term photostability on copolymers containing them than do benzothiadiazole or thienopyrazine units, while the most photostable donor units appeared to be dithienosilole units (Figure 8.25, **91**) [694]. The alkoxy-BDT units found in the high performing polymers **63–66** described earlier in this section were not studied, but their alkoxy groups may reduce their stability as electron-rich rings such as dialkoxybenzenes are usually more susceptible towards aerial oxidation than unsubstituted benzenes are. How useful the insights gained by this study will be for development of future high performing polymers is as yet unknown. Recently, 6% efficiency was reported from a device using a polymer **92** containing a photostable dibenzosilole donor [695]. The desired goal of combining high efficiency with good photostability seems achievable, which is supported by this high efficiency (6.74%) with good stability (>95% of PCE after 1,000 h) very recently reported from devices using the polymer **93** [696]. The high stability has

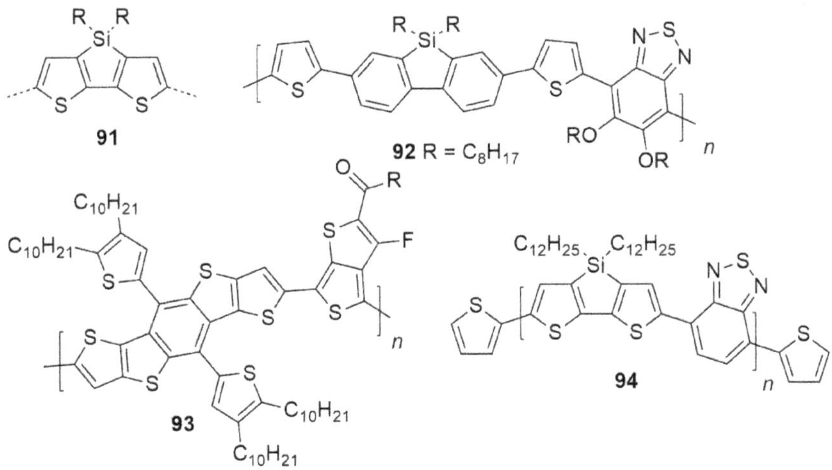

Figure 8.25 Highly photostable donor units and polymers

been attributed to the crystalline nature of the blend film due to the highly conjugated planar nature of the donor unit in the polymer. There have been some studies on degradation in devices of dithienylcyclopentadiene-benzothiadiazole copolymers related to **24** (Figure 8.7), which suggest that the presence of PCBM increases the rate of degradation of the cyclopentadiene unit by oxygen more than that of the benzothiadiazole unit [697] and that attachment of ester or alcohol groups to the ends of the alkyl substituents may increase their thermal stability [698]. Whether these findings have more general application remains to be seen.

A number of other factors that are relevant when designing the synthesis of a polymer for high efficiency OPVs are polymer purity, molecular weight, polydispersity and also end groups. The addition of even traces of palladium catalyst to the polymer film has been shown to lower the efficiency of devices using PTB7 (**63**), indicating that catalyst residues from the Stille coupling used to make it, which can be shown to be present in the films, may be detrimental to device performance [699]. This suggests that improving the synthetic procedure to use less catalyst and/or to more efficiently remove it may be necessary to enhance device performance. A study of the performance of devices using PTB7 of different molar masses found that the efficiencies were highest for samples with molar masses of >100 kDa, which was attributed to the presence of homo-coupled oligomers in the lower molar mass samples [700]. Similarly, it has been shown that the efficiency of devices using PTB7 with a molar mass of ca. 100 kDa is reduced for samples with higher polydispersity [701]. Thus, ensuring the reproducible production of high molar masses and low polydispersities by careful control of polymerisation conditions and polymer isolation and purification will be needed to obtain reliable high efficiency devices. With respect to the role of end-groups,

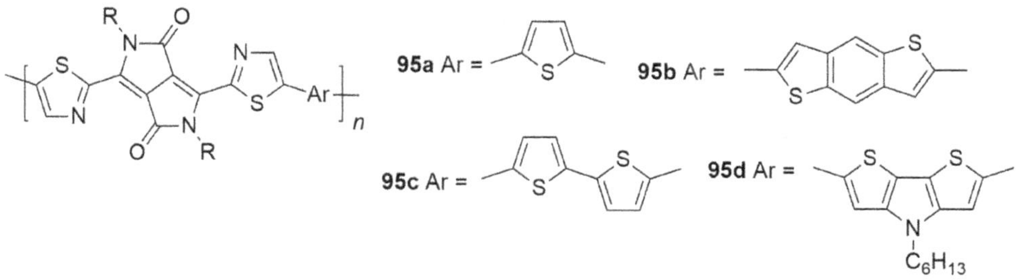

Figure 8.26 Diketopyrrolopyrrole copolymers with low photon energy losses

there has been a study showing that the efficiency of devices using the silole-copolymer **94** made by Stille coupling can be increased from 4.2 to 4.7% when the polymer is treated with excess thienyl reagents to ensure that all residual bromine and tin groups are replaced to give thiophene end groups, as shown [702].

Two new areas that are just beginning to be explored as ways to increase the efficiency of polymer solar cells are changing the energy of triplet states in the donor, and the reduction of photon energy losses. It has been demonstrated that triplet states may contribute to loss of energy through recombination and that this may be minimised by raising triplet exciton energy levels though as yet no design rules have been proposed [703]. There has been a report that incorporation of low (up to 1.1 mol%) phosphorescent iridium complexes into PTB7 can enhance the efficiency of unoptimised solar cells from 4 to 5% [704]. It remains to be seen whether similar enhancement will be seen in the optimised PTB7 (**63**) cells with efficiencies of 10% discussed earlier in this section. In most polymer solar cells there is an energy loss of at least 0.6 eV compared to the optical bandgap of the donor, due to energy losses involved with the charge transfer state and exciton splitting processes. Recently, it has been reported that diketopyrrolopyrrole-based polymers **95** (Figure 8.26) show energy losses of less than 0.6 eV in solar cells with PC71BM as an acceptor [705]. The best performing polymer (**95d**) had an efficiency of 5.6%. Whether such reductions can be obtained in other classes of polymers remains unknown but, as diketopyrrolopyrrole polymers are among the classes of polymers which have shown high charge carrier mobilities in FETs (see Section 8.3), further exploration of their potential as solar cell materials seems promising.

Clearly there still exists considerable scope for synthetic chemists to design and make new polymers which will produce enhanced efficiency in OPVs. As yet, we still do not fully understand all the factors that affect the performance nor do we have reliable ways of predicting solid state properties, particularly of blends of materials such as are used in BHJ devices, which is an even more serious gap in our knowledge and barrier to potential rational designing of suitable materials.

Nonetheless we now have some useful hints which can help to narrow down the possibilities and the goal of producing solar cells with both efficiency $>15\%$ and long lifetimes looks eminently attainable within the near future.

We have now discussed all the major classes of one-dimensional linear conjugated polymers and in Chapter 9 we will move on to discuss hyperbranched systems and dendrimers which may be thought of as two- or three-dimensional conjugated systems.

9

Hyperbranched Polymers, Star Polymers and Dendrimers

So far, we have covered polymers in which there is a quasi-linear conjugated π-system, making them essentially one-dimensional objects. However, emissive hyperbranched polymers and dendrimers can also be constructed containing conjugated units. These are two- or three-dimensional objects but the conjugation extends only along (often short) linear conjugated pathways, though there is considerable scope for energy or charge transfer between conjugated segments within them. The difference between dendrimers and hyperbranched polymers is that the former have discrete, well-defined structures, so that they are usually considered to be macromolecules rather than polymers, whereas the latter are polydisperse with often ill-defined structures containing many defects. A third class of materials to be covered in this chapter are star polymers, in which a number of linear polymer chains are linked to a central core. They are somewhat analogous to dendrimers but with linear rather than branched chains.

9.1 Hyperbranched Emissive Polymers

Hyperbranched polymers (HBPs) [706] have not been much explored as functional materials for organic electronic applications, except in LEDs [150]. They are usually rather ill-defined materials whose main advantages are their good solubility and efficient intra-molecular energy transfer. As well as emissive materials, they may have utility as charge-transporting/blocking layer materials, as their lack of crystallinity due to their branched structures can be advantageous for such purposes in LEDs. There are two general methods for making hyperbranched polymers containing conjugated chromophores. The first is the self-condensation of AB_2 molecules in which the A and B units react to form a bond. Here the bond forming reaction can be a condensation reaction such as the Wittig reaction or a metal-halide cross-coupling reaction like the Suzuki coupling. As an example of the former, Figure 9.1 shows how a hyperbranched PPV **1** can be made by Wittig

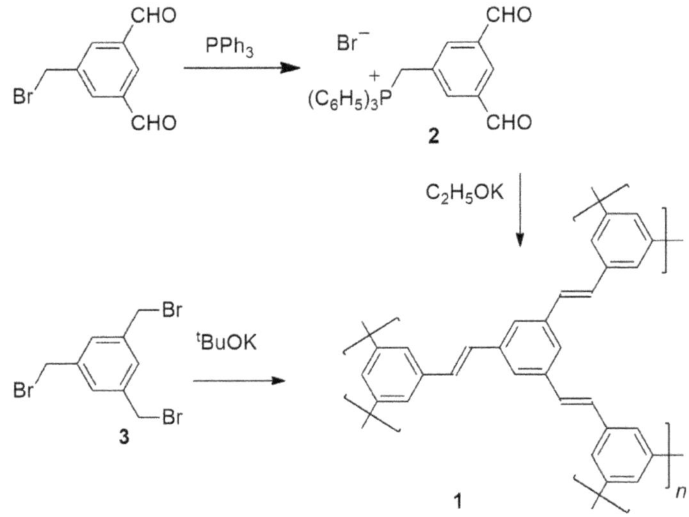

Figure 9.1 Synthesis of 1,3,5-hyperbranched PPV

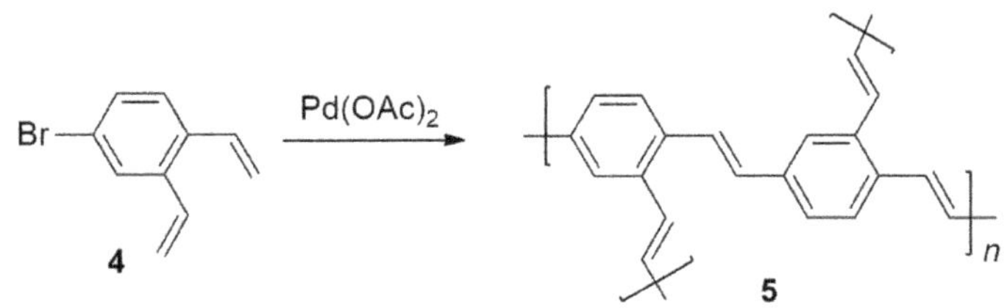

Figure 9.2 Synthesis of a 1,2,4-hyperbranched PPV

coupling of the monomer **2**. In this case the Gilch method can also be used to make **1** by treatment of the trihalide **3** with base [707]. The latter route is reported to produce higher molar mass material. As might be expected from the short conjugation lengths, this polymer is a blue-emitting material ($\lambda_{max} = 446$ nm).

As an example of the second type of reaction, Heck coupling of the diene **4** produces a 1,2,4-linked PPV **5** (Figure 9.2). Due to there being longer linear conjugated units in **4** and so a greater effective conjugation length than in the 1,3,5-linked **1**, this polymer is a green emitter whose emission maximum red-shifts slightly from 524 nm to 550 nm as the molar mass increases [708].

A variant of this approach using a branched A_3 and a linear B_2 monomer is illustrated by the preparation of the copolymers **6** and **7** (Figure 9.3) prepared by Wittig condensation of the trialdehydes **8** and **9**, respectively, with the bisphosphonium salt **10** [709].

While these approaches employ condensation reactions to link the units, oxidative coupling of branched oligothiophenes **11** has been used to form

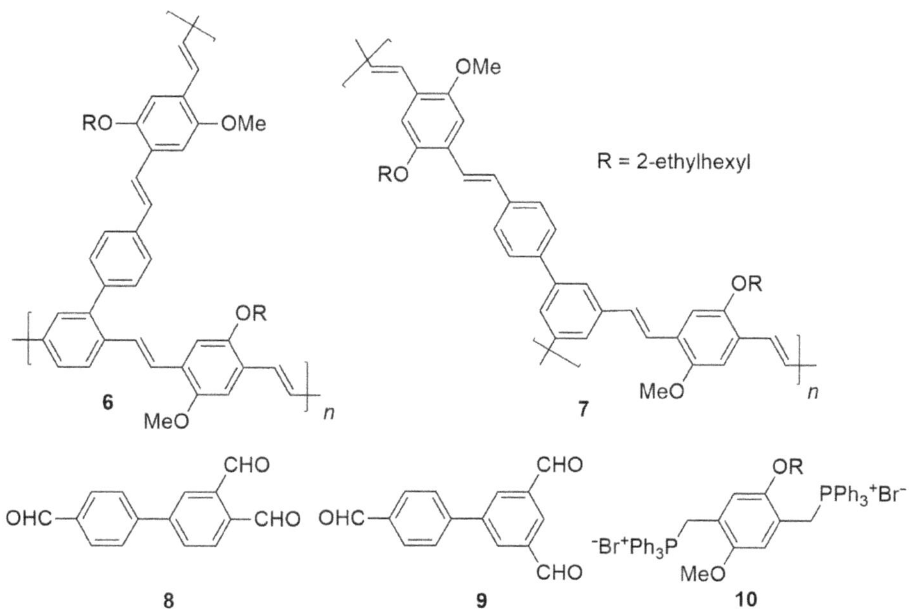

Figure 9.3 Hyperbranched ROPPVs and the units used to prepare them

hyperbranched polythiophenes **12** (Figure 9.4) [710]. In a variation of this approach, the bis(trimethylsilyl)terthiophene **13** can be used as a precursor to **11a** or to the diiodo-adduct **14** and boronate **15**, which undergo Suzuki coupling to give **16**, which can be converted by desilylation and oxidative coupling to the dendrimer **17** [711, 712]. Higher order dendrimers are obtainable by conversion of **16** into suitable halides and boronates in an iterative process. This illustrates the relationship between the synthesis of HBPs and dendrimers. By means of selective protection and deprotection steps, the routes used to make HBPs can be used in a controlled iterative manner to make dendrimers, which may be thought of as well-defined hyperbranched oligomers.

The second approach is the cyclotrimerisation of alkynes, as developed by Tang and co-workers [98, 713]. As illustrated in Figure 9.5, a wide range of hyperbranched polyarylenes **18** have been made by the co-cyclotrimerisation of arylene diynes with aryl- or alkyl-acetylenes as end groups using the same types of catalysts used for preparing polyacetylenes (see Chapter 2). As can be seen, the method is compatible with a wide range of arenes (**18a**, **b**), and heteroarenes (**18c**, **d**) and can also be used to incorporate arylene vinylene (**18e**) [714] or arylene ethynylene units (**18f**) [715, 716]. Other units that have been polycyclotrimerised in this way include silanes and triphenylamines bearing two or three alkyne substituents. By suitable choice of chromophores, emission colours ranging across the visible spectrum can be obtained, as exemplified by polymers **18g**, **18h** and

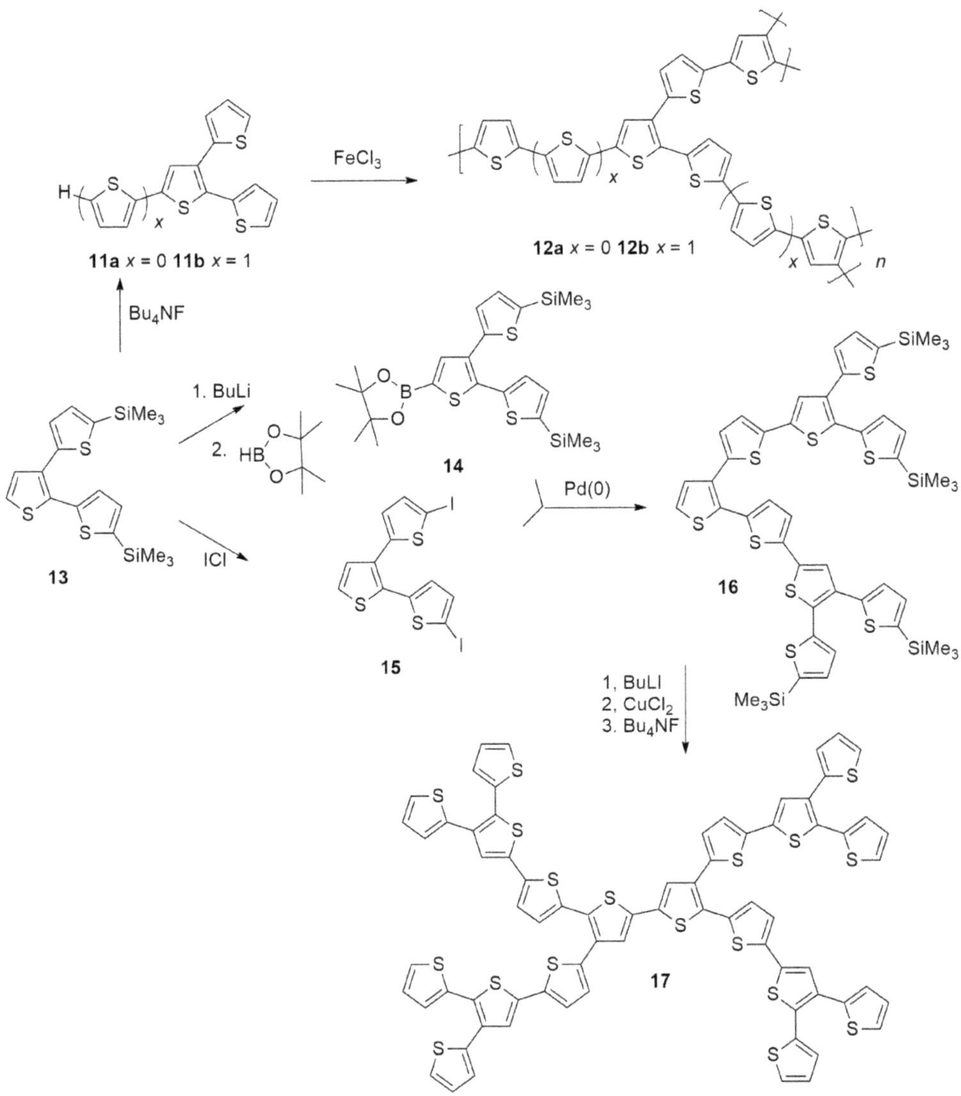

Figure 9.4 Synthesis of hyperbranched polythiophenes and thiophene dendrimers

18i, which display, respectively, blue ($\lambda_{max} = 438$ nm), green ($\lambda_{max} = 545$ nm), and red ($\lambda_{max} = 665$ nm) emission [717].

More recently it has been found that milder conditions can be used to cyclotrimerise electron deficient monomers such as **19** and **20** to form hyperbranched polymers **21** (Figure 9.6) [98]. Whereas hyperbranched polymers **18** formed by transition-metal catalysed cyclisations contain some 1,2,4-benzene linkages, the polymers **21** made by base-induced cyclisation are completely 1,3,5-linked. It remains to be seen if this method can be used to incorporate as wide a

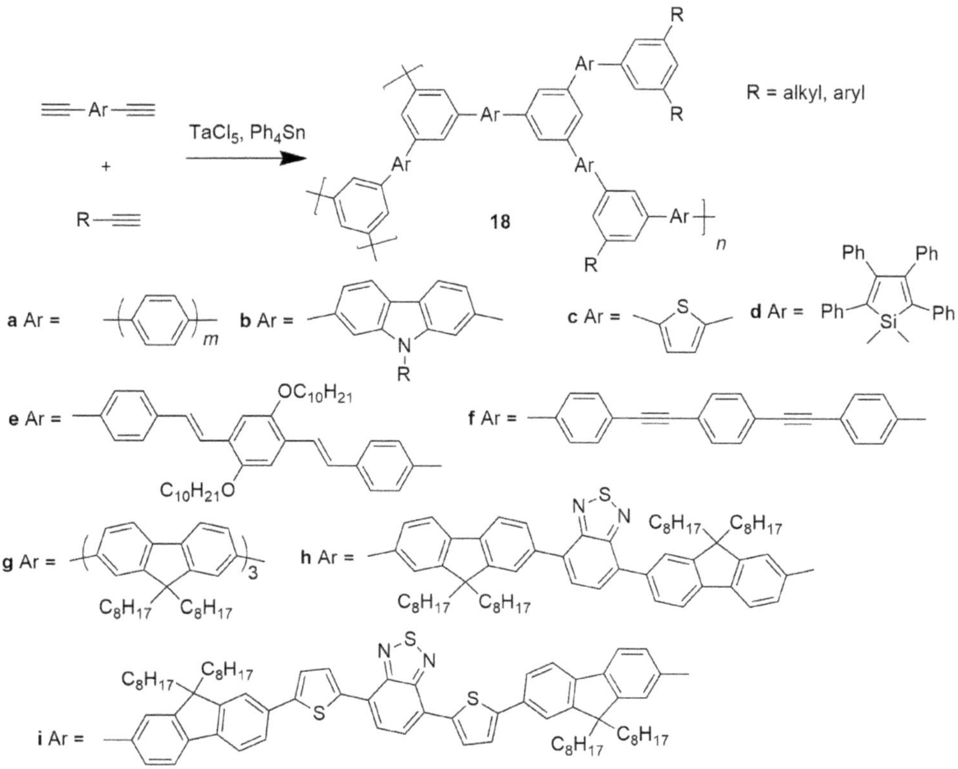

Figure 9.5 Synthesis of hyperbranched polyarylenes by cyclotrimerisation

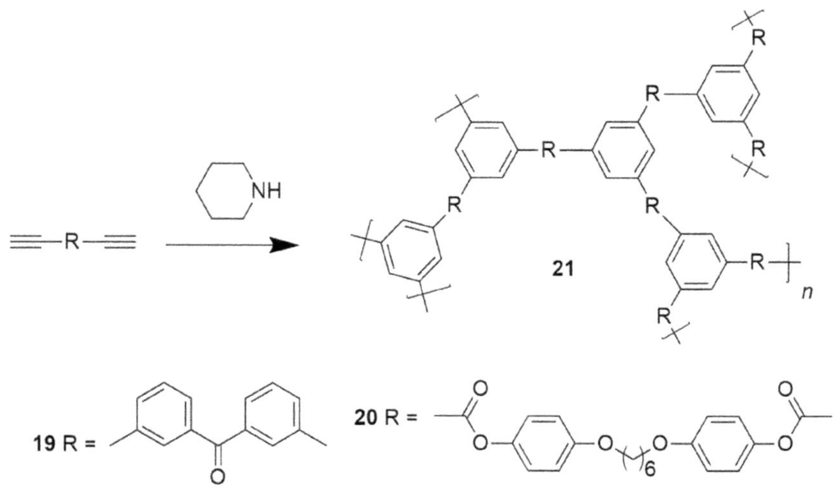

Figure 9.6 Base-induced cyclotrimerisation of electron-poor diynes

range of possible chromophores as the previous method, but its milder synthetic conditions and enhanced regioregularity make it a potentially attractive method for the synthesis of emissive polymers.

In the case of the hyperbranched RO-PPVs **6** and **7**, it is reported that the 1,2,4-substituted polymer **6** displays much more efficient EL emission than the 1,3,5-substituted polymer **7** [718]. As would be expected, given its longer conjugation segments, the green emission from **6** ($\lambda_{max} = 533$ nm) is notably red-shifted compared to the blue-green emission ($\lambda_{max} = 498$ nm) observed from **7**. What is striking is that the emission efficiency is an order of magnitude higher for **6** than **7** (28 cd A^{-1} and 2.1 cd A^{-1}, respectively), even though the PL quantum yields for both materials are similarly high for both materials. It is interesting that the quantum efficiency of **6** in solid state is much higher than in solution (0.64 vs 0.35) [709], which is the opposite of the behaviour usually seen for linear conjugated polymers where, as we have previously noted, aggregation tends to quench luminescence. Such behaviour has been observed for a number of other hyperbranched polymers and also some branched small molecules [98]. Indeed, there exist molecules which show emission only in the solid state or when aggregated in solution. Such materials have potential applications in sensors as binding an analyte to them should produce quenching – the opposite of the enhancement seen in some conjugated polymer-based sensors described in Chapter 7.

For example, recent work by Ma et al. has shown that novel 2D HCBs based on pyrazine (poly(2,3,5,6-styrylpyrazine)) show promise as fluorescent sensors for TNT-based explosives [719]. In particular, enhanced exciton quenching in the presence of TNT was observed, facilitated by the multi-dimensional transport pathways provided by the HCP structure.

HCPs also have a reasonably long history as an active layer component in solar cells. However, the challenge in developing OPV devices from branched polymers in general is that these materials need to maintain their conjugation in all three dimensions. A lack of conjugation in any direction acts as a defect to charge conduction and, hence, limits device performance. In addition, unlike linear polymers, π–π stacking is hindered by the branched nature of the polymer and thus most hyperbranched polymers are amorphous [706]. Branched polythiophenes have been explored as solar cell materials in the literature but in general have delivered lower conversion efficiencies (~0.6%) in bulk heterojunction structures than their linear chain counterparts [710]. More recently, solution-processable C_{3h}-symmetric benzotrithiophene (C_{3h}-BTT)-based hyperbranched conjugated polymer nanoparticles (BTT-HCPNs) with tuneable particle size via Stille mini-emulsion polymerisation have been used in solar cells delivering efficiencies of around 1.5% [720].

9.2 Emissive Dendrimers

Dendrimers containing fluorescent chromophores have not been extensively investigated as materials for use in LEDs, though they have shown some promise as emissive materials [721]. Their well-defined sizes and shapes have made them especially suitable materials for studying the fundamental photophysics of multi-chromophoric luminescent materials, especially the nature and dynamics of energy or charge transfer processes between chromophores [722, 723]. The chromophores may be incorporated as core units, as substituents on the periphery or, especially for arylene vinylene or arylene ethynylene dendrimers, the branching units (dendrons) themselves may be luminescent. Dendrimers are prepared either by attachment of the preformed dendrons to a core (convergent approach) or by building the dendrons stepwise from the core (divergent approach). In both cases there is a limit to the size at which the dendrimers can grow and still maintain structural regularity, especially when relatively rigid dendrons are used. The third generation is the limit for most classes of emissive dendrimers.

One advantage of dendrimer structures with emissive cores is that the large dendrons prevent interaction between the chromophores, thus suppressing the effects of excimer or aggregate formation, which tend to red-shift emission and/or reduce efficiency. Here larger dendrons (higher dendrimer generations) are more effective. The disadvantage is that the dendrons are usually not very good charge carriers, which may reduce the efficiency of LEDs using them, especially for higher generation dendrimers. This can be overcome to some extent by incorporating charge transporting units at the periphery of the dendrimer. The effect of dendrimers size on EL efficiency is determined by the extent to which the effects of core isolation and the partially insulating nature of the dendrons counterbalance each other.

The emission from most emissive dendrimers comes from the core and so it is possible to obtain all emission colours by choosing appropriate core molecules as exemplified by the molecules **22–24** (Figure 9.7) with phenylene vinylene dendrons [724–726]. These were prepared by the divergent approach using iterative Heck coupling to synthesise the dendrons which were then attached to the cores by either condensation with pyrrole (**24**) or Horner coupling (**22** and **23**). The synthesis of **24** is an unusual case in which the core is prepared in situ rather than dendrons being attached to a pre-formed core. These show, respectively, blue ($\lambda_{max} = 480$ nm), yellow-green ($\lambda_{max} = 550$ nm) and red ($\lambda_{max} = 660$ nm) emission. The PL and EL efficiencies for **22** have been found to be dependent on the dendrimer generation, with the highest efficiencies being seen for the second generation dendrimer [727]. The emission colour of **22** was reported to undergo a rapid change from blue to white by an undetermined process which was not thermally induced, with the higher generation materials showing greater stability,

22 Ar =

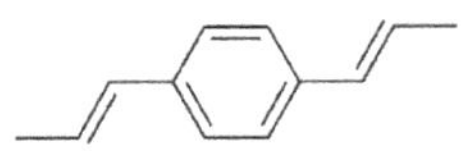

23 Ar =

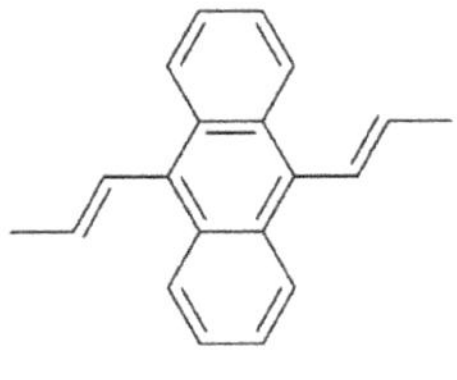

24 Ar =

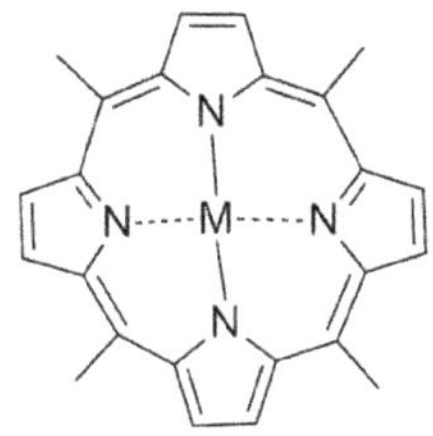

M = 2H, Pt

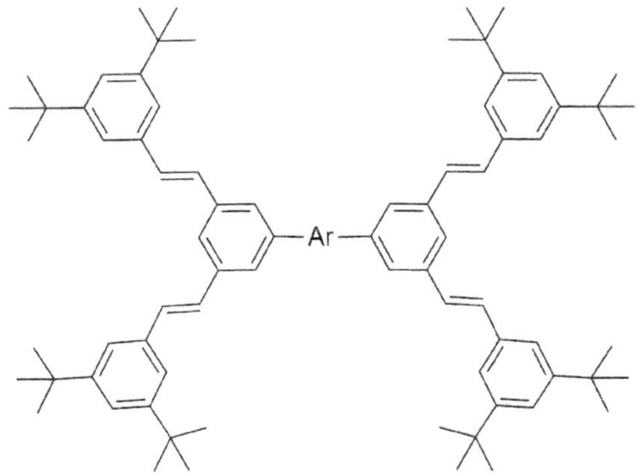

Figure 9.7 Phenylene-vinylene dendrimers with blue, green and red emission

suggesting it may arise from some sort of inter-molecular interaction. The most efficient emissive dendrimers are those containing phosphorescent metal complexes as cores, which are discussed in Chapter 11.

A similar divergent strategy involving iterative Diels–Alder addition of functionalised tetraarylcyclopentadienones to alkynes has been developed by the Müllen group to produce emissive polyphenylene dendrimers with rylene dyes as cores [728]. One attractive feature of this method is that it is possible to incorporate a variety of functional groups into successive layers of the dendrimer. The ultimate example of this is the second generation dendrimer **25** (Figure 9.8) which contains blue-emitting chromophores on the periphery, green-yellow emitting chromophores on an inner layer and a red emitter at the core. Due to efficient energy transfer, excitation of any of the chromophores produces PL emission mainly from the core ($\lambda_{max} = 708$ nm) [729]. This molecule thus acts like a light-harvesting antenna, though further structural design, for example, incorporation of a group for binding to an electron-accepting surface, is required to obtain something from which the energy might be directly harvested in a usable form. Red EL has been obtained from a blend with the charge transporting polymer PVK [728].

Dendrimers **26** (Figure 9.9) illustrate how dendrimers can be prepared with emissive chromophores at the core and charge-transporting moieties at the periphery. These non-rigid dendrimers show bright PL and EL due to efficient energy transfer between the periphery and the core. The emission colours are, respectively, blue ($\lambda_{max} = 480$ nm) for the coumarin **26a** and green ($\lambda_{max} = 550$ nm) for the quinquethienyl **26b**

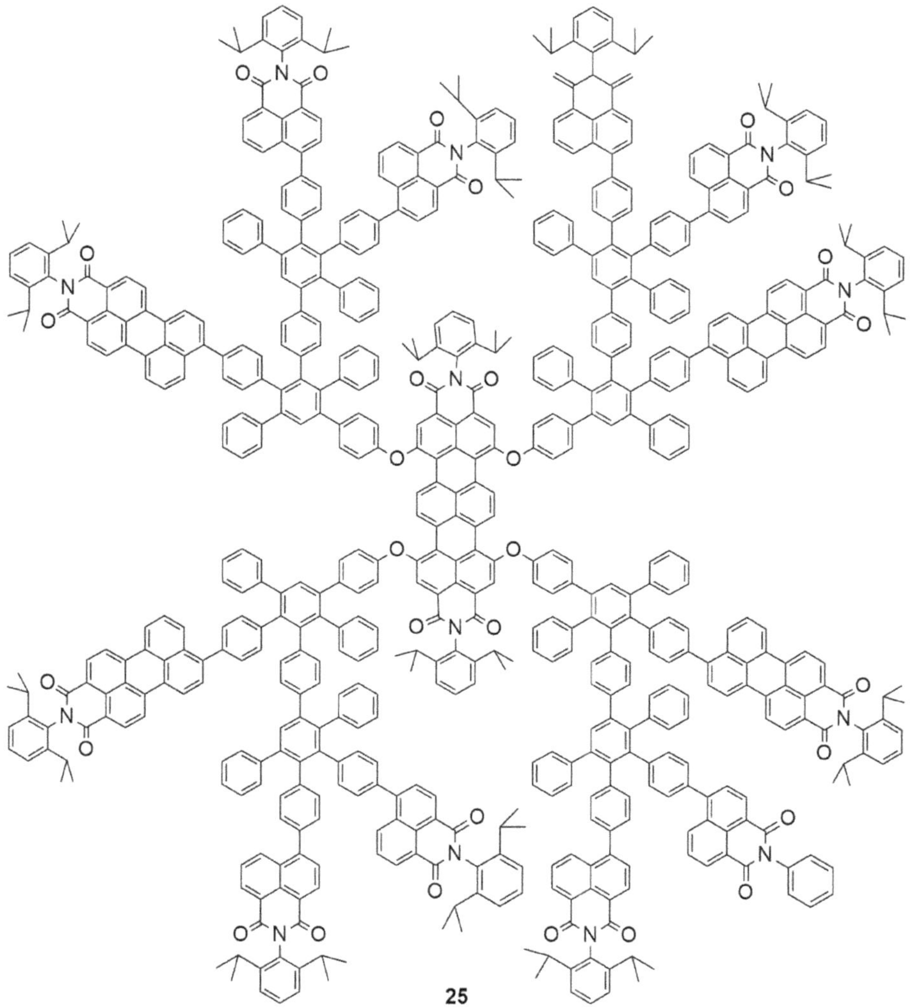

25

Figure 9.8 Polychromophoric dendrimer showing light harvesting properties

cored-dendrimer [730]. The flexibility of the dendrons in these materials may assist the groups to get closer enough for efficient charge transfer between them, but the same approach has also been used successfully for dendrimers with more rigid dendrons based on phenylene vinylenes [731] or polyphenylenes [732].

Again, the enhanced emissive properties exhibited by dendrimers also means that they have potential application as sensors based on emission quenching produced by the analyte species. Indeed, dendrimers can offer advantages over conventional conjugated polymers since they can be monodisperse with the number of chromophores available for emission quenching determined by the dendrimer generation [733].

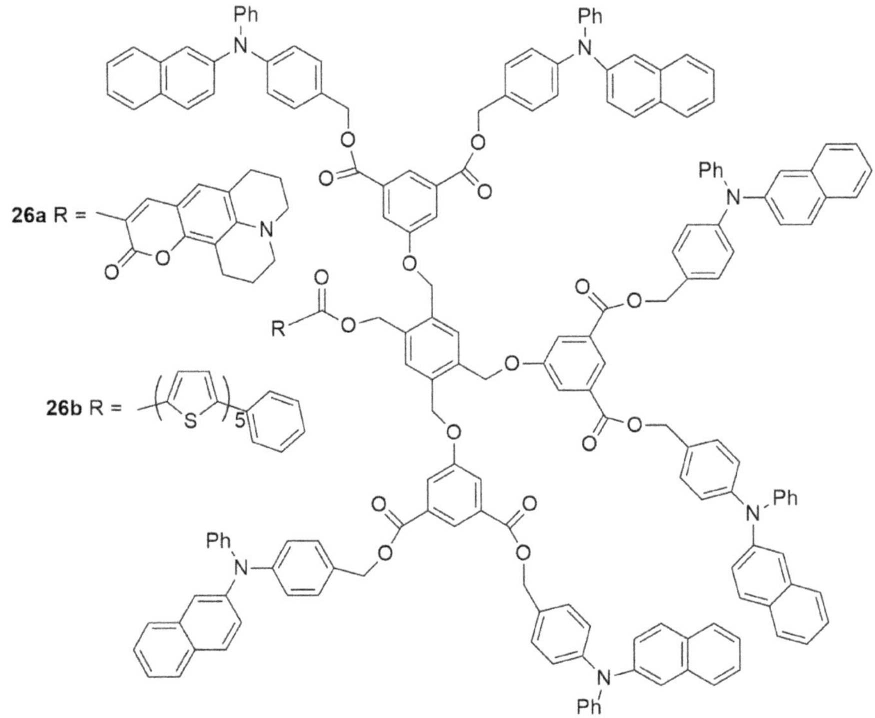

Figure 9.9 Emissive dendrimers with charge transporting groups on the periphery

Early work on solid state films focussed on dendrimers based on carbazole and showed that the quenching in the presence of 2,4-dinitrotoluene (DNT) increased with dendrimer generation in solution but remained essentially constant in the solid state [734]. Subsequent work on the structure–function properties of dendrimer films revealed that sorption of the analyte is driven by film swelling and resulting in an exponentially increasing diffusion constant [735]. Most recently, the sensitivity limits of carbazole dendrimers has been measured; confirming a parts per billion level sensitivity to TNT analogue analytes that is insensitive to dendrimer generation [736].

Dendrimers have been investigated in solar cells over many years, with these materials offering a more ordered branched structure than HCPs [733]. Early work on blending phenyl-cored thiophene dendrimers with PCBM delivered efficiencies as high as 1.3%, even though the band gap at 2.1 eV was relatively high [737]. Efficiencies of up to 1.72% were initially obtained with thiophene dendrimers such as **17** [711, 738]. Recent work has shown that lower concentrations of thiophene-based dendrimers can deliver PCEs of around 3.5% when blended with $PC_{70}BM$ and when the morphology has been optimised via a judicious choice of blend ratio and careful solvent annealing [739].

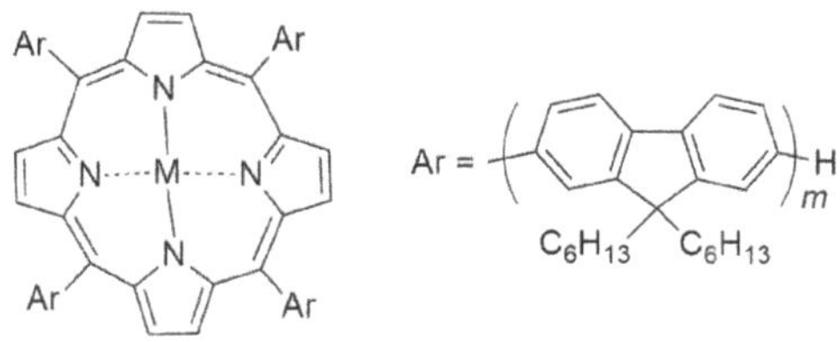

M = 2H, Pt

27

Figure 9.10 Red-emitting star molecule with blue-emitting arms

9.3 Emissive Star Polymers

Star-polymers in which a number of linear polymer arms radiate from a central core are related to dendrimers, but, due to their not having branched arms, have much less dense structures so that interactions between the arms are much more possible than are seen in dendrimers. While both divergent and convergent approaches could be used to make such materials generally the convergent approach in which polymer or oligomer units are attached to the core is used. As emissive star polymers display emission from the conjugated arms, they are thus more susceptible to the effects of chromophore aggregation or excimer formation, though less so than standard linear polymers. A variety of emissive star polymers have been made, mainly with blue-emitting chains such as oligofluorenes [150]. A wide range of cores have been investigated, including 1,3,5-trisubstituted benzenes, truxenes, tetraarylsilanes and tetraarylmethanes, and also inorganic materials such as polysilsesquioxanes. If the core is emissive then energy transfer can produce emission from the core, for example, the porphyrin-based molecule **27** (Figure 9.10) shows red emission ($\lambda_{max} = 662$ nm) [740]. Such molecules may have applications as light-harvesting antennas. To date very little work has been completed in this area. One study has constructed multi-armed P3HT star polymers with a gold nanoparticle core. However, the device efficiency was reduced compared to a P3HT:PCBM control OPV device [741]. Alternatively, if the energy transfer from arms to core is incomplete, such star polymer structures might be white emitters (see Chapter 12).

One feature of dendrimers and star polymers is that the chromophores are of fixed sizes. It is also possible to prepare polymers in which the chromophores are of uniform size, by attaching the chromophores as side-chains or by interrupting the conjugation between them with non-conjugated spacers. It is to such materials that we now turn.

10

Polymers with Molecular-like Chromophores

One problem with the sorts of systems we have discussed so far is that they contain chromophores with a wide range of conjugation lengths and, thus, potentially different absorption and emission spectra. Indeed, single molecule spectroscopy has demonstrated that a single polymer chain can contain a number of emissive chromophores of different sizes and emission spectra [343]. In the bulk, efficient energy transfer generally leads to emission, primarily from the longest and thus lowest energy chromophores. One way around the problems inherent in the presence of multiple chromophores of differing properties is to produce materials in which all the chromophores are identical and so have the same predictable properties. This arrangement can be achieved by either attaching the chromophores as substituents onto a non-emissive chain – this situation is somewhat reminiscent of the case of polyacetylenes discussed in Chapter 2 where the unsubstituted polymer is non-emissive but substituted polyacetylenes can be strongly luminescent – or by inserting non-conjugative spacers into the polymer chain at regular intervals, so producing a polymer consisting of a series of linked chromophores whose π-systems cannot overlap. In both cases the hope is that the materials will show the sort of well-defined spectral properties observed for molecular chromophores but combined with the good materials properties (e.g. solution processability and good film-forming) of polymers. The potential disadvantage is that the non-conjugated units are insulators which may reduce charge transport and, thus, device efficiency.

10.1 Polymers with Emissive Side Groups

Polymers can be prepared in which luminescent moieties are pendant onto the polymer chain, either by polymerisation of monomers containing the chromophore or by attaching chromophore units to an existing polymer chain. These approaches enable the utilisation of the vast body of expertise developed in classical polymer

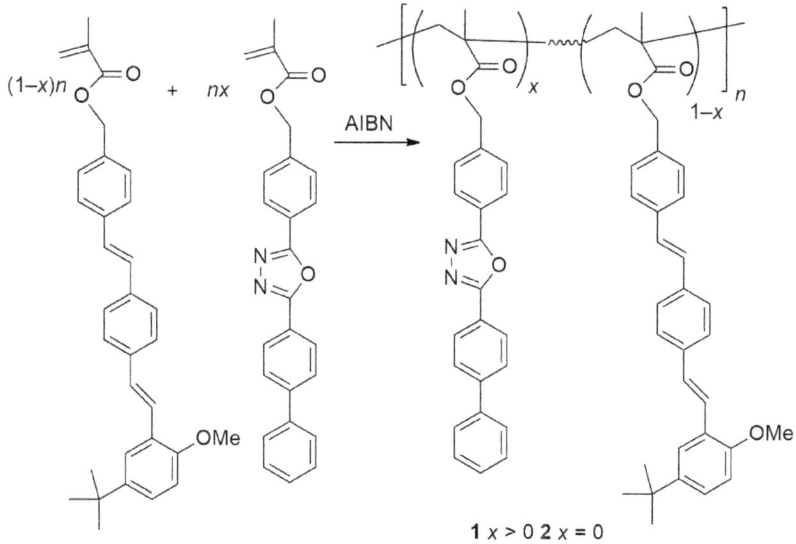

1 $x > 0$ **2** $x = 0$

Figure 10.1 Synthesis of a copolymer with pendant functional groups

synthesis, for example, in synthesis of commercially important classes of polymers such as polystyrenes or polyacrylates, to control the physical properties of the polymer, for example, molecular mass and polydispersity. As the chromophores used are usually of fairly short conjugation length this method is particularly attractive for the synthesis of blue emitting polymers.

The first method is exemplified by the synthesis of the copolymer **1** bearing a mixture of diaryloxadiazole charge transport and distyrylbenzene emissive functional groups (Figure 10.1) [742].

This synthesis illustrates one important point about this method – it is possible to attach more than one functional group onto the polymer in a one step synthesis. Most often a combination of an emissive and a charge transporting unit are attached onto the polymer as this improves emission efficiency. In this case, whereas the copolymer **1** produces blue EL ($\lambda_{max} = 457$ nm) the corresponding homopolymer **2** without oxadiazole units displays only short-lived non-uniform emission as it has very poor conductivity. Simple radical polymerisation, as shown here, gives random copolymers controlled radical polymerisation methods to enable block copolymers to be produced. While such structures have no obvious advantages for LEDs, with poor results reported from them to date [150], they may have potential as materials for OPVs, though so far no good results have been reported from such materials [743].

In addition to radical polymerisation methods such as shown here, metathesis has been used to prepare polymers with functional side-chains. For example, ROMP of the norbornene monomer **3** produces a blue-emitting ($\lambda_{max} = 475$ nm)

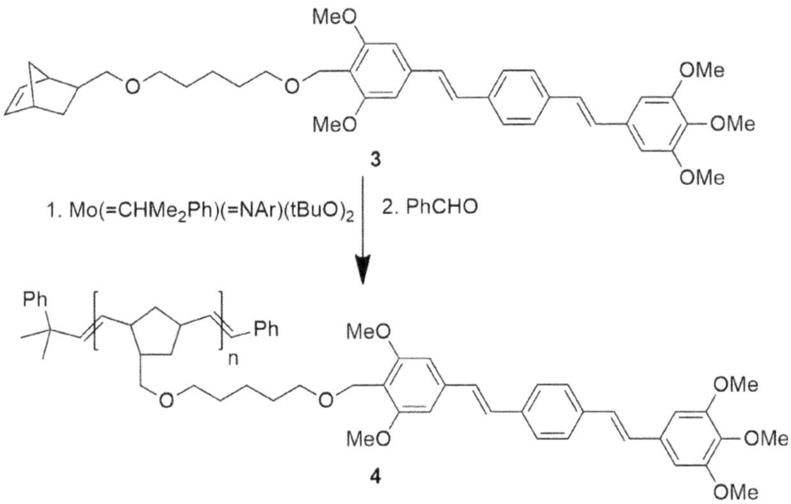

Figure 10.2 Synthesis of a polynorbornene with chromophore side-chains by ROMP

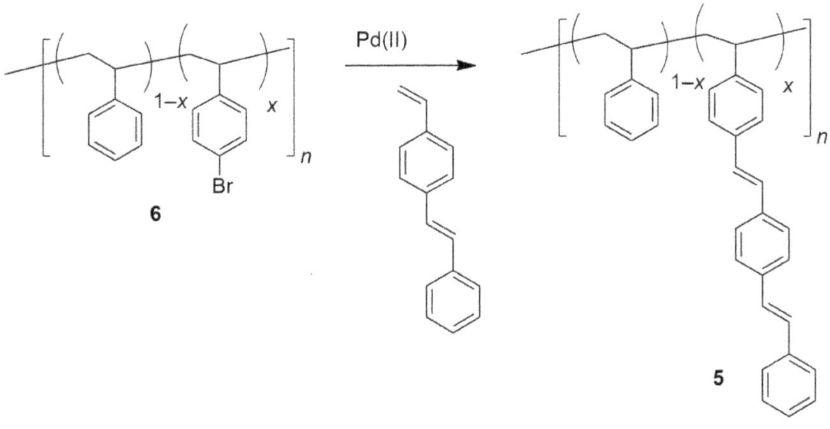

Figure 10.3 Attachment of a chromophore to a styrene copolymer

polymer **4** (Figure 10.2) [744]. One great advantage of metathesis methods is that they are usually living in nature, allowing block copolymer synthesis.

The second method is exemplified by the preparation of the blue-emitting ($\lambda_{max} = 460$ nm) copolymer **5** (Figure 10.3) in which the emissive chromophore is attached by Heck coupling to a (commercially available) copolymer of styrene and bromostyrene **6** [745]. A number of other methods, including Williamson etherification, Wittig coupling and Suzuki coupling, have been used to attach chromophores to polymers based on siloxanes, styrenes or acrylates [150].

By these methods, a wide range of polymers bearing emissive chromophores as side-chains have been prepared and used in LEDs [150]. Depending upon the

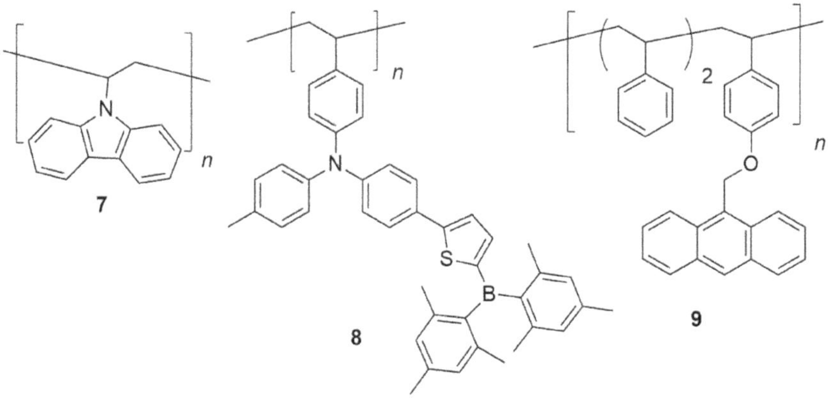

Figure 10.4 Representative side-chain polymers with blue, green and red emission

chromophore chosen, the emission can be any colour from violet-blue for poly(*N*-vinylcarbazole) (PVK, Figure 10.4, **7**) ($\lambda_{max} = 410$ nm), through green for polymer **8** ($\lambda_{max} = 521$ nm), to red for polymer **9** ($\lambda_{max} = 590$ nm). A variety of charge transporting units have also been investigated, with carbazoles being the most widely used hole-transporting and oxadiazoles the most widely used electron transporting units chosen to date. Commercially available PVK is commonly used as a hole-transporting polymer in LEDs, including as a host materal for molecular emitters. Similar EL properties have been observed for copolymers with a mixture of carbazole and emissive substituents [746] and for blends of PVK with the emissive homopolymer [747]. This suggests that, by putting both units on the same polymer, we can get the benefits of a blend without the disadvantage of phase separation often seen for blends. Device efficiencies for materials with fluorescent chromophores (phosphorescent ones are discussed in Chapter 11) have generally been disappointing and the device lifetimes are also reported to be short in many cases, for example, for compounds **1** and **2**. In their case, incorporation of a cross-linkable cinnamoyl unit followed by light-induced cross-linking to produce an insoluble film was reported to enhance device lifetimes [742], but this is not a generally applicable method as some chromophores will be susceptible to degradation during cross-linking.

Because the polymer backbones in these materials are usually flexible, there is considerable scope for the emissive side-chains to form excimers and emissive aggregates, especially when there are high amounts of chromophores present.

In general, excimers (excited dimer states) or exciplexes (excited complex state) are detrimental to both OLEDs and OPV devices. In the case of OLEDs, the formation of excimers can lead to fluorescence quenching of the chromophore emission. A reduction of the luminance is the main drawback, which is also accompanied with a broadening of the emission spectrum [748]. This broadened

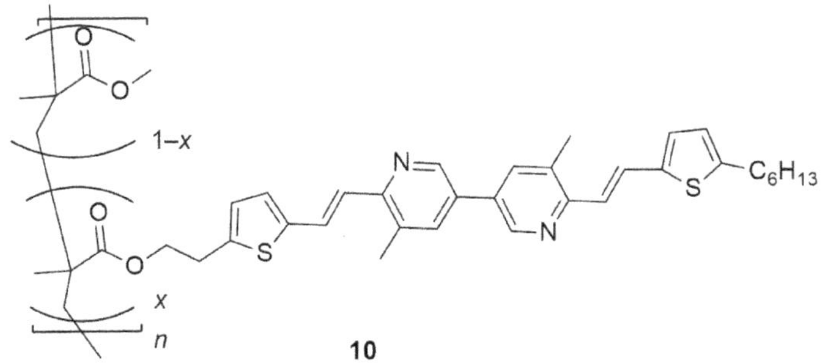

10

Figure 10.5 Copolymers displaying variable emission colour

emission can have applications for developing broad spectral emitters but, as will be discussed in Chapter 11, has most application in phosphorescent emission. Historically, most OPV devices used fullerene derivatives as acceptors but with the advent of non-fullerene acceptor (NFA) materials, the role of excimers or excimer-like states in OPVs has been investigated by several research groups [749]. In general, excimer states do not participate in charge generation but can act as trap sites for fast recombination, which reduces photocurrent generation in certain NFA blend materials. In the case of pendant chromophore polymers, recent work has shown that polymers with freely moving pendants are not suitable for use as upconverters in photovoltaic or electroluminescent devices due to transient intra-chain excimer formation [750].

An unusually dramatic example of the effects of side-chain aggregation is provided by the copolymers **10** (Figure 10.5) whose solid-state PL is dependent upon the amount of the emissive chromophore attached [751]. Thus, copolymers with 2 wt%, 8 wt% and 49 wt% of chromophore, respectively, display blue ($\lambda_{max} = 455$ nm), blue-green ($\lambda_{max} = 476$ nm) and yellow ($\lambda_{max} = 528$ nm) emission, due to the increasing amount of aggregation of the emissive side-chains.

10.2 Copolymers with Isolated Chromophores in the Main Chain

These types of polymers share the advantages and disadvantages of the just discussed polymers with emissive side-chains, and here also two general synthetic approaches exist to prepare them. Either the chromophore is prepared as part of the polymerisation step (e.g. by a Wittig polycondensation) or a substituted chromophore is polycondensed with a suitable linking agent (e.g. by a polyesterification) [150]. The first approach is well exemplified by the synthesis of polyethers (e.g. **11**) with distyrylbenzene segments separated by flexible alkyl chains, by a Wittig polycondensation method (Figure 10.6) first developed by Karasz and co-workers [752, 753] and subsequently applied by other groups.

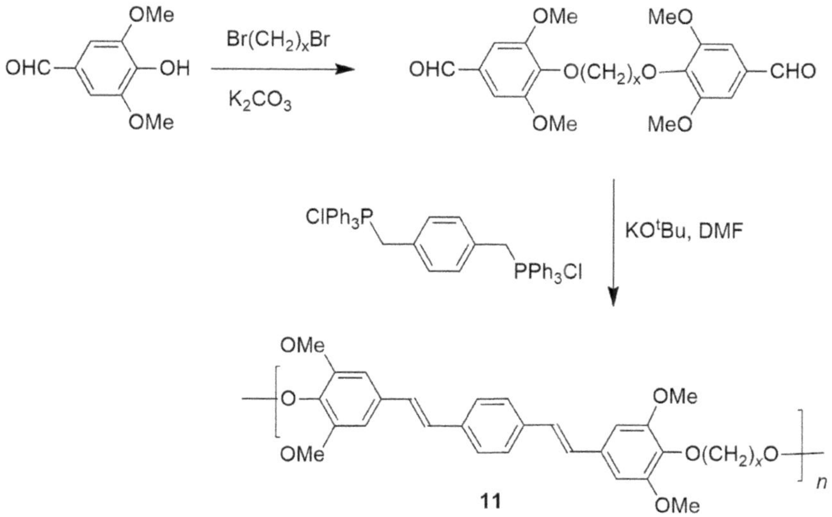

Figure 10.6 Wittig route to copolymers with isolated chromophores

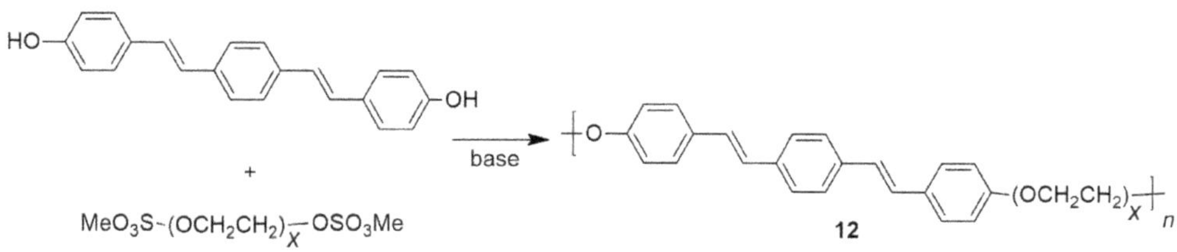

Figure 10.7 Condensation route to copolymer with all-trans chromophores

A disadvantage of this method is that a high proportion of the double bonds formed have *cis*-geometry, and to get all-*trans* configuration the polymers must be treated with iodine, which leads to a loss of material during the subsequent purification step. This problem is surmountable by using Heck coupling to prepare the arylene vinylene chromophore [754]. Similar copolymers with oligoarylene chromophores can be made by using Suzuki or other metal-catalysed cross coupling reactions as the polymerisation method, while polymers with isolated cyano-PPV units can be prepared by Knoevenagel condensation [150].

The second approach of condensing a suitable chromophore with a linking agent also has the advantage for arylene vinylene chromophores of avoiding the problem of *cis-trans* isomerism. Thus, condensation of an all-*trans* bis(4-hydroxystyryl) benzene with polyethylene glycol dimesylate gives an all-*trans* polymer **12** (x = *ca.* 20) (Figure 10.7) [755]. Other reactions for preparing emissive copolymers by condensation of suitably functionalised chromophores with spacers include etherification, esterification, amidation, hydrosilylation and hydroboration [150].

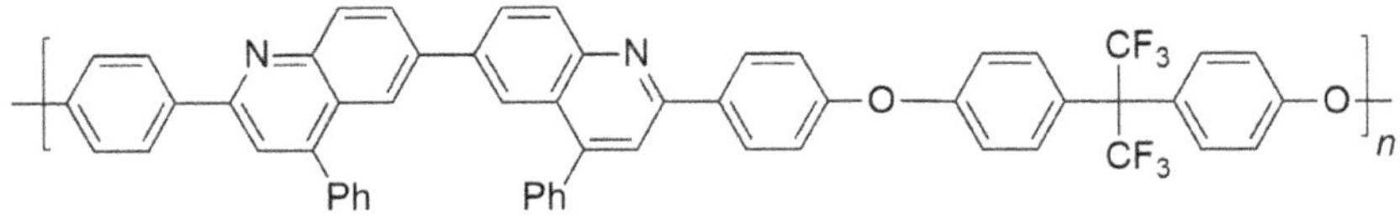

13

Figure 10.8 Polymer with isolated chromophores showing high EL efficiency

A very large number of emissive polymers with isolated chromophores have been prepared by these methods for use in LEDs [150]. By appropriate choice of chromophores, emission colours ranging across the entire visible spectrum have been obtained but, as with the side-chain polymers discussed, the device performances have generally not been good. This is in part due to poor charge transport due to the presence of the insulating spacer units. To overcome this charge transporting units can be incorporated into the spacers. The most widely used units have been triarylamines for hole transport and oxadiazoles for electron transport [150]. The best performing LED using a polymer with isolated chromophores is one reported using the blue-emitting ($\lambda_{max} = 450$ nm) copolymer **13** (Figure 10.8) with quinoline chromophores from which 4% efficiency was obtained, though the lifetime has not been reported [756].

Generally, the nature of the spacer has little effect upon the absorption and emission of the polymer, though (usually small) changes in emission spectra have been observed in some cases. Where the spacer tends to have an effect is upon the ability of the chromophores to interact to form excimers or emissive aggregates, which is due to its length.

In other words, the potential disadvantage of this approach is that, even with a spacer, any fluorophores that are not sufficiently separated may form excimers with nearest neighbours. These interactions are more likely if the polymer can bend or fold. Recent work to prevent excimer formation has involved developing polymers with both intra-chain and inter-chain spacers, thereby reducing the formation of emissive aggregates whilst maintaining efficient energy transfer [757].

The mechanical and optical properties of the polymers **11** and **12**, for example, are dependent on the length of the alkyl spacer. Thus, longer spacers produce a drop in the T_g of **11**, so that for $x = 12$ it is about 50°C, which eventually causes a drop in efficiency of EL devices as the T_g drops too close to the operating temperature. This is counterbalanced by better phase separation occurring with longer alkyl spacers leading to increased confinement of the exciton on the conjugated segments and so improved EL efficiency. The effect of these two contrary effects is that the efficiency reaches a peak for $x = 10$ and then starts to drop again for longer spacers [758]. A similar effect is seen for copolymer **12**, where a polymer with a long ethylene glycol spacer ($x = 20$) shows only short-lived

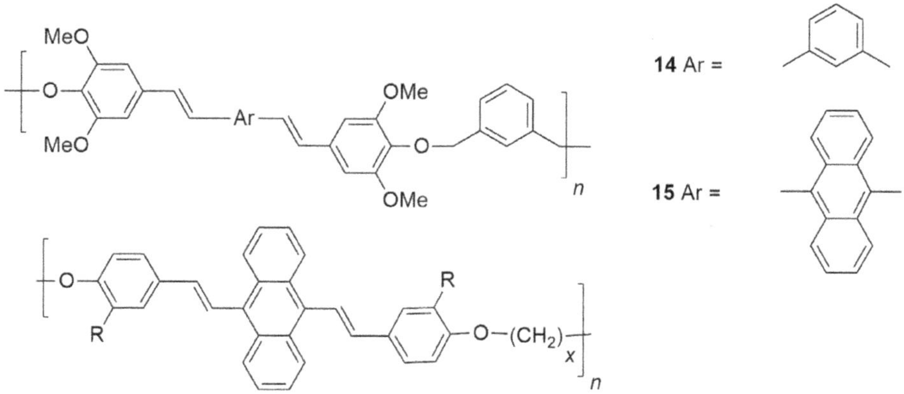

14 Ar =

15 Ar =

16 R = H
17 R = OMe

Figure 10.9 Copolymers illustrating effects of spacer on properties

blue EL [755] whereas materials with shorter spacers (x = 3 or 6) show stable blue emission (λ_{max} = 466 nm) [759, 760]. More recently, the same trends of decreasing charge mobility and rising PL efficiency with increasing spacer chain length have been reported for other polymers made by metathesis, which contain terfluorene chromophores linked by alkyl chains [761]. Curiously improving the regioregularity of these polymers was found to show the opposite effects – it raised charge carrier density but lowered luminescence efficiency. The T_gs of such polymers can be increased by using a 1,3-phenylene spacer unit as found in polymers such as **14** (λ_{max} = 413 nm) or **15** (λ_{max} = 533 nm) (Figure 10.9) instead of a polyalkyl spacer [762]. The effect of aggregation upon emission colour is exemplified by copolymers **16**. These show yellow PL (λ_{max} = 580 nm) in solution with, depending upon the chain-length, a slight red (x = 7) or blue shift (x = 12) in the solid-state spectrum [763]. This variance can be explained by differing degrees of polymer inter-chain interactions due to variable amounts of phase separation. The PL maximum of the methoxy-substituted polymers **17** shows a slight blue-shift with increasing chain-length from x = 4 (λ_{max} = 583 nm) to x = 12 (λ_{max} = 578 nm), presumably for similar reasons [764]. The EL maximum for the latter is at 574 nm, showing a close match between the two spectra.

A good example of where all of these effects have been combined together effectively was reported by Gather et al., who developed a electroluminescent copolymer of electron and hole-transporting units with red-, green- and blue-emitting chromophores [765]. This successful development of main chain multichromophore structure was used to build efficient and highly stable polymeric white organic light-emitting diodes (WOLEDs). We will return to the topic of WOLEDs in Chapter 11 on phosphorescent materials.

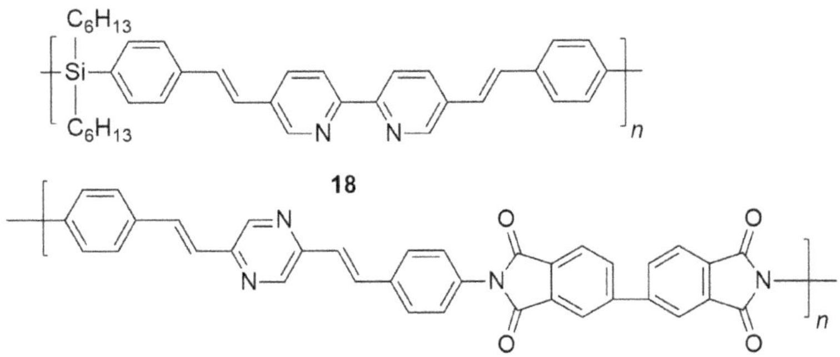

18

19

Figure 10.10 Copolymers displaying aggregate or excimer emission

Excimer-like emission has been seen from polymers with isolated chromophores. An unusual example of this, the molar mass dependent colour of emission from the silane **18** (Figure 10.10), which can be made either by Horner or Wittig polycondensations, with the latter producing lower molecular weights. Both samples showed blue PL in solution (λ_{max} = 442 nm), with a strong shoulder at 418 nm for the lower molar mass material, but their solid-state PL spectra were quite different, with maxima for the higher mass material being markedly red-shifted (λ_{max} = 524 nm, cf. 479 nm) compared to the lower mass material [766]. It was suggested that this arose from the longer molecular chains allowing more highly ordered molecular alignment in the solid state, resulting in stronger interactions between chromophores, producing emissive aggregates or excimers. Another example of materials showing excimer-like emission are polyimides (e.g. **19**) prepared by polycondensation of 2,5-bis(4-aminostyryl)pyrazine with alicyclic or aromatic tetracarboxylic dianhydrides, which show weak orange-red EL (λ_{max} = 550–565 nm) [767]. Since the emission from molecular distyrylpyrazines is green or blue [768], this emission must come from an aggregate or excimer, suggesting that phase separation leading to stacking of the π-systems is occurring within the polymer films.

It is possible to include two different chromophores within the same copolymer. In this case efficient energy transfer processes will cause the emission to come primarily or even solely from the lowest bandgap chromophores. For example, in polymer **20** (Figure 10.11) the indenofluorenes spacer is a violet-blue emitter, but the solid-state emission from the copolymer comes solely from the arylene vinylene moieties so that the PL is blue (λ_{max} = 440 nm) and the EL blue-green (λ_{max} = 470 nm) [769]. Similarly, the copolymers **21a–c** show yellow PL (λ_{max} = 538–550 nm) in solution due to efficient Förster energy transfer from the other chromophores to the distyrylanthracene chromophores. More complex behaviour is seen in the solid-state, with the PL depending upon various factors,

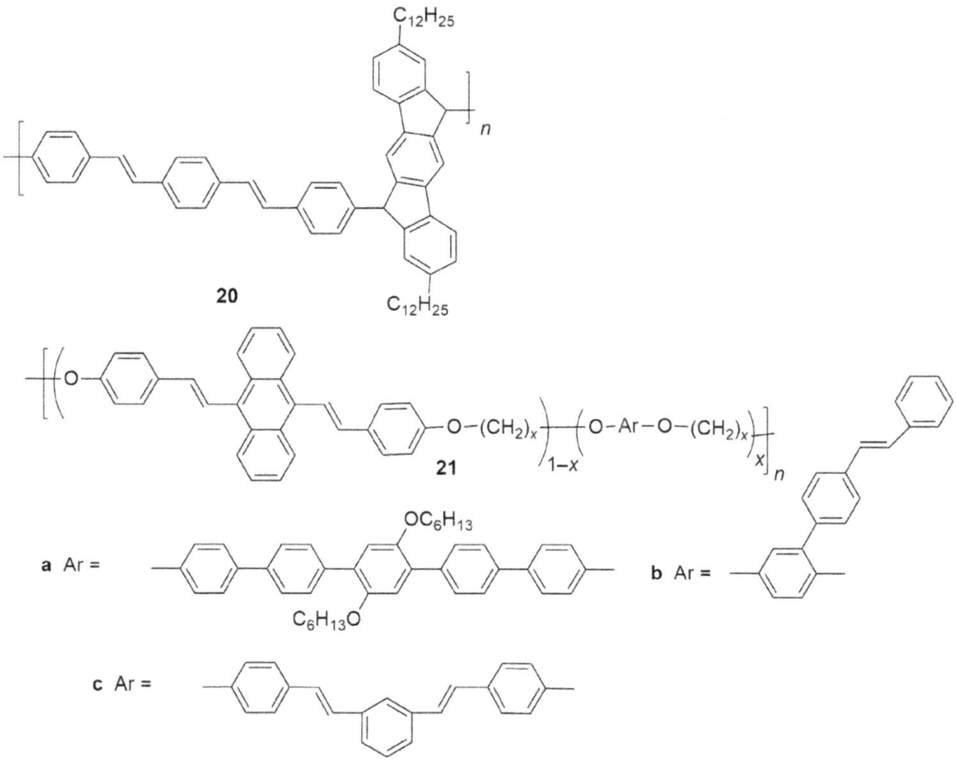

Figure 10.11 Copolymers with two different chromophores in the chain

such as the relative compositions, length of spacer units and excitation wavelength. In some cases, effects like red-shifting of maxima and changes in shape of the spectra have been observed [770].

So far, we have discussed polymers which are covalently linked. Polymers can also be produced of which luminescent monomer units are held together by non-covalent bonds such as metal-ligand coordination or hydrogen bonds – these are known as luminescent supramolecular polymers. The first such materials reported were the luminescent polymers **22** formed by self-assembly of the emissive units through complexation between the terpyridine ligands on chromophore and zinc ions (Figure 10.12) [771]. This method is compatible with a wide range of functional groups and so materials can be made with emission ranging over the whole visible range [772]. Polymers **22a** and **22b** have been used to make, respectively, blue (λ_{max} = 450 nm) and green (λ_{max} = 572 nm) emitting LEDs [771].

If zinc is used as the metal ion in such polymers the complexation is readily reversible and since zinc(II) has a full *d*-shell, the metal complex does not contribute to the electronic transitions within the polymers. If other transition metals (e.g. ruthenium) are used, the resulting complexes are less labile and are

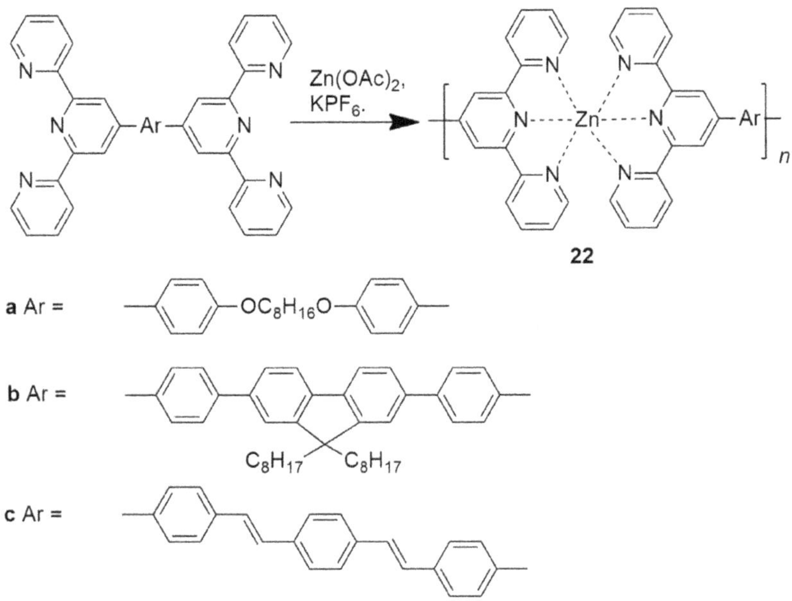

Figure 10.12 Self-assembled EL polymers formed by metal complexation

themselves emissive, which complicates the optical behaviour. The reversibility of the zinc-terypyridine complexation enables the complexes to be broken up by addition of stoichiometric zinc, and the resulting zinc complexes can then be reassembled with other terpyridine-substituted chromophores to form alternating copolymers (e.g. **23**) (Figure 10.13) [773].

The emission from the copolymer **23** is red ($\lambda_{max}=590$ nm) and matches that of the homopolymer **24**, whereas the emission from **25** is blue-green ($\lambda_{max}=466$ nm), suggesting that efficient energy transfer is occurring between chromophores within the copolymer. While the emission from **24** is very similar to that of the corresponding monomer ($\lambda_{max}=581$ nm), that from **25** is red-shifted and broader than that of the monomer ($\lambda_{max}=434$ nm), even in solution, suggesting that aggregation of chromophores may be occurring within the latter polymer. It has been demonstrated that this disassembly–reassembly method can be used to make luminescent three-dimensional networks which may have potential applications as light-harvesting systems if they can be made porous so as to allow materials to act as energy or charge acceptors [773]. There is to date one report of a metallopolymer linked ruthenium–terpyridine complex being used as electron donating material in an OPV but the efficiency was very low [774].

An alternative method for forming luminescent polymers and/or networks is selective hydrogen bonding. There are few reports of such systems being used in LEDs. Schenning, Meijer and co-workers have suggested that assemblies of

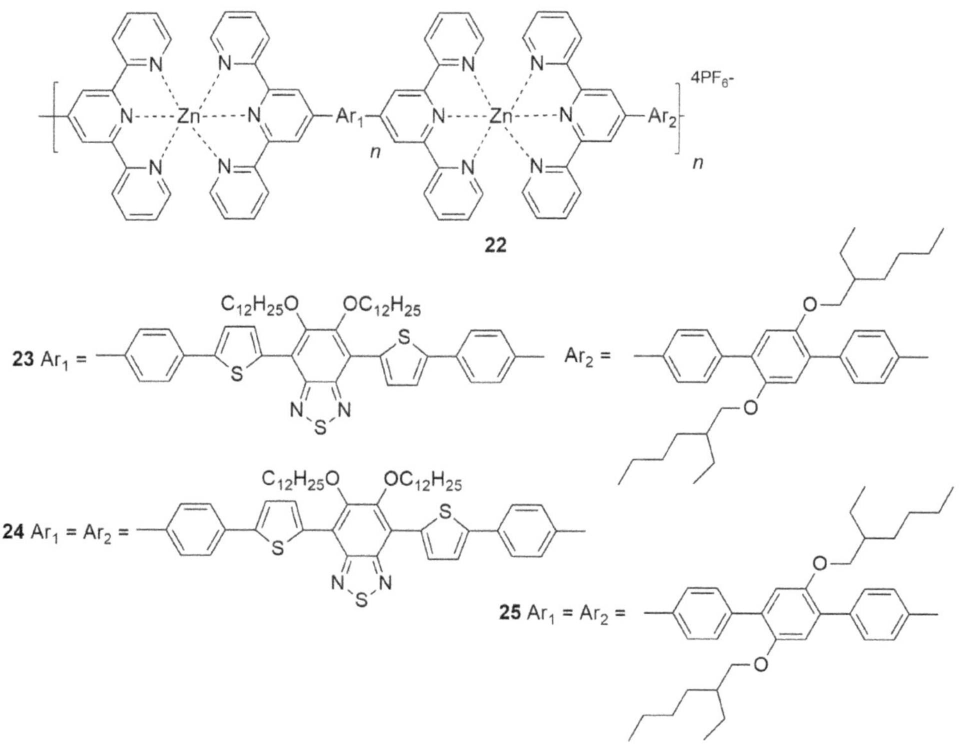

Figure 10.13 Metallo-homopolymers and an alternating copolymer

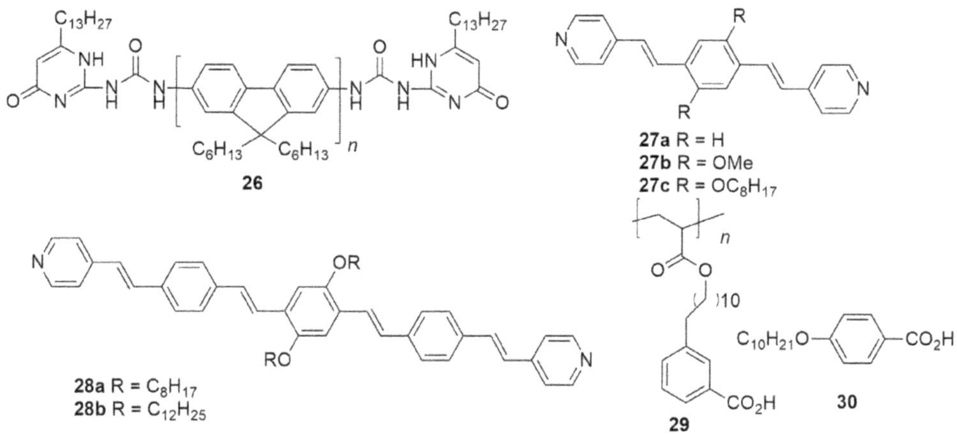

Figure 10.14 Molecules forming luminescent hydrogen-bonded complexes

oligofluorenes **26** ($n = 1$–7) (Figure 10.14) which display blue PL ($\lambda_{max} = ca.$ 410–420 nm) in the solid-state, might be promising candidates for LED applications, but to date their EL has not been reported [775]. There has been a report of LEDs using two-component luminescent supramolecular polymers.

A Taiwanese group found that networks formed by hydrogen bonding between pyridine-terminated oligomers such as **27** or **28** and polyacids such as **29** were luminescent with emission colours ranging from violet-blue ($\lambda_{max} = 396$ nm) for **27a**:**29** to red ($\lambda_{max} = 642$ nm) for **28a**:**29** [776]. These complexes, however, did not form good films, so LEDs had to be constructed from blends of **27c** in PVK complexed with the molecular acid **30**. The complex produced green EL ($\lambda_{max} = 510$ nm) with a narrower emission peak and higher brightness than was obtained from uncomplexed **27c** in PVK. This shows that hydrogen-bonded complexes of chromophores may display enhanced emission properties and further investigations of such materials is warranted.

While the device performance of materials with isolated chromophores has to date generally been disappointing, there is no doubt the concept has potential. Much further study is needed, however, to realise this potential. In particular, the use of non-covalent interactions to form assemblies or networks of chromophores is still in its infancy. Since nature uses this approach in forming light-harvesting systems it has particular utility in unlocking the secrets of how photosynthetic systems work.

So far, we have looked at materials which display fluorescence, which is to say they emit only from a singlet excited state. Phosphorescent materials in which the emission from a triplet excited state is allowed are of great interest as devices using them are potentially much more efficient as all the excitation energy might be usable. So, in Chapter 11, we will consider the synthesis of phosphorescent polymers.

11

Polymers for Phosphorescent LEDs

So far, we have been discussing materials in which the emission comes from a singlet excited state – that is, *fluorescent* materials. In these materials the triplet states have generally been non-emissive – that is, they are not *phosphorescent* (Figure 11.1).

By contrast, phosphorescence is a spin forbidden process, which requires the electron to change its spin during the transition from the high to the low energy orbital through spin-orbit coupling (Figure 11.2).

This coupling is much stronger in high atomic number elements, so the most efficient phosphors contain heavy metals such as palladium, ruthenium, platinum, osmium or iridium, while standard conjugated polymers which consist mainly of light elements such as carbon, hydrogen, nitrogen and oxygen are, therefore, intrinsically weakly phosphorescent at best. To obtain phosphorescence from LEDs based on conjugated polymers thus requires incorporation of phosphors either as dopants within a polymer host or built into the polymer structure, either as side-chain substituents or as co-monomers within the main chain. In both cases the polymer should have sufficiently higher triplet energy than the phosphor to permit efficient energy transfer via Dexter transfer to the latter. Since the lowest triplet energies in materials are lower than the lowest singlet energies, this requires polymer hosts to have significantly larger bandgaps than the dopants, so that the fluorescent emission of the polymer is considerably blue-shifted compared to the phosphor emission. This requirement is a major problem in designing hosts or co-monomers for blue phosphors. Blue emitting polymers will work well as hosts for red or green phosphors but violet-emitting materials with high triplet energies are needed as hosts or co-monomers for blue phosphors. In addition, as discussed in Chapter 10, the incorporation of phosphors can result in emission quenching and/or broadening, either through the formation of aggregates resulting in excimers (dopant–dopant) or exciplexes (dopant–host) [748]. However, it is also possible to exploit the excimer emission itself to complement or enhance the intrinsic monomer emission of the host system. While this approach has been shown to be

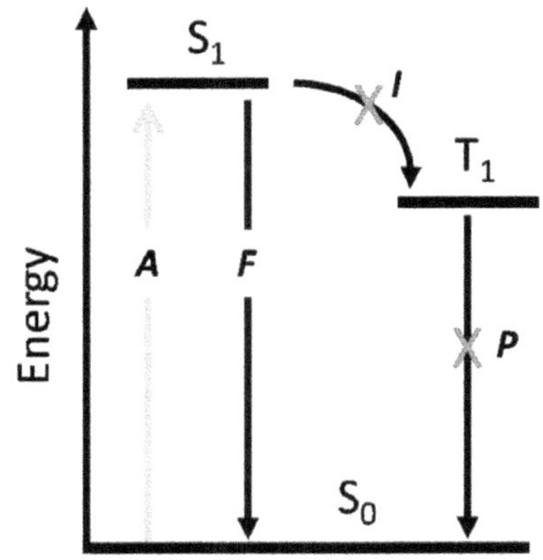

Figure 11.1 Photophysics of fluorescence highlighting: light absorption (*A*), prompt fluorescence (*F*), inter-system crossing (*I*) and phosphorescent decay (*P*). The grey crosses indicate spin forbidden processes

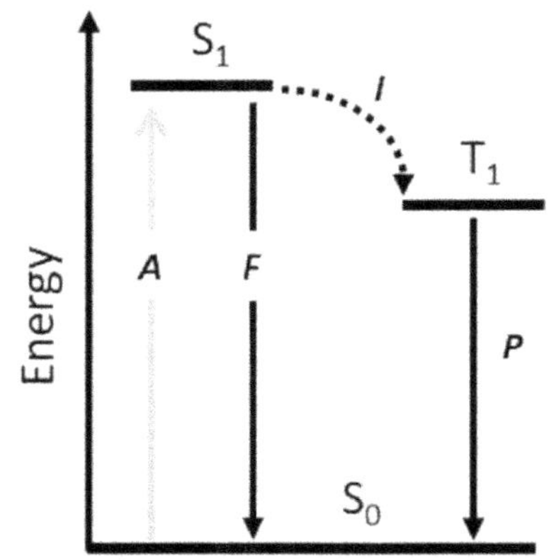

Figure 11.2 Photophysics of phosphorescence highlighting: light absorption (*A*), prompt fluorescence (*F*), inter-system crossing (*I*) and phosphorescent decay (*P*). The dotted line represents a partially allowed process

especially useful in the case of platinum and iridium dopant complexes in OLEDs based on small molecules [777, 778], there are fewer studies of the role of excimer emission for Pt complex dopants in conducting polymer hosts [779].

More recently, an alternative to the use of costly and toxicologically uncertain heavy metal complexes (such as the platinum and iridium complexes) has risen to the fore. These so-called third generation emitters are based on the thermally activated delayed fluorescence (TADF) process (Figure 11.3). Similar to phosphorescence, these molecules are capable of harvesting the triplet states with triplet-harvesting yields close to 100% [780, 781]. In a third-generation emitter, there are two fluorescent pathways. First, a fraction of the singlet excited state

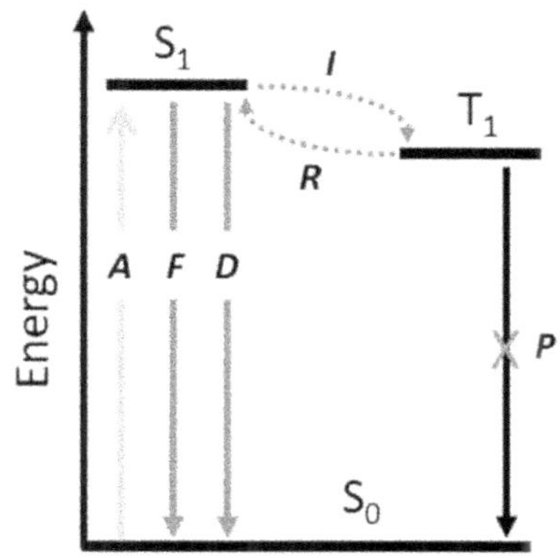

Figure 11.3 Photophysics of thermally activated delayed fluorescence (TADF) highlighting: light absorption (*A*), prompt fluorescence (*F*), delayed fluorescence (*D*), inter-system crossing (*I*), reverse inter-system crossing (*R*) and phosphorescent decay (*P*). The dotted lines represent partially allowed processes. The grey cross indicates a spin forbidden process

decays rapidly via immediate fluorescence to the ground state. Second, a portion of the singlet excited state that undergoes inter-system crossing can return back to the higher energy singlet state via a thermally excited process, known as reverse inter-system crossing (RISC) and can then subsequently decay more slowly to the ground state via fluorescence [782]. As such, the time-resolved fluorescence spectrum consists of both a fast (prompt fluorescence (PF)) and a slow (delayed fluorescence (DF)) component [783].

A number of TADF polymer emitters have been reported in the literature. However, achieving highly efficient TADF polymers has proved more challenging due to the difficulties in achieving small enough energy gaps between the S_1 and T_1 states, especially for red-orange emitting polymers [784]. The methods by which TADF units can be incorporated into polymers are generally the same as those used to introduce phosphors [785, 786] but, as is discussed in Section 11.2, there is an extra design possibility in introducing TADF units into polymers, due to such units containing separate donor and acceptor subunits. Initially, TADF polymers showed low device efficiencies but recently some very promising results have been obtained and this is one area of research into emissive polymers in which really significant advances remain possible.

11.1 Emissive Polymers as Hosts for Phosphors

Blending phosphorescent molecules such as porphyrins or metal complexes with polymers offers a way to harvest the triplet energy of both the polymer and dopant for obtaining emission and, thus, potentially to combine the useful physical properties of the polymer (good film-forming properties) with the desirable optical properties (high emission efficiency) of the phosphor. The obvious problem is that the blend may be unstable over time; leading to phase separation and loss of emission purity and/or efficiency. There has been a lot of work done into developing this approach using both metallo-porphyrins and metal complexes (mainly platinum- and iridium-based) as dopants, with emission colours ranging from the blue to the near infrared [150]. For red or green emitting phosphors, blue-emitting polymers such as polyfluorenes [787], poly(3,6-carbazole)s [788] or LPPPs [789] have proved good hosts with high device efficiencies obtained from devices using such blends with low (up to 4 wt%) amounts of dopants. Poly(N-vinylcarbazole) (PVK) has also proven to be a very good host for red or green emitting phosphors, though not so good for red as for green emitters due to poorer spectral overlap with the former, and the only report of NIR emission from a blend in PVK shows quite low efficiency [790]. Nonetheless, efficiencies of over 10% have been reported for both red and green emitting blends [791]. Blue-emitting phosphors, as previously mentioned within this chapter, present something of a problem in view of the high triplet energy and large bandgap needed by any host, and here carbazole-based materials such as PVK have produced the best results, though the efficiencies are lower than for the other colours [791]. The high efficiencies obtained using PVK as a host appears to be a combination of its large bandgap, high triplet energy and good hole transport properties, though the addition of molecular oxadiazoles to the blend to improve electron transport is needed to obtain the highest device efficiencies, which may have implications for the long-term stability of the device performances. Other materials, shown in Figure 11.4, with built-in charge transporting units such as the carbazole

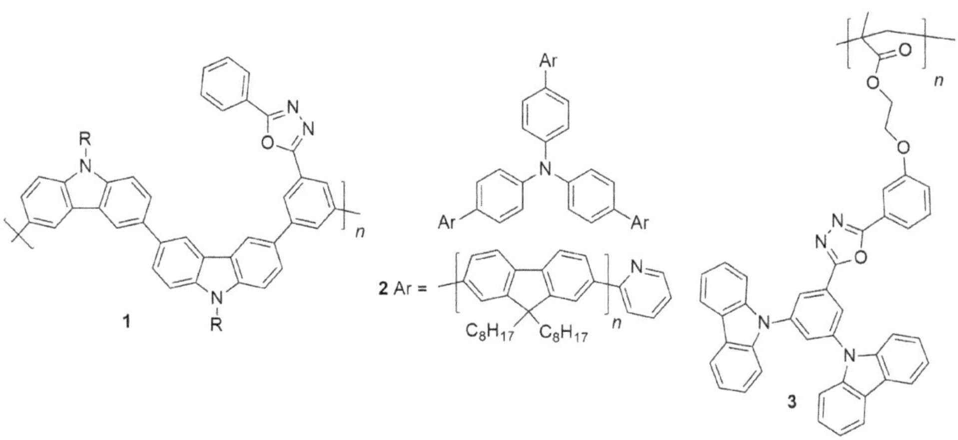

Figure 11.4 Polymers with charge transporting units designed as host materials for phosphors

copolymer **1** [788], the triarylamine-based star polymer **2** [792] and the copolymer **3** [793] have also proven to be good hosts for phosphors, though reported device results to date are not as good as those from PVK blends. The incorporation of oxadiazoles units into polymers **1** and **3** is an attempt to replicate the high efficiencies obtained using molecular oxadiazoles in PVK hosts while avoiding the possible phase separation problems associated with such blends. The lower device results obtained using them presumably reflects that they have different ratios and distributions of the two charge transporting units from those found in the best performing blends, but with improved design of copolymers it may be possible to replicate the results from the ternary blends.

11.2 Phosphorescent Polymers

To overcome the phase separation problems inherent in blends described in Section 11.1, phosphorescent polymers have been investigated in which the phosphors are covalently bound to the polymer structure either within the main chain or as side-chains. As was mentioned in Chapter 5, phosphorescence has been observed from Ph-LPPP, which arose when an unidentified phosphorescent impurity was introduced during the synthesis, presumably through a reaction between a precursor polymer, the palladium catalyst residues from a Suzuki polycoupling and a lithium reagent [356]. Such accidental introduction of phosphorescent moieties remains unique in the literature but is a potential side-reaction in other polymer syntheses and shows the need to be careful in removing residues of catalysts from materials. The first conjugated polymers with a phosphorescent unit deliberately incorporated into the chain were the fluorene copolymers **4** and **5** (Figure 11.5) which were prepared by incorporating the green (**4**) or red (**5**) emitting iridium complexes bearing monobrominated aryl ligands into a Suzuki polycondensation between dibromofluorene and a fluorene bisboronate [794]. For low molar mass polymers, emission was seen only from the phosphors, but for longer polymer chains with lower iridium content some blue emission from the polyfluorenes was also discernible. This approach has since been copied by other workers who incorporated other phosphors such as the red and green-emitting salen complexes found, respectively, in **6** [795] and **7** [796] or red-emitting porphyrins as in **8** [797, 798]. The highest efficiency yet reported is 26.3 cd A^{-1} (18%) from a polymer **9** containing a red-emitting osmium complex [799]. Besides polyfluorenes, this method has also been used to make phosphorescent polymers based on phenylenes and carbazoles containing red or green-emitting phosphors, with efficiencies of over 5 cd A^{-1} (4%) obtained in some cases [150]. To date, there has been no report of a blue emitting phosphor being incorporated into a conjugated polymer chain, but there is one report of such

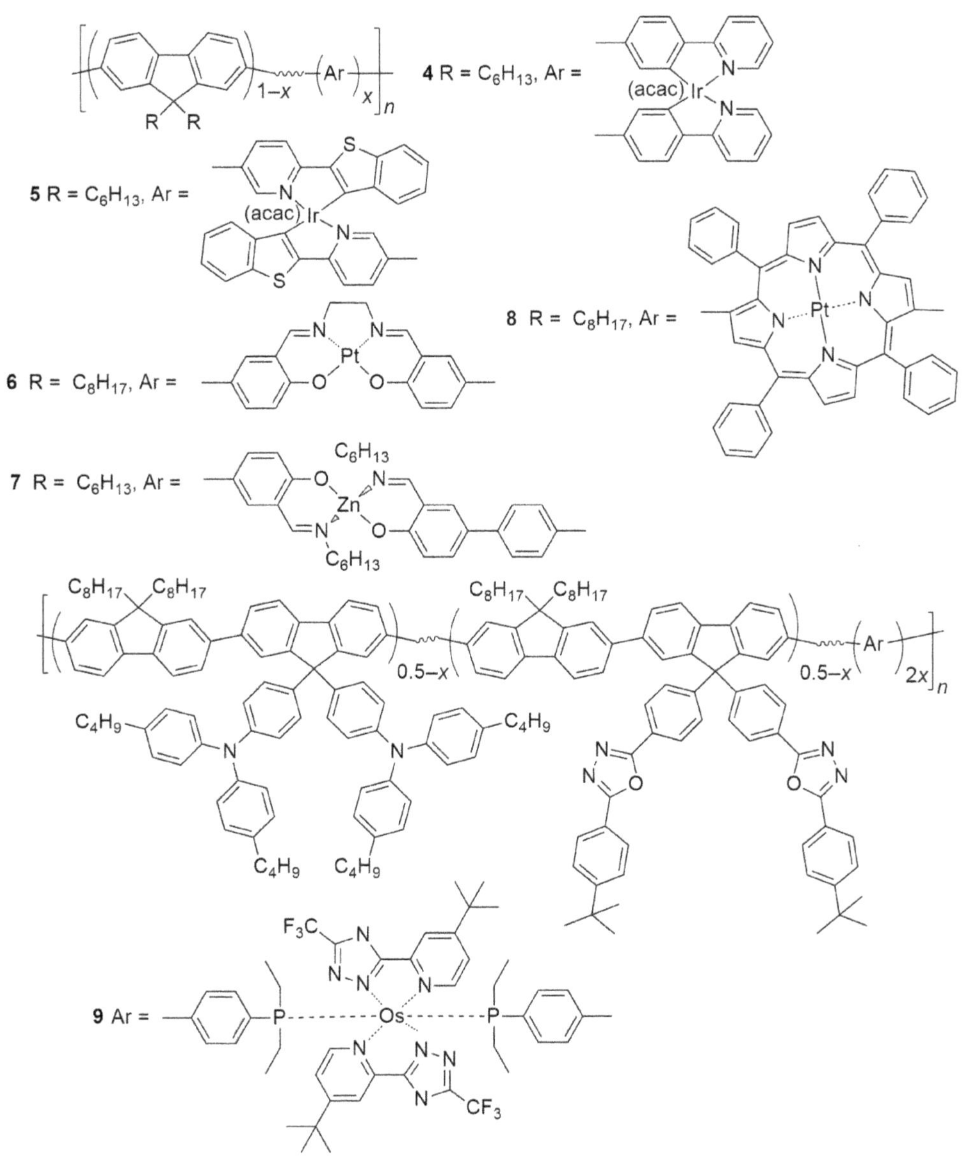

Figure 11.5 Phosphorescent fluorene copolymers

a group being attached as a side-chain to a polyfluorene [800]. The emission spectra was not affected but the EL efficiency was enhanced by the presence of the phosphor.

Hyperbranched phosphorescent polymers **10** (Figure 11.6) have been made whose green EL ($\lambda_{max} = 520–530$ nm) comes from the iridium complexes [801].

The best performing phosphorescent macromolecules are dendrimers such as **11–15** (Figure 11.7) containing phosphorescent iridium complexes as cores with

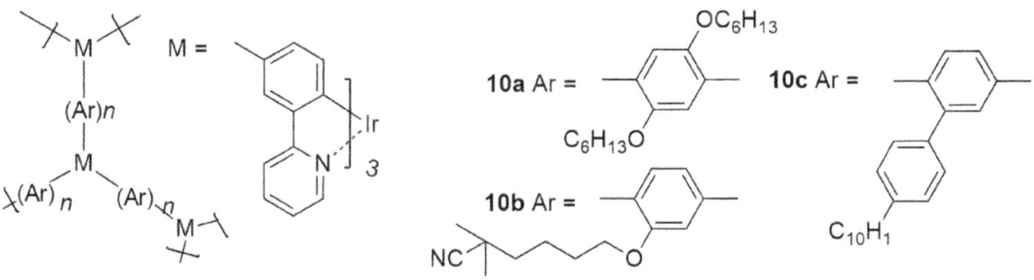

Figure 11.6 Phosphorescent hyperbranched polymers

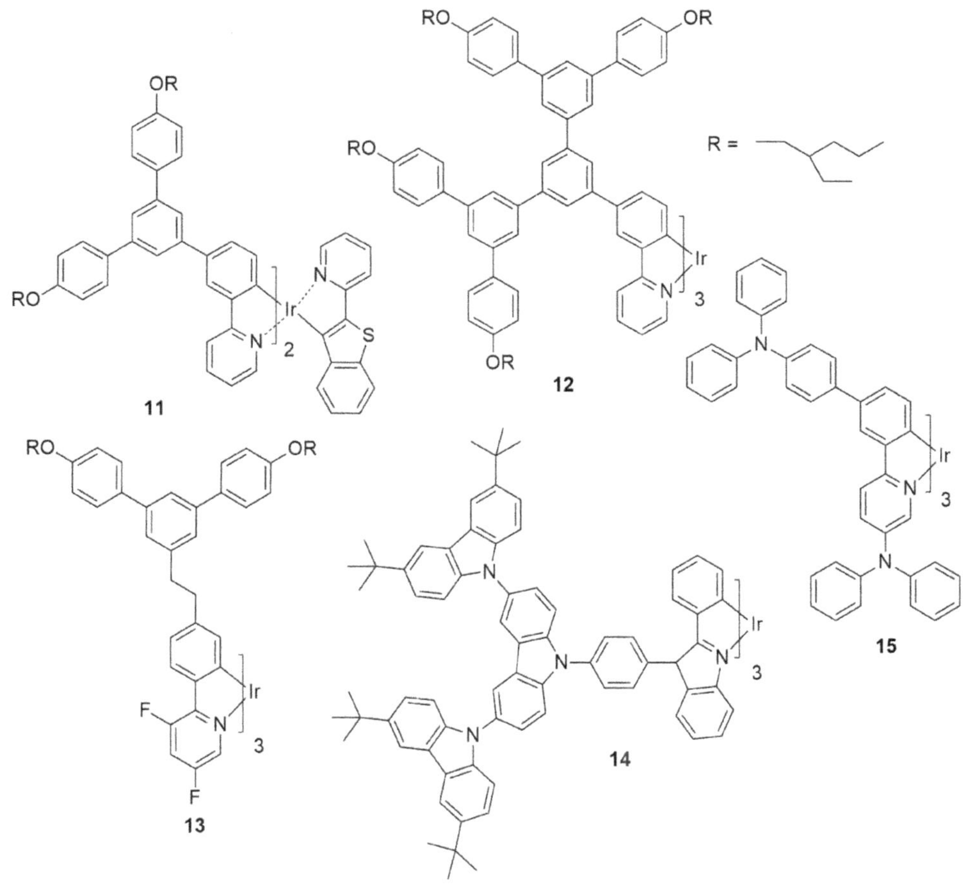

Figure 11.7 Phosphorescent dendrimers

phenylene dendrimers. By changing the ligands it has been possible to obtain high efficiency red (**11**, $\lambda_{max} = 634, 693$ nm, 5.7% efficiency), green (**12**, $\lambda_{max} = 518$ nm, 16% efficiency) or blue (**13**, $\lambda_{max} = 460$ nm, 10.4% efficiency) [721, 802]. The highest efficiencies were obtained by blending the dendrimers with carbazole-based host materials to improve the charge injection and transport. One obvious

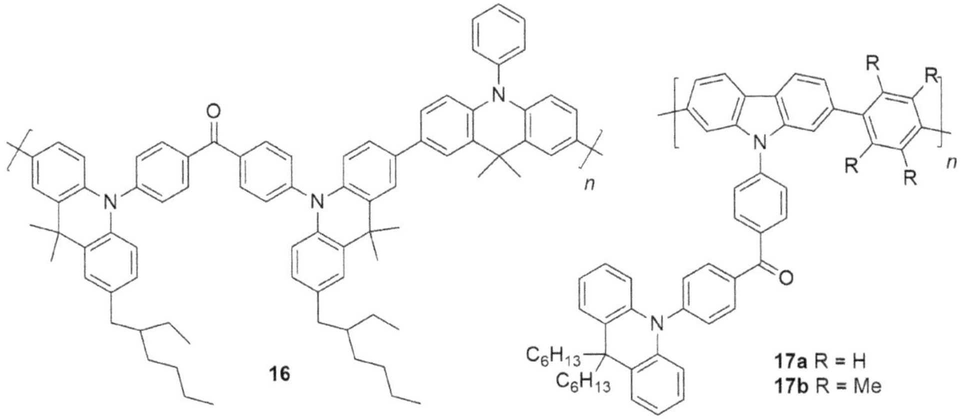

Figure 11.8 Conjugated polymers displaying TADF

way to improve charge transport in dendrimers is to incorporate charge-transporting units into the dendrimers. This approach has been successfully demonstrated by the dendrimer **14** with carbazole-based dendrons which produces green EL with an efficiency of up to 16.6% (57.9 cd A^{-1}) [803]. Similar high efficiencies (21%, 52.4 cd A^{-1}) have also recently been obtained from similar orange-emitting dendrimers such as **15**, with triphenylamine instead of carbazole hole-transporting units incorporated into the dendrons [804].

Conjugated polymers have been made containing TADF units in the main chain but here an extra design aspect becomes important. TADF units contain a donor and an acceptor sub-unit. The polymers can thus be made either with both sub-units attached together within the chain or with the donor sub-unit in the chain and the acceptor pendant to it. The former presents greater problems in obtaining efficient emission as the orbitals tend to become delocalised across multiple units which may result in loss of TADF properties, so that the latter approach has been more widely adopted [785, 786] but some polymers with good efficiencies have been made by the former approach (e.g. **16**, Figure 11.8), from which green EL with an efficiency of 9.3% has been reported [805]. The efficiency of both designs of polymers can often be further enhanced by incorporating meta-linkages or substituted phenylene linkers to disrupt the long-distance conjugation along the chain, so that, for example, for polymers **17**, which exemplify the pendant acceptor design, the external quantum efficiency of **17a** with a phenylene linker is only 1.4% but that of **17b** with a tetramethylphenylene linker is 23.5% [806].

Dendrimers showing TADF have been made which contain acceptor units such as benzophenone at their core and donor units such as carbazoles in the branches, but those containing conjugated branches have to date shown only modest device performance [786]. Much better efficiencies have been seen from dendrimers with

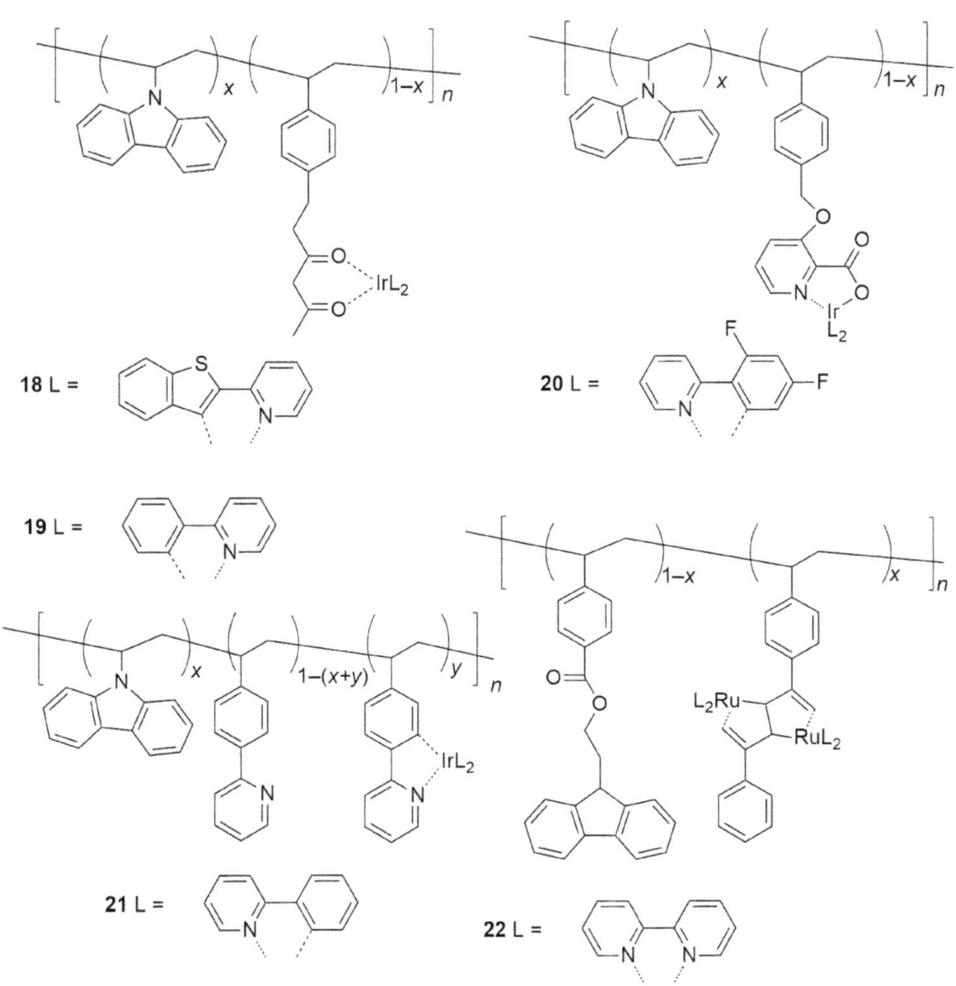

Figure 11.9 PVK copolymers with phosphorescent side-chains

donors on the periphery linked to acceptor cores by non-conjugated branches, but those lie outside the scope of this book.

The other way to incorporate phosphors into luminescent polymers is as side-chains. Here the problem of charge transport might be overcome by making copolymers with a combination of charge-transporting moieties and emissive units as side-chains. This approach has been successfully demonstrated for PVK copolymers such as **18**–**21** (Figure 11.9) [807, 808]. By varying the ligands, relatively efficient red (**18**, $\lambda_{max} = 620$ nm, 5.5% efficiency), green (**19**, $\lambda_{max} = 520$ nm, 9%, **21**, $\lambda_{max} = 512$ nm, 4.4%) and blue (**20**, $\lambda_{max} = 475$ nm, 3.5%) EL can be obtained. Near-infrared EL ($\lambda_{max} = 790$ nm) has been obtained from the structurally related polystyrene **21**-bearing ruthenium complexes [809].

Even higher efficiency has been obtained from copolymers **23** (Figure 11.10), bearing a mixture of hole and electron-transporting units, made by living radical polymerisation [810]. The efficiency of the green EL ($\lambda_{max} = 512$ nm) was highest (10.0%) when the iridium loading was 8 mol%. Curiously, this efficiency is lower than for a blend of the copolymer **24** with 8 mol% of the untethered iridium complex. Increasing the iridium loading to 13% lowered the EL efficiency to 7.3% due to self-quenching, which indicates that great care must be taken in getting the balance between emissive and charge transporting units correct in such copolymers.

Whereas the above polymers are all made by radical polymerisation methods, other materials with phosphorescent side-chains have been made by ROMP, for example, polynorbornenes **25** (Figure 11.11), bearing green or yellow-emitting

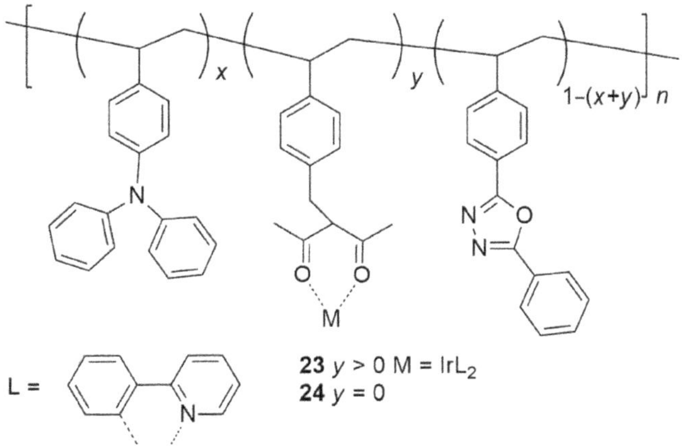

Figure 11.10 Copolymers bearing phosphors and electron- and hole-transporting units

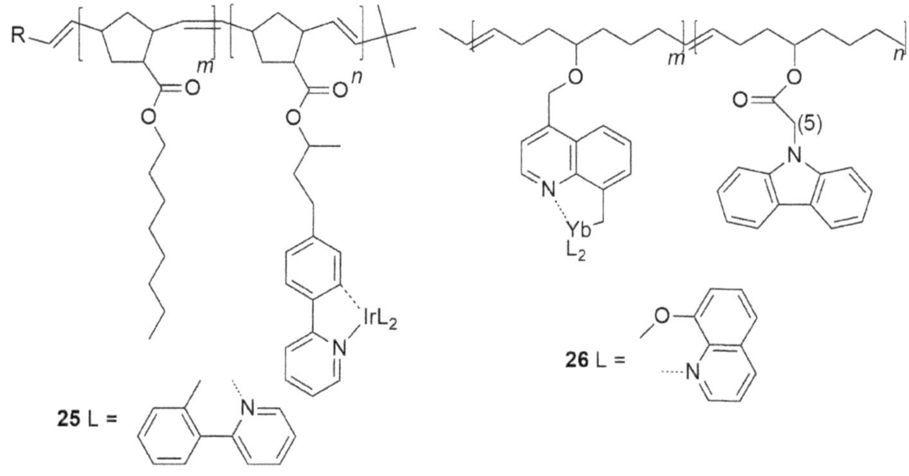

Figure 11.11 Phosphorescent copolymers prepared by ROMP

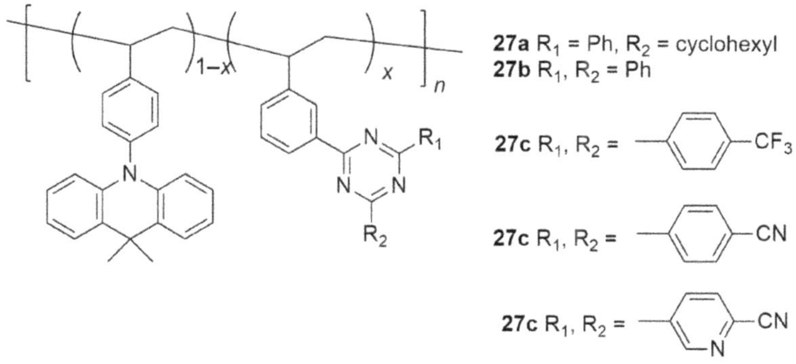

Figure 11.12 TADF copolymers utilising through-space charge transfer

iridium complexes [811]. ROMP has also been used to make polynorbornenes and polycyclooctenes **26** bearing ytterbium complexes which emit in the infrared (λ_{max} = 971 nm) [812].

These approaches have also been used to make polymers bearing TADF units [785, 786]. In the case of TADF polymers, the TADF effect can also be achieved by attaching donor and acceptor units separately, in which case the TADF is induced by through-space charge transfer, as in polymers **27** (Figure 11.12) in which the emission colour can be tuned between blue for **27a** (λ_{max} 453 nm) to red for **27e** (λ_{max} 616 nm) by altering the substituents on the triazine acceptor [813].

One other possibility in attaching phosphors or TADF units as side-chains is that one can put more than one different emissive unit on. As we shall see in Chapter 12, this is a useful route towards obtaining white emission from LEDs.

12

Polymers for White-Emitting PLEDs

White light emission from low-power LEDs is highly desirable in many applications, particularly lighting, which currently accounts for 15–20% of global energy consumption and, consequently, around 5% of global greenhouse gas emissions [814]. As such, there has been considerable recent research effort towards developing low-power white organic light-emitting diodes (WOLEDs) [815]. The difficulty in obtaining white light from organic materials is that no chromophore is known which emits it. White emission thus usually arises by simultaneous emission from a number of chromophores with complementary emission. This usually means either a blue emitter combined with an orange emitter or a combination of a red and a green emitter. A particular problem with white emission is that of colour stability as, if the chromophores bleach at even very slightly different rates, then, over time, the emission colour will cease to be white. Also, the different electrical properties of different chromophores mean that the emission may be voltage-sensitive, as the various chromophores may display different turn-on thresholds and/or respond differently to voltage changes. Obtaining reasonably pure white emission also usually requires getting the balance between the chromophores right within a relatively small tolerance zone. At this point, one must take into account the accepted definitions of white emission, based upon CIE coordinates (x, y), which define the colour with respect to the components of red, green and blue on a two-dimensional colour surface and colour rendering indices (CRI), that define how well the colour of objects illuminated by it match those illuminated by a standard light source – natural daylight in the case of white light. More specifically, the CRI index determines the accuracy with which the light source renders eight particular pastel colours. Thus, a CRI of 100 corresponds to a light source that exactly matches natural daylight.

The CIE co-ordinates for pure white emission are (0.33, 0.33) but light with co-ordinates close to this is still white to our eyes – the term 'near white' we use to describe some materials' emission refers to light that our eyes do not perceive as

Table 12.1 *CRI requirements for typical outdoor and indoor lighting applications*

Environment	Application	Minimum CRI
Outdoor	Street lighting	65
Indoor	High-bay lighting	70
Indoor	Ceiling lighting	80
Indoor	HDTV	90

being quite white (i.e. it has a faintly perceptible colour). The terms 'cool white' or 'warm white' are used for near white emission with, respectively, yellow and blue tones. The conventional wisdom is that a CRI difference of more than five points is required before there is a perceptible difference to the eye and that, generally speaking, a CRI of above 80 is required for a high-quality white light source [816]. However, the acceptable colour tolerance for a given light source is highly dependent upon the application. For example, white LED lighting for ultra-high-definition television production requires a CRI of over 90 [817]. Table 12.1 highlights some typical CRIs for a range of popular lighting applications [814].

Two approaches are possible towards such systems: blends of different molecules with complementary emission or molecules containing a mixture of different chromophores. An obvious problem with the former approach is that of phase separation, leading to unstable emission, which the latter approach avoids, but both systems have the difficulty that, when the spectra of the chromophores overlap, then energy transfer will inevitably lead to emission from the lower bandgap chromophores occurring to a disproportionate degree. We have already seen this approach used as a way to tune the colour of the emission of high-bandgap materials (e.g. polyfluorenes) by incorporating a few mol% of either fluorescent or phosphorescent chromophores. In both approaches, therefore, it is necessary to have inefficient or incomplete energy transfer occurring to the lower bandgap chromophore. This can happen in blends if there is incomplete mixing as the energy transfer mechanisms require the two chromophores to be in close proximity (within a few nanometres of each other) or if the low-bandgap material is in insufficient quantities to allow complete quenching of the high-bandgap material, that is, there are simply too few acceptor chromophores for all the prospective donors to transfer their energy to within the lifetime of their excited states. This latter case is also the only way in which white emission can be obtained from a single emissive molecule.

12.1 White EL from Blends

There have been a number of reports of white or near-white EL obtained from blends of blue-emitting and red-orange-emitting conjugated polymers [150].

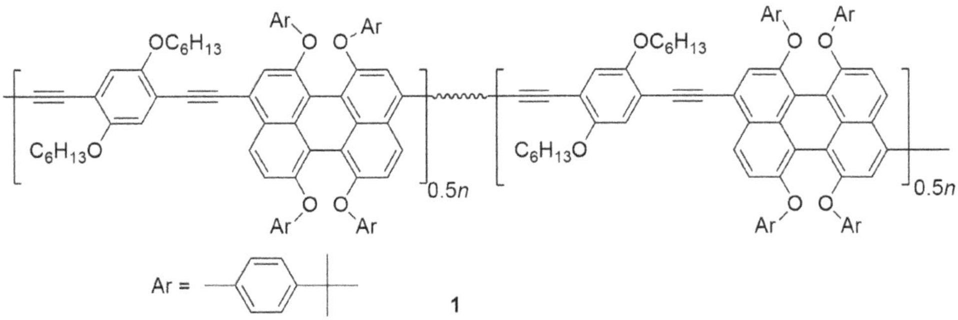

1

Figure 12.1 Red-orange-emitting polymer which produces white EL in blends with MeLPPP

An example of how the obtaining of white EL depends upon the blend composition is shown by the behaviour of blends of blue-green-emitting Me-LPPP (Chapter 5) and the red-orange-emitting polymer **1** (Figure 12.1) [818–820]. The emission from the blends is predominantly red-orange, even when the amount of **1** present is as low as 0.2 wt%. Interestingly, the efficiency for the orange EL emission from this blend is higher (1.6%) than from pure Me-LPPP (1%) or **1** (0.01%) using the same device configuration. Higher concentrations of **1** cause a rapid drop in device efficiency. Similarly, the optimal PL efficiency for the blend with 0.7 wt% **1** is higher (41%) than for Me-LPPP 30% or **1** (11%) and again drops as the concentration increases. These effects indicate how blending can increase device efficiency by separating polymer chains and reducing non-emissive decay pathways. When the content of **1** is reduced to 0.05 wt%, white EL is produced, due to simultaneous emission from both polymers with reasonable (0.8%) efficiency [819–821]. Adding the insulating polymer poly(methyl methacrylate) (PMMA) to the blend leads to separation of the emissive polymer chains by the PMMA chains and less efficient energy transfer so that, while a higher amount of **1** (0.08 wt%) is required for obtaining white emission, the EL efficiency goes up to 1.2%. These effects help illustrate the subtle effects of changing blend composition, which must be considered when attempting to optimise performance from blends, and some of the techniques one may employ in doing so.

White emission has also been observed from blends of blue-emitting polyfluorenes containing a few wt% of green and/or red-orange-emitting PPV derivatives (Figure 12.2). The highest reported efficiency (6%) is from a blend of the PDAF **2a** with 2 wt% of red-orange-emitting MEH-PPV (Chapter 3) [822]. White EL is also seen from ternary blends of blue-emitting **2a** (95 wt%), with the green-emitting **3** (4 wt%), and either MEH-PPV or the CN-PPV **4** as a red-emitting component (1 wt%) [823]. An example of efficient (1.64%, 4.08 cd A^{-1}), voltage stable white emission is that reported from a blend of **5** (9 wt%) in poly

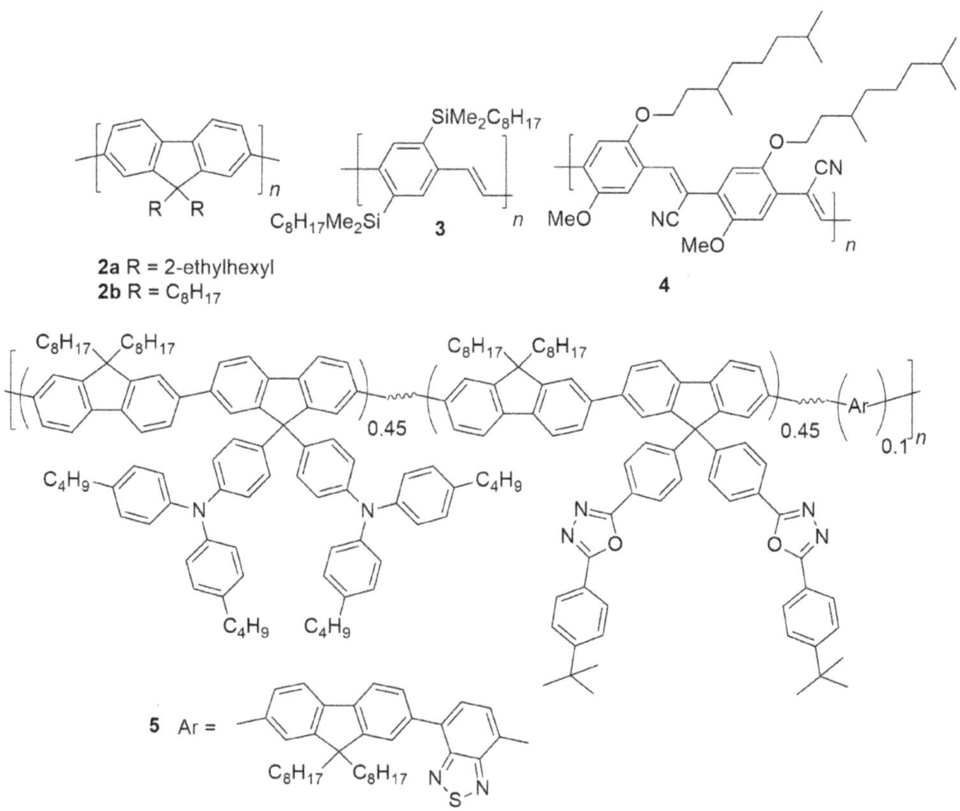

Figure 12.2 Other polymers used to produce white-emitting blends

(dioctylfluorene) **2b** [824]. A 10:1 blend of PVK copolymers **6** and **7** (Figure 12.3) bearing, respectively, red- and blue-emitting iridium phosphors produced white EL with high (4.5%) efficiency [807].

White EL can also be obtained by blending red, yellow and/or green emitting dyes or phosphors into blue-emitting polymer hosts [150]. Suitable host materials include poly(N-vinyl carbazole) (PVK, Chapter 10), polyfluorenes, and fluorene-based copolymers. Again, ternary blends have been used. For example, a high efficiency (21 cd A^{-1}) has been observed from a ternary blend of a yellow-orange-emitting dye with a mixture of a polyfluorene and a green-emitting fluorene–fluorenone copolymer [825]. One can also use a mixture of phosphorescent and fluorescent dopants in the same host. This has been used to make an efficient (6.12%, 13.2 cd A^{-1}) white LED from a blend of a blue fluorescent dye and an orange phosphorescent osmium complex in PVK [826]. A blend of PVK with a mixture of blue and orange emitting iridium-cored dendrimers also produces highly efficient white EL (37 cd A^{-1}, 18.5%) [804]. It has also been reported that doping **8** with a red-emitting porphyrin produced white EL with a high efficiency

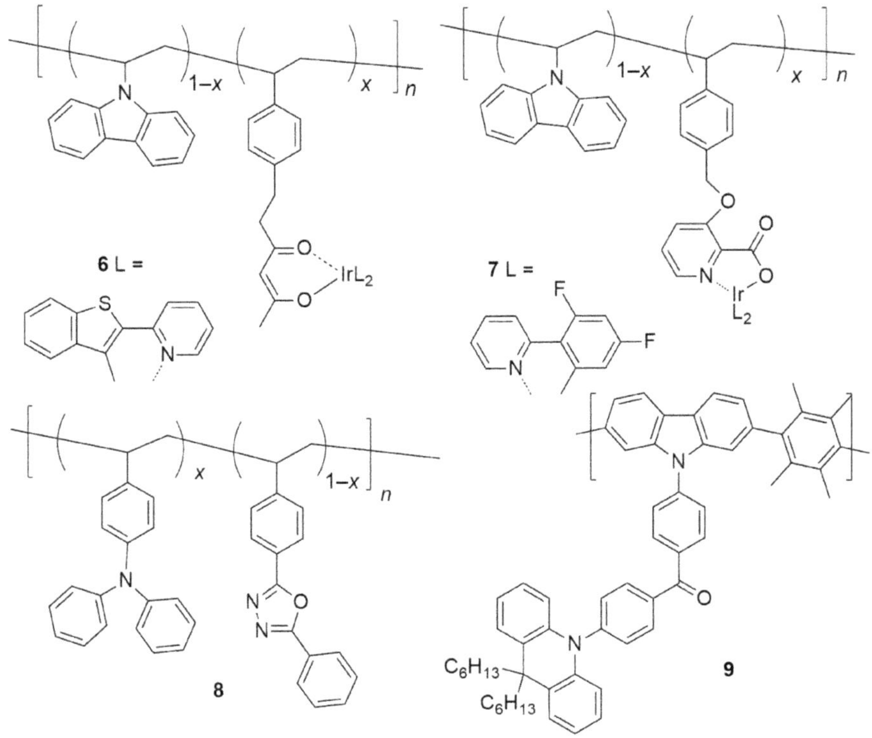

Figure 12.3 Emissive polymers used in blends for white EL

of 4.9% [810]. The highest efficiency to date from a blend of an emissive polymer and a red-orange dopant (20.9%, 61.1 cd A^{-1}, 56.4 lm W^{-1}) has been obtained by doping a highly efficient thermally activated delayed fluorescence (TADF) red-orange-emitter into a blue-green-emitting copolymer **9** which itself contains TADF units [806]. The stability of these blends and their emission has not to the best of our knowledge been reported, but it is likely to be the major obstacle to their use, and the approach of obtaining white emission from a single material is likely to prove superior.

12.2 White EL from Single Polymers

There are a number of examples in the literature in which blue-emitting polymers can display white emission, arising from the formation of long wavelength emission from aggregates or excimers [150]. An example of this is the white EL seen from the anthracene-substituted polynorbornenes **10a**, **b** (Figure 12.4) [827]. These display blue-green PL in the solid-state but an extra emission band in the red appears in the EL spectrum, which is attributed to aggregates.

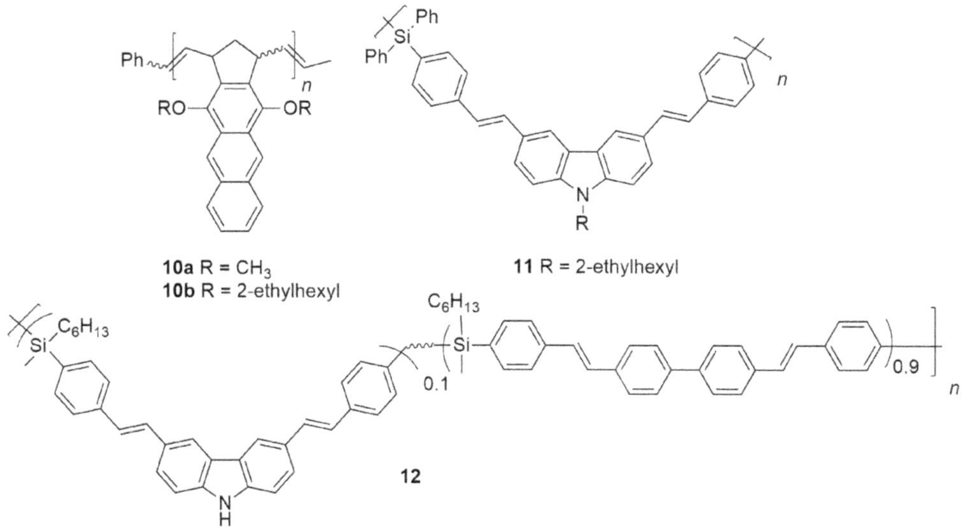

Figure 12.4 Polymers showing white EL due to exciton emission

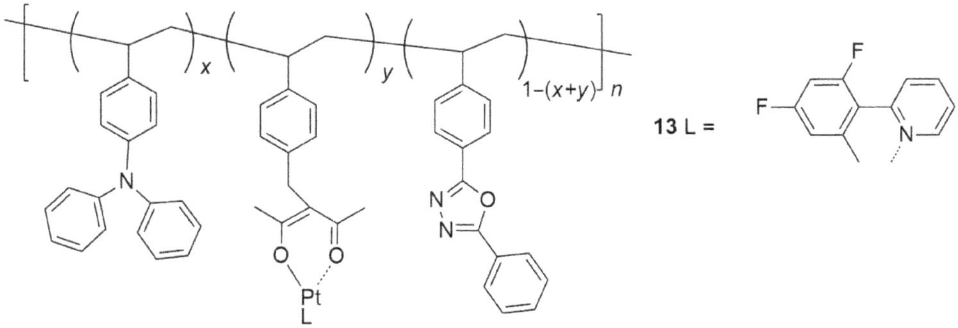

Figure 12.5 Side-chain copolymer for near-white EL

Sometimes this white light emission is a voltage-dependent phenomenon as, for example, in the polymer **11** whose emission shows a blue band around 420 nm and a yellow band around 580 nm, which has been attributed to excitons. As the driving voltage increases the latter becomes stronger so that the EL becomes near white at driving voltages above 12 V [828, 829]. The similar copolymer **12** produces white EL at higher operating voltages due to the appearance of an extra band in the red part of the spectrum, which was attributed to charge transfer complexes since it was not seen in the PL spectrum [830].

There are other white- or near-white-emitting polymers for which the long wavelength emission is due to well-defined emissive chromophores. For example, the polymer **13** (Figure 12.5) produces efficient (4.6%) near-white EL due to a

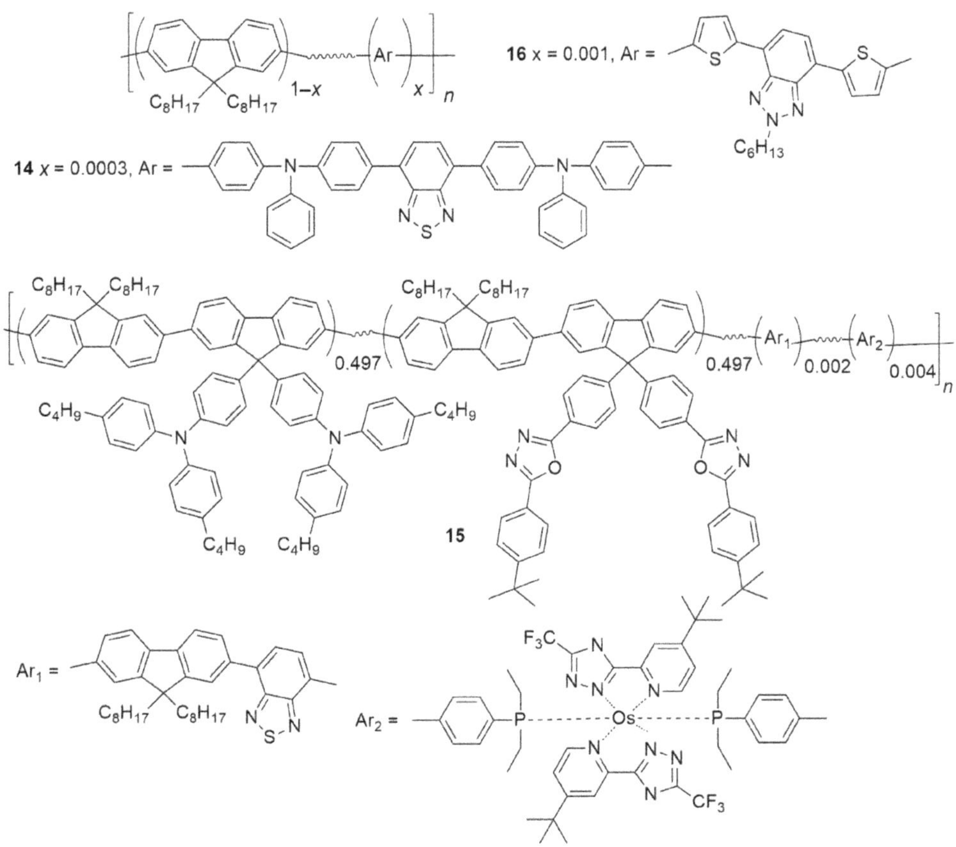

Figure 12.6 Fluorene-copolymers showing white EL

combination of blue emission from the charge transporting units and long wavelength emission from the platinum complex [831].

To date, the most effective method for obtaining efficient white EL from single polymers has been to incorporate low mole fractions of red, orange, or green emitting chromophores into blue-emitting polymers such as polyfluorenes. These copolymer structures can be achieved using either fluorescent or phosphorescent chromophores. For example, the copolymer **14** (Figure 12.6) containing 0.03 mol % of an orange-emitting benzothiadiazole-based co-monomer produces white EL with a reported efficiency of up to 8.99 cd A^{-1} [832]. The copolymer **15** containing a red-emitting osmium-based phosphor displays even better efficiency (10 cd A^{-1}, 5.1%) [799]. One should note that this performance is much better than the figure of 4 cd A^{-1} mentioned in Section 12.1 using the similar copolymer **5** blended with a phosphor; suggesting that attaching a dopant to a polymer is a better strategy than bending it with the polymer. Until recently, the highest reported efficiency to date (11 cd A^{-1}) for a white LED based on this type of copolymer came from the

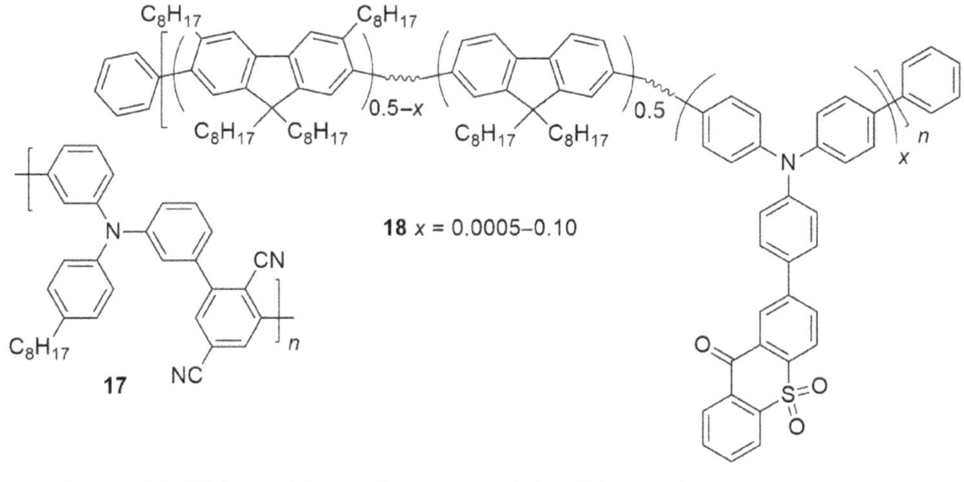

Figure 12.7 White-emitting polymers containing TADF units

copolymer **16** with a benzotriazole-based red-emitter as comonomer [833]. In all these cases a very low (typically 0.01–0.05 mol%) of the red or orange emitter is needed if one is to obtain white light. At higher concentrations, the emission becomes dominated by the dopant emission and so is not white. Only at these low concentrations is the energy transfer from the fluorene backbone sufficiently inefficient that one gets enough blue emission for our eyes to perceive the light as being white.

An advance in the design of such copolymers which has pushed their external quantum efficiencies up to around 15% has been the incorporation of units which show TADF. As mentioned in Chapter 11, the TADF chromophores typically consist of a linked donor and acceptor moiety, and it is possible to incorporate the TADF unit either with both moieties in the chain as in copolymer **17** [834] (Figure 12.7) or with one of these moieties within the chain and the other pendant to it as in copolymers **18** [835], but there is as yet too little data to assess the relative merits of these two approaches. An efficiency of 15.4% (51.9 cd A^{-1}, 50.9 lm W^{-1}) has been obtained from a device using **17** with CIE coordinates of (0.34, 0.57). The efficiency and emission colour from copolymers **18** was found to be dependent upon the amount of the TADF chromophore incorporated into the copolymer. The purest white emission with coordinates of (0.32, 0.31) and (0.36, 0.36) was obtained when x was 0.0005–0.001 mol%, whereas the polymers with higher amounts of the dopant showed orange emission (λ_{max} 602–622 nm). The EL efficiencies were modest (0.3–0.8%) for the pure white-emitting materials, but 4.8% was obtained for the copolymer with 0.02 mol% of the TADF unit. When this latter polymer was doped into a blue-emitting host (1,3-bis(*N*-carbozlyl)benzene) at a ratio of 1:19, warm white EL (0.42, 0.45) was obtained with 9.9% efficiency.

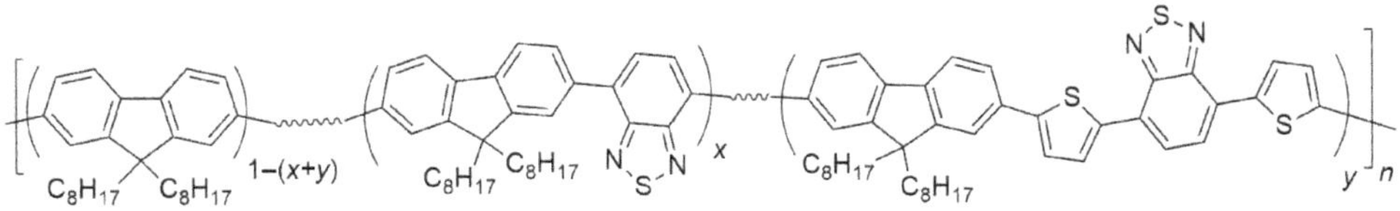

19 $x = 0.00005$, $y = 0.0001$

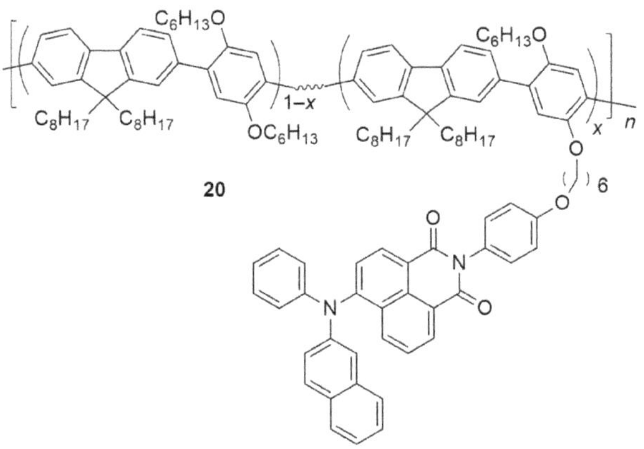

20

Figure 12.8 Other copolymers for efficient white EL

A more complicated approach is to introduce two dopants onto the chain. This approach makes getting the balance of the emission right harder but can give purer white (i.e. with CIE coordinates closer to (0.33, 0.33)), than for the polymers described earlier in this section, whose emission, while white to the eyes, is generally not strictly pure white. Pure white emission has been obtained with good efficiency (6.34 cd A^{-1}) from the ternary copolymer **19** (Figure 12.8) containing both green- (benzothiadiazole) and red-emitting (dithienylbenzothiadiazole) chromophores [836]. It is also possible to attach the dopants to the conjugated backbone as side-chains. An efficient (22 cd A^{-1}) white LED that has been obtained by this approach used a fluorene-phenylene alternating copolymer **20** with an orange-emitting naphthalimide dye as a side-chain on some phenylene units [837]. The copolymer with 0.05 mol% of the dye produced white EL, while orange EL was obtained for copolymers with higher dye content.

Obtaining white emission from polymers with non-emissive backbones is also possible. The block copolymer **21** (Figure 12.9) bearing blue and red phosphors has been made by controlled radical polymerisation [838]. Since the two emissive units are on different blocks, phase separation of the blocks leads to a morphology containing red and blue-emitting domains, so that, by varying the ratio of the chromophores, the emission colour can be tuned from red to white. White EL has been obtained from a copolymer with a ratio of 10 blue to 1 red phosphor (n:m = 10:1). The reported efficiency is modest (1.5%) but there is considerable scope for device optimisation.

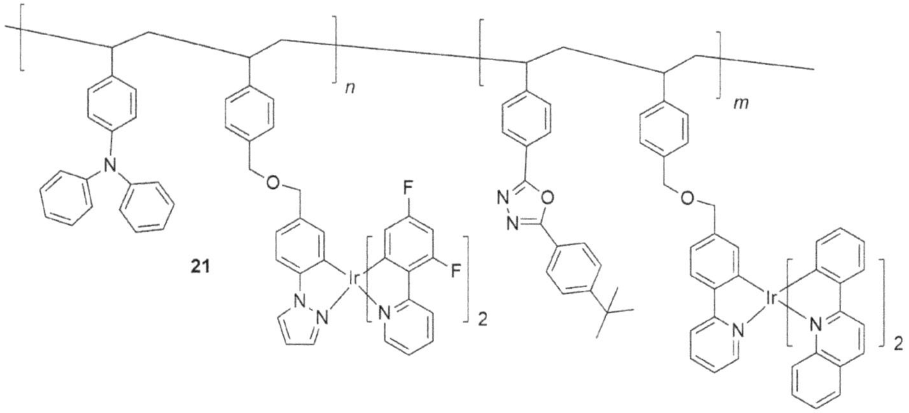

Figure 12.9 White-emitting block copolymer with emissive sidechains

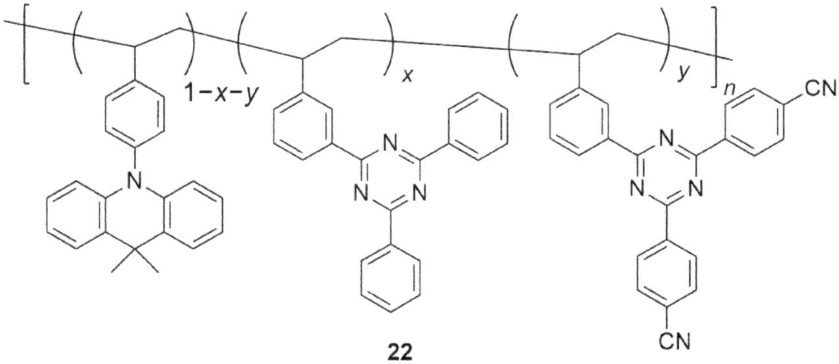

Figure 12.10 White-emitting TADF side-chain copolymers

This approach can be used to make white emission via TADF as exemplified by copolymer **22** (Figure 12.10) in which through-space charge-transfer between a donor and two acceptor units produces a mixture of blue and yellow light [813]. The amount of the blue-emitting acceptor was held at $x = 0.05$, while the amount of the yellow emitter was varied from $y = 0.004$ to $y = 0.008$. The CIE coordinates changed from (0.31, 0.42) to (0.35, 0.47), with efficiencies of up to 14.1% being obtained (37.9 cd A^{-1}, 34.8 lm W^{-1}).

White emission from hyperbranched systems would be hard to design given their irregular structures, but it is not hard to imagine that a star polymer with blue-emitting arms and an orange- or red-emitting core might produce white emission as it would not be possible for all the arms to transfer their energy to the core at the same time. Indeed, this approach has proven successful with white emission obtained from star polymers **23** (Figure 12.11) with an orange-emitting core bearing 6 blue-emitting polyfluorene arms [839]. These polymers were made by

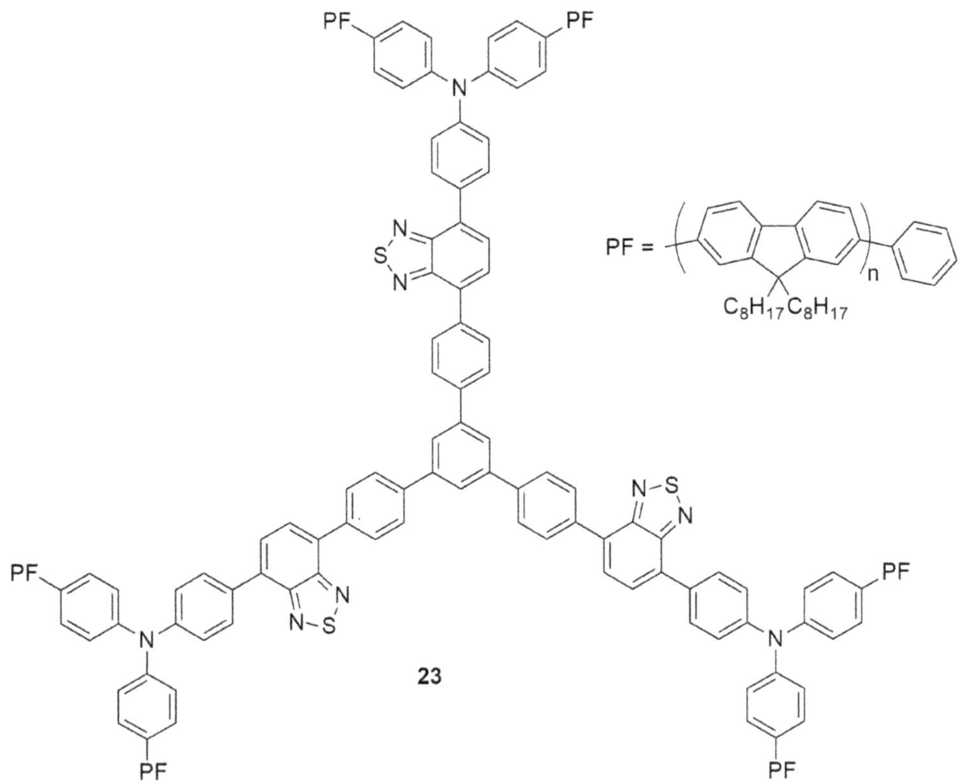

23

Figure 12.11 Star polymer for white EL

adding the hexabrominated core as a comonomer into an AB-type Suzuki polymerisation of a bromofluorene monoboronate. To get white EL, the core was kept to 0.01–0.03 mol%. For example, almost pure white EL was obtained with high efficiency (18 cd A^{-1}, 6.36%) from the polymer with 0.02 mol% of core.

There remains one final strategy for obtaining white emission, and that is to combine different chromophores together by non-covalent linkages to form a white-emitting supramolecular copolymer. Such a structure has been obtained by mixing together the oligomers **24–26** (Figure 12.12) which, respectively, produce blue, green and red emission [840]. Complementary hydrogen-bonding between the terminal ureidopyrimidinone units produces random copolymers whose emission colour could be tuned by varying the ratio of the components. Pure white PL was obtained from a blend of the three in a ratio of **23**:**24**:**25** = 84:10:6 and near white EL was reported, but the efficiency was low. Since the EL efficiency of devices using each of the individual components alone was also low, this result may suggest a generic charge transport problem in the complexes, but further work would be needed to determine if there is a fault in the concept or merely in the specific materials and/or device setup used.

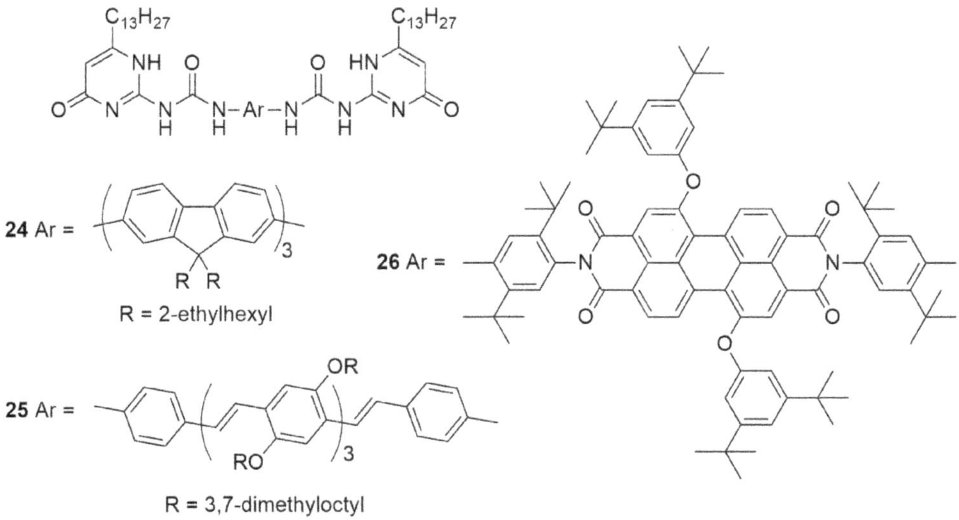

Figure 12.12 Oligomers for making a white-emitting supramolecular complex

As can be seen there are a variety of strategies that may be adopted to obtain efficient white emission. While data has not been published for the stability of the emission from most of these materials, there is good reason to be optimistic that devices with acceptable lifetimes will be obtainable after further optimisation of molecular and device design. We have now covered all the various types of conjugated (and in some cases not-so-conjugated) polymers which have been developed for use in organic electronic applications and have shown how they can be optimised to maximise their effectiveness as functional components. However, we have not covered all of the types of devices such materials might be used in. There are a few other device types which have not been mentioned so far, some of which require extra performance criteria and, thus, extra design parameters. It is to these we now turn in Chapter 13.

13

Polymers for Other Luminescent Devices

So far, we have discussed the use of emissive polymers in standard LEDs which contain a single emissive layer with possibly one or more charge transporting layers sandwiched between two electrodes. These are not the only type of emissive device that can be made using emissive conjugated polymers. However, most of these do not require any special extra features in the polymers concerned and so the design principles we have discussed in preceding chapters remain sufficient for producing materials for efficient devices.

13.1 Light-Emitting Devices with Non-standard Configurations

There have, for example, been reported devices using multiple emissive layers. The only extra design feature one has to consider here for the polymers is that the deposition of one layer must not disturb the underlying layer, so one needs either to render the lower layer insoluble through cross-linking (thermal or photo-induced) or make use of orthogonal solubility, that is, the solvent for the new layer must be one the underlying layer is not soluble in. Even with this condition, some mixing can occur at the interface between the layers which may produce new emission colours. The emission from these devices may also be voltage-dependent, with the recombination zone moving from one layer into another [150, 841]. Devices have also been constructed with the symmetrical structure electrode/ 'insulating' polymer/emissive polymer/'insulating' polymer/electrode. These can be operated using alternating current, that is, directly from the mains, and were styled symmetrically configured alternating-current light-emitting (SCALE) devices by their inventors [842–844]. The emissive polymers used were mainly pyridine-based – either polypyridine or poly(pyridine vinylene) derivatives [150]. The so-called insulating polymers used have been emeraldine-base polyaniline (PANI), sulphonated polyaniline (SPAN) and P3HT, which are actually used as charge-injecting polymers. The position of the recombination zone may vary with bias so they are not truly symmetrical in behaviour. There have been no recent reports on these devices, suggesting that they offer no real advantage over LEDs, since the need to use a transformer to convert AC to DC is already a feature of devices such as laptops and is hardly onerous, and their device performance was not better than standard LEDs nor is their structure intrinsically cheaper to make. There appear to be no special requirements for emissive polymers to be used within them save that the solvents used to deposit should not be good solvents for the 'insulating' polymer layers.

By contrast, there are other classes of emissive devices in which the mechanism of the device operation requires new aspects to be taken into consideration when designing a polymer for use within them. Here we will discuss two sub-classes of device: (a) light-emitting transistor devices and (b) light-emitting electrochemical cells (LECs).

13.2 Light-Emitting Transistors

Organic electronic light-emitting transistors (OLETs) have been under development for approximately the past 15 years [845]. These devices combine transistor switching with light emission and have been foreshadowed as a simpler approach for flexible active matrix full colour electroluminescent display technology, thereby potentially enabling high aperture ratios (ratio of light-emitting area over total pixel area) [846]. Further applications of these devices include optical communication systems, electrically pumped organic lasing and solid-state lighting sources [847]. As such, there has been a significant interest in the properties of the organic and inorganic materials that would go into these versatile devices [848].

Both unipolar and ambipolar (balanced charge carrier characteristics) OLET devices have been developed [849]. In the case of unipolar OLET devices, there is a dominant charge carrier (electron or hole) and consequently light emission occurs close to the electrode responsible for injecting the lower mobility carrier. By contrast, in an ambipolar OLET device, the electron and hole mobilities are more balanced and the location of the emissive region can be controlled by the gate voltage. As might be expected, the materials evolution for OLET applications has closely mirrored the development of materials for OLED applications, initially exploring direct emission, incorporating fluorescent emitters and more recently TADF materials. However, the integrated nature of the OLET device architecture means that there is also the challenging requirement for both high mobility (to ensure fast switching) and high external quantum efficiency (to ensure high external quantum efficiency). Arguably, this two-fold requirement means that to-date the performance of OLET devices continues to lag that of conventional OLED

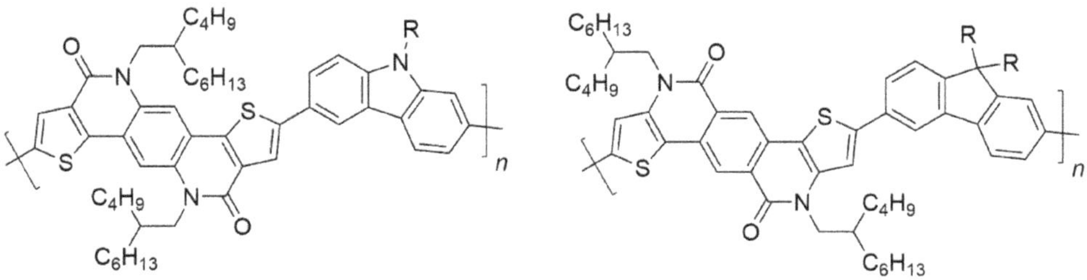

1 R = 2-ethylhexyl **2** R = 2-ethylhexyl

Figure 13.1 Semi-ladder copolymers used to make efficient OLETs

devices and further developments in organic electronic materials physics and chemistry have been required before OLETs can reach their full potential. In considering the ideal light-emitting polymer semiconductors for OLETs we need to design materials that have the appropriate energy levels to minimise charge injection barriers from metal contacts, high ambipolar charge carrier mobility to centralise channel light emission and high fluorescence quantum yield [848].

Excitingly, recent progress in combining these properties has been made. In particular, a series of semi-ladder polymers such as **1** and **2** (Figure 13.1) made by copolymerisation of weak acceptors (TPTQ or TPTI) and weak donors (fluorene (F) or carbazole (C)) have been developed that combine efficient luminescent and charge transporting properties [850, 851]. Despite these advances, there are still many challenges that need to be overcome before OLETs become a key optoelectronic technology. In particular, new materials with improved emission and mobility characteristics need to be developed with a focus on n-type and ambipolar materials [852].

13.3 Light-Emitting Electrochemical Cells (LECs)

Light-emitting electrochemical cells or LECs differ from conventional LEDs in one important respect. The emissive layer consists of a blend of an emissive polymer with an electrolyte, which makes these layers much better ionic conductors. These devices therefore show some marked differences in their behaviour from LEDs, including efficiencies which are relatively insensitive to electrode work functions, non-rectifying current voltage characteristics and emission with similarly low threshold voltages under both forward and reverse bias, which are in turn nearly independent of the film thicknesses. The very first LECs consisted of mixtures of conjugated polymers such as MEH-PPV, mixed with poly(ethylene oxide) (PEO), a well-known ion-transporting polymer, and lithium triflate (LiOTf), sandwiched between ITO and aluminium electrodes [853–858]. Later, a device using a surface configuration with inter-digitated electrodes was developed [855, 859]. Multilayer LECs or LECs containing blends of more than one emissive polymer are also possible. By appropriate choice of emissive polymers, red-, green-, blue- or white-emitting LECs have been fabricated [150].

Much of the early work on LECs was devoted to understanding the mechanism of their operation and how it differed from that of conventional LEDs or from electrogenerated chemiluminescence (ECL) (*vide infra*) which was initially thought to be a possible mechanism for the generation of light in LECs [841, 860]. This showed that, in these devices, p- and n-doped regions are produced at the anode and cathode, which then extend towards the centre of the emissive layer to form an internal p–n junction where recombination and emission occur. In surface-configured devices this junction can even be seen directly [861]. Interestingly, when emissive polymers are blended with low concentrations of ions the device behaviour is intermediate between that of standard LEDs and LECs, with low threshold voltages as in LECs, but retaining the rectifying behaviour seen for LEDs [862]. Later another type of LEC was developed in which the p–n junction is formed at a temperature well above the intended operating value and then cooled down to freeze in the junction as, when the device is run at a lower temperature, the lower ion mobilities prevent it moving. These devices retain most of the advantages of the original dynamic junction LECs, such as high efficiency and low threshold voltages, but without their drawbacks of low response times (junction is already formed) and over-doping. Like LEDs, these devices show rectifying behaviour and emit only under forwards bias. The first examples of these devices using MEH-PPV involved junctions stable only below 200 K [863–866], but later devices using other PPV derivatives were developed where the junction could be formed at 60–80 °C and then frozen at room temperature [867]. Later it was shown that, by using ionic liquids as the electrolytes, such junctions could also be produced at similar temperatures in LECs using MEH-PPV [868]. It has been reported that briefly heating MEH-PPV-based frozen-junction PLECs above the T_g of the polymer enhances the device efficiency by converting the frozen p–n junction to a p–i–n junction [869].

An obvious problem with using a blend as an emitting layer is that of phase separation, as the hydrophilic PEO is not compatible with most conjugated polymers which are usually somewhat hydrophobic in nature. To overcome this, polymers, mainly RO-PPV derivatives such as **3** or **4** (Figure 13.2), but also polyfluorenes such as **5** or polythiophenes like **6**, have been prepared which bear ionophoric oligo(ethylene oxide) or crown ether units as the side-chains, specifically for use in LECs (they can also be used of course in LEDs) [150]. One can also use polymers such as **7** with interrupted chromophores linked by ionophoric units. Such polymers can be used in LECs blended with suitable lithium salts without the need for PEO.

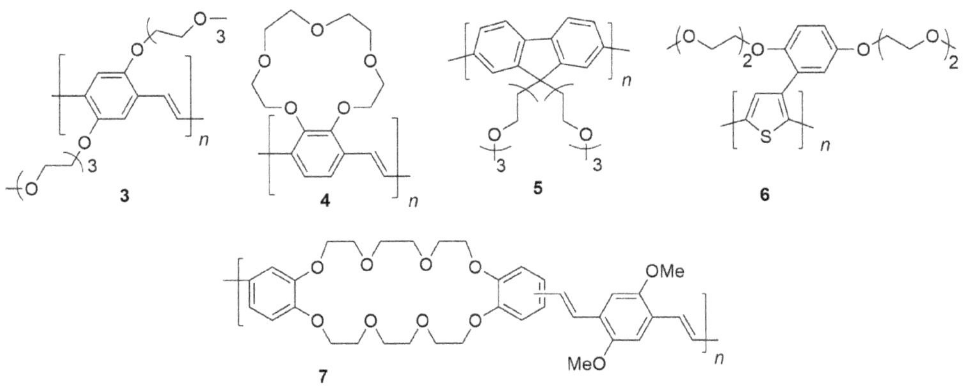

Figure 13.2 Some polymers bearing ion-transporting units for use in LECs

The main claimed advantages of the LEC compared to conventional LEDs were the low threshold voltages and the sometimes-higher device efficiencies. However, advances in LED design, including use of better charge transporting layers have since eroded these advantages and the shelf and device lifetimes for LEDs are reportedly better than for LECs, so that these devices seem to have become of little interest to academic and industrial researchers, with few recent reports on them.

13.4 Electrogenerated Chemiluminescence Cells

Electrogenerated chemiluminescence (ECL) involves the generation of light-emitting species by electron transfer in solution between radical ions formed by electron transfer reactions with electrodes, so that an ECL cell is essentially an electrolytic cell. As mentioned in Section 13.3, ECL was at one time thought to be a possible explanation for the emission seen from LECs, but, as has been described in Section 13.3, it has been shown that this is not correct. There are a few reports of ECL being observed from films of conjugated polymers such as MEH-PPV [870] or P3HT [871], in solution or as a blend with PEO for the PPV derivative **8** (Figure 13.3) [872]. The ECL spectra generally closely match the PL spectra. Devices similar to LEDs have been made with polymer solutions or gels sandwiched between the electrodes, from which the emission is generated by ECL [873]. For a device using a gel of MEH-PPV, the red-orange-emission has an emission maximum ($\lambda_{max} = 575$ nm) intermediate between the emission maxima for the polymer in solution and the solid state ($\lambda_{max} = 560$ nm and 592 nm, respectively) [874]. Blue-green ECL ($\lambda_{max} = 430$, 450, 482 nm) has been observed from a solution of a polydialkylfluorene sandwiched between two transparent electrodes [875]. Thus, in principle, any emission colour is attainable from an ECL cell. Given the encapsulation problems inherent in such devices (leakage of

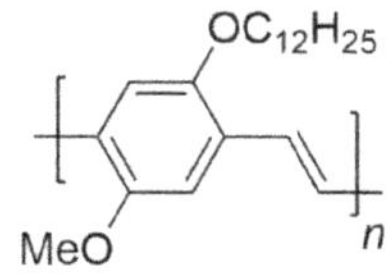

8

Figure 13.3 PPV derivative used for ECL

solutions must obviously be avoided) and the lack of any reported enhancement in their device performance over LEDs, these devices seem likely to remain scientific curiosities, rather than serious competitors to standard LEDs.

While ECL cells, unlike LECs, do not introduce any specific new design requirements for conjugated polymers to be used within them, there is one class of device which probably does require new requirements but to date it is not clear what they should be; that is, an LED which emits laser light.

13.5 Polymer Microcavities and Lasers

Lasing, that is the emission of coherent light, can be produced in two ways. It can be obtained by exciting the emissive material, either in solution or as a thin film by laser irradiation. This is referred to as optically pumped lasing. This has been observed for many organic materials, including conjugated polymers, and is the basis for dye lasers, in which the emission from an inorganic semiconductor laser is converted into emission of a different colour by excitation of organic dyes in solution. While this is very useful as a way to obtain laser colours not available from the limited number of inorganic laser materials, the ultimate goal of research into polymer-based lasers is to obtain electrically pumped lasing – that is to obtain laser emission by electrical stimulation of a film of the polymer. To date, production of such a device – a laser diode – has proven elusive, but it is thought to be not impossible, suggesting that with advances in device design and/or improvements in materials it could be attainable. There have been a number of reviews on the development of lasing from conjugated polymers, which cover the history of the field and discuss the strategies being investigated towards producing a laser diode [150, 876–879]. Here we will give only a brief overview of the field.

The first observation of lasing from a conjugated polymer was from a solution of MEH-PPV in 1992 [880]. Four years later, lasing from a blend of MEH-PPV and titania nanoparticles in polystyrene was reported [881]. The same year also saw the report of optically pumped lasing behaviour from a thin film of PPV in a microcavity device [882] followed soon afterwards by similar reports of optically-pumped super-radiance and lasing phenomena from films of other conjugated polymers [883–885]. A microcavity structure is one in which the electroluminescent material is sandwiched between two planar, highly reflecting mirrors. Such a device acts as a resonator for a standing electromagnetic wave producing a marked narrowing in the emission spectrum, when the polymer film thickness is of the order of the wavelength of the emitted light. This device structure is thus useful for obtaining enhanced emission, with narrower emission bands, and the colour can even be tuned by changing the device parameters, as has been demonstrated by several groups [886–890]. A microcavity device has been constructed using ladder-type PPP in which the emission colour could even be tuned from red to blue simply by altering the thickness of the polymer layer [891]. Microcavities are just one of the device architectures from which electrically pumped lasing has been sought. Lasing has now been reported from many of the classes of conjugated polymers discussed in this book, with emission colours covering the whole spectral range, so that conjugated polymers potentially offer a way to tune laser emission colours to produce almost any visible wavelength desired. One material that has attracted particular interest due to its extremely good optical properties has been Me-LPPP, which has been used to make a flexible optically pumped polymer laser [892, 893].

While considerable progress has been made towards understanding the material and device requirements for producing a polymer laser diode, we are still far from being able to define what properties a putative material would need to be a serious candidate for making such a device. Indeed, optically pumped polymer laser diodes have been in existence for over 20 years; however, despite extensive research, an electrically pumped polymer laser remains elusive. The key barriers to electrical stimulation of a lasing response in organic materials arise from their intrinsic properties. In particular, there are a number of issues that any new polymer lasing materials will need to overcome. First, the typically low and unbalanced carrier mobilities of semiconducting polymers limits population inversion and introduces substantial charge-induced absorption losses if the charges cannot be sufficiently cleared from the lasing region. Second, triplet and/or polaron absorption and associated singlet–triplet annihilation result in losses that require complex synthetic strategies such as polymers with low inter-system-crossing rates and triplet scavenging moieties to mitigate. As discussed in Chapters 11 and 12, TADF materials offer considerable promise as a solution to the issues associated with triplet generation due to their ability to recycle triplet excitons via reverse inter-system crossing. However, the anticipated high current density thresholds for electrically pumped lasing [894] are beyond current TADF materials [895]. Third, current electrodes and electrode–organic interfaces typically produce absorption losses and unbalanced injection, increasing the lasing threshold significantly, placing further demands for more detailed molecular design rules [896].

Obviously, the high quantum efficiency and small Stokes shift seen in rigid systems such as bridged phenylenes, high charge carrier mobilities and high thermal and photostability appear to be among them. In addition, more work needs to be undertaken regarding electrode and cavity design and fabrication. The advances in design and synthesis of conjugated polymers and in device design may well lead to the fabrication of a prototype laser diode within the quite near future, but for now this remains a 'Holy Grail' of organic electronics research.

13.6 Integrated Polymer Devices and Other Devices

One area still to be fully investigated is the combination of a variety of polymer-based organic electronic devices. For example, polymer LEDs might be coupled with polymer-based photovoltaic devices to make optocouplers [855, 897], which might be useful in optically based systems in which light replaces electricity as the medium for transmission of data. Integrated devices in which a polymer-based FET drives a polymer LED have already been made [898–900], representing a major step towards the development of all-polymer integrated circuits. It may also be possible to combine LEDs, OPVs or FETs with other devices using conjugated polymers, for example so-called plastic retinas, in which a combination of MEH-PPV and polyaniline has been used to make polymer grid triodes for image enhancement [901]. Given that it is now possible to fabricate all the components of an electronic device by printing techniques [902], the possibility has thus opened up of fabricating complicated electronic devices entirely by inexpensive solution-based techniques.

14
Conclusion and Outlook

As we have shown in the previous chapters, the field of conjugated polymers has developed a lot since the initial pioneering work on polyacetylene. A very wide range of materials have been designed and prepared for use in a range of devices, including transistors, light-emitting diodes and solar cells. Thanks to simultaneous advances in device design and fabrication techniques, in the understanding of the physics of organic electronic materials in and outside devices, and in the methods for making and purifying organic materials, devices can now be made using conjugated polymers whose performance is in many, though not all, respects comparable to the well-known commercial devices based upon inorganic semiconductors. Since the polymers are usually cheaper to process and are capable of forming large, robust, flexible films suitable for use on non-standard shaped surfaces, their devices possess potential advantages of cost over existing ones. As a result, such polymer-based electronic devices are now entering the marketplace and all current projections show them taking an increasing share of the market over the next few decades.

One reason why polymers will not totally displace conventional inorganic materials is that, as just stated, they cannot match all aspects of the latter's performance. For example, polymers have been made which, when doped, display conductivities comparable to metals, but they lack metals' mechanical strength and robustness, especially as wires, and more importantly the best conducting materials are unstable towards air in their conductive forms. So, while thin films of conductive polymers will have uses, especially where good conductivity is combined with optical translucence and flexibility, they cannot be expected to replace metals (e.g. for electrical wiring or for inter-connects in silicon chips). Thus, the future of organic materials lies as semiconductors not conductors.

Polymer-based light-emitting devices have already successfully entered the commercial marketplace and, as their performance improves, more sophisticated display devices such as PLED-based televisions will become viable – prototypes have already appeared and commercial release is imminent. This is because efficient and reasonably long-lived red, green and blue-emitting materials have been developed. Green remains the most efficiently produced colour, but the problems originally found for other colours have been satisfactorily overcome. While blue emission has faced some problems due to colour instability, advances in our understanding of the causes of the problem, such as formation of defects by oxidative degradation have, enables imaginative solutions to be found, including design both of new structures with greater stability towards oxidation and also new synthetic routes which prevent formation of defects. Thus, the lifetimes reported for blue PLEDs are now satisfactory though still have room for improvement. For red PLEDs, the use of red-emitting dyes or phosphors as co-monomers in blue hosts seems to be a good way around the problems of poor colour purity and lower efficiency previously seen for red-emitting conjugated polymers. Efficient white emission has also now been obtained through some imaginative designs, which take advantage of energy transfer processes within the materials. While issues of colour stability still need addressing, there is good reason to believe polymer-based white LEDs will soon be a possible option for applications such as lighting requiring white emission.

Conjugated polymers have been made which display charge carrier mobilities comparable or even better than amorphous silicon, though still well short of crystalline silicon or graphene. Thus, printable organic-based transistors for use in a range of applications are possible and are likely to prove commercially viable in some applications. Prototype devices using them are under development as we write this, and commercial products should appear within the next few years.

Commercialisation of printed transistor devices has followed closely on the heels of that of OLED technology. One of the earliest examples is that of Plastic Logic, which originally started as a spin-out from Cambridge University. Ultimately, this venture evolved into two companies: Plastic Logic Germany, specialising in flexible displays, and FlexEnable, specialising in flexible organic electronics technologies and OTFT materials. In the United States, E-Ink Corporation was set up in 1997 to commercialise electronic ink applications based on technology from MIT Media Labs. Subsequently, it merged with Prime View International (established in 1992 to produce thin film transistor liquid crystal display technology in Taiwan) to form E Ink, specialising in e-readers and e-paper displays based on OTFT technology. More recently, there have been an increasing number of companies working in the OTFT space with a strong focus on wearables, especially for health-monitoring applications. For example, SmartKem aims to deliver specific materials chemistry for wearable OTFT display applications based on microLED technology.

The relative slowness in development of OPV compared to OLED and OFET technology does in part reflect the more difficult problems involved in that one has had to optimise the performance of a blend of two materials, when the critical parameters for each were only vaguely understood and their behaviour when blended was often not understood or even predictable. However, the development of what look like robust models for predicting the maximum efficiency of devices from the predictable orbital energy levels of polymer structures has greatly aided our capacity to pre-screen possible structures and select the most promising ones. What is still lacking is the ability to predict solid state packing of even relatively simple materials, let alone the behaviour of blends of them.

Consequently, of all the main classes of organic electronic devices, organic photovoltaic devices have traditionally been viewed as being the greatest distance away from serious commercial applicability. Historically, however, commercialisation featured early in the development of the organic photovoltaic field, with several start-up companies (encompassing materials and devices) emerging, growing, merging, being acquired or going bankrupt. An early mover in this space was Quantum Solar Energy Linz (QSEL), founded in 1997 from the Linz Institute for Organic Solar Cells (LIOS) and subsequently acquired by Konarka Technologies, Inc. that was founded by, amongst others, the Nobel laureate Alan Heeger, in 2001, as a spin-off from the University of Massachusetts, Lowell. Plextronics was founded in 2002 in Pittsburgh as a spin-off company from Carnegie Mellon University, primarily as a materials supply company based on the regioregular P3HT synthesis developed by Richard McCullough. Solarmer Energy was founded in California in 2006, licensing OPV technology developed at the University of California, Los Angeles and new semiconducting material technology developed at the University of Chicago. In the United Kingdom, Ossila was founded in 2009 at the University of Sheffield, focussing on supplying the materials and equipment for organic electronics research. In Cambridge, the company Eight19 was founded in 2010 to commercialise organic solar cell technology developed at the Cavendish Laboratory. However, it became increasingly clear that commercial scale OPV was challenging in the short-term and a number of these initial start-up companies were forced to file for bankruptcy, including Konarka (May 2012) and Plextronics (January 2014). Subsequently, Belectric and Solvay acquired the rights to the assets and IP to Konarka and Plextronics, respectively. In 2016, the French company ARMOR launched industrial production of a new generation of photovoltaic material, designed and manufactured in France.

Devices with PCEs in excess of 15% were reported during the writing of this book, far in excess of the 5% efficiency threshold proposed by Chamberlain in 1983 [903] and double the 8–10% efficiencies that were state of the art as recently as 2012 [904], so why is OPV not yet a commercially viable technology? Previous attempts have highlighted that the successful commercialisation of OPVs is governed by three key parameters: device efficiency, lifetime and cost. As described in previous chapters, especially Chapter 8, the OPV research community has primarily been focussed on improving device efficiency. However, to date, reducing OPV material synthetic cost has received far less attention. Indeed, scientific research at the large scale is currently limited and advances in the OPV material cost and scalability is urgently required for the rapid industrial development of printed solar technologies [905].

Looking ahead, OPV is currently emerging from the classic 'Valley of Death' commercialisation phase, with a number of restructured and consolidated companies developing large scale OPV products. Interestingly, the companies that withstood the 'OPV crash' most successfully focused on supplying materials and tools to the research community. Future materials advances focussing on inexpensive materials synthesis at scale are required to enable OPV's promise of a low-cost sustainable energy technology for everyone.

Two areas that seem to be crucial for determining the efficiency and lifetime of devices using organic molecules in general and conjugated polymers in particular are the effects of light and passage of current through them on the molecular structures, that is, their susceptibility to electrochemical or photo-oxidation, and the interactions between the materials and electrodes, that is, effects of doping and the nature of barriers to charge injection. Much work has been done towards understanding these problems for conjugated polymers [150], which in turn has enabled their (partial) solution through improved synthetic design or better handling, especially during device fabrication.

While some problems still remain to be solved before organic electronic devices such as OFETs, OLEDs and OPVs attain their full potential, not to mention the fundamental problems still being faced by those seeking to produce a polymer laser diode, the remarkable progress made through chemists, physicists and device engineers working together allows up to be (not too) cautiously optimistic that these can and will be overcome, and that the twenty-first century may truly become as some have boldly proclaimed it will be: 'the age of plastic'.

References

[1] R. Thompson, S. Swan, C. Moore, and F. vom Saal, 'Our plastic age', *Philosophical Transactions of the Royal Society B: Biological Sciences*, vol. 364, pp. 1973–1976, 2009.

[2] 'Patents for inventions', UK Patent Office, 1857, p. 255.

[3] L. H. Baekeland, 'Method of making insoluble products of phenol and formaldehyde', US Patent 942,699, 1909.

[4] S. C. Rasmussen, 'Electrically conducting plastics: Revising the history of conjugated organic polymers', in *100+ Years of Plastics. Leo Baekeland and Beyond*, E. T. Strom and S. C. Rasmussen (eds.) (ACS Symposium Series). Washington, DC: American Chemical Society, 2011.

[5] D. E. Weiss and B. A. Bolto, *Physics and Chemistry of the Organic Solid State*, vol. II. New York: Interscience Publishers, 1965, ch. 2.

[6] S. B. Mainthia, P. L. Kronick, and M. M. J. Labes, 'Electrical measurements on polyvinylene and polyphenylene', *Journal of Chemical Physics*, vol. 37, no. 10, pp. 2509–2510, 1962.

[7] R. McNeill, R. Siudak, J. H. Wardlaw, and D. E. Weiss, 'Electronic conduction in polymers. I. The chemical structure of polypyrrole', *Australian Journal of Chemistry*, vol. 16, pp. 1056–1075, 1963.

[8] B. A. Bolto and D. E. Weiss, 'Electronic conduction in polymers. II. The electrochemical reduction of polypyrrole at controlled potential', *Australian Journal of Chemistry*, vol. 16, pp. 1076–1089, 1963.

[9] B. A. Bolto, R. McNeill, and D. E. Weiss, 'Electronic conduction in polymers. III. Electronic properties of polypyrrole', *Australian Journal of Chemistry*, vol. 16, pp. 1090–1103, 1963.

[10] C. K. Chiang, Y. W. Park, A. J. Heeger, H. Shirakawa, E. J. Louis, and A. G. MacDiarmid, 'Conducting polymers: Halogen doped polyacetylene', *Journal of Chemical Physics*, vol. 69, pp. 5098–5104, 1978.

[11] C. J. Brabec, V. Dyakonov, and U. Scherf (eds.), *Organic Photovoltaics*. Weinheim: Wiley-VCH, 2008.

[12] A. Saeki, Y. Koizumi, T. Aida, and S. Seki, 'Comprehensive approach to intrinsic charge carrier mobility in conjugated organic molecules, macromolecules, and supramolecular architectures', *Accounts of Chemical Research*, vol. 45, pp. 1193–1202, 2012.

[13] H. Sirringhaus, 'Reliability of organic field-effect transistors', *Advanced Materials*, vol. 21, pp. 3859–3873, 2009.

[14] H. G. O. Sandberg, T. G. Bäcklund, R. Österbacka, and H. Stubb, 'High-performance all-polymer transistor utilizing a hygroscopic insulator', *Advanced Materials*, vol. 16, pp. 1112–1115, 2004.
[15] D. Elkington, D. Darwis, X. Zhou, W. Belcher, and P. C. Dastoor, 'The fabrication and characterization of poly(4-vinylpyridine)-based thin film transistors exhibiting enhanced ion modulation', *Organic Electronics*, vol. 13, pp. 153–158, 2012.
[16] J. Shinar and R. Shinar, 'Organic light-emitting devices (OLEDs) and OLED-based chemical and biological sensors: An overview', *Journal of Physics D: Applied Physics*, vol. 41, p. 133001, 2008.
[17] M. Pope, H. P. Kallmann, and P. Magnante, 'Electroluminescence in organic crystals', *Journal of Chemical Physics*, vol. 38, pp. 2042–2043, 1963.
[18] C. W. Tang and S. A. VanSlyke, 'Organic electroluminescent diodes', *Applied Physics Letters*, vol. 51, pp. 913–915, 1987.
[19] J. H. Burroughes, D. D. C. Bradley, A. R. Brown, R. N. Marks, K. Mackay, R. H. Friend, et al., 'Light-emitting diodes based on conjugated polymers', *Nature*, vol. 347, pp. 539–541, 1990.
[20] T. Förster, 'Zwischenmolekulare energiewanderung und fluoreszenz', *Annalen der Physik*, vol. 437, pp. 55–75, 1948.
[21] S. Kappaun, C. Slugovc, and E. J. W. List, 'Phosphorescent organic light-emitting devices: Working principle and iridium based emitter materials', *International Journal of Molecular Sciences*, vol. 9, pp. 1527–1547, 2008.
[22] B. Geffroy, P. le Roy, and C. Prat, 'Organic light-emitting diode (OLED) technology: Materials, devices and display technologies', *Polymer International*, vol. 44, pp. 572–582, 2006.
[23] C. W. Tang, S. A. VanSlyke, and C. H. Chen, 'Electroluminescence of doped organic thin films', *Journal of Applied Physics*, vol. 65, pp. 3610–3616, 1989.
[24] Y. D. Jin, X. B. Ding, J. Reynaert, V. I. Arkhipov, G. Borghs, P. L. Heremans, and M. Van der Auweraer, 'Role of LiF in polymer light-emitting diodes with LiF-modified cathodes', *Organic Electronics*, vol. 5, no. 6, pp. 271–281, 2004.
[25] G. E. Jabbour, B. Kippelen, N. R. Armstrong, and N. Peyghambarian, 'Aluminum based cathode structure for enhanced electron injection in electroluminescent organic devices', *Applied Physics Letters*, vol. 73, pp. 1185–1187, 1998.
[26] L. S. Hung, C. W. Tang, and M. G. Mason, 'Enhanced electron injection in organic electroluminescence devices using an Al/LiF electrode', *Applied Physics Letters*, vol. 70, pp. 152–154, 1997.
[27] M. Wohlgenannt, K. Tandon, S. Mazumdar, S. Ramasesha, and Z. V. Vardeny, 'Singlet and triplet exciton ratios in conjugated polymers', *Nature*, vol. 409, pp. 494–497, 2001.
[28] B. H. Wallikewitz, D. Kabra, S. Gélinas, and R. H. Friend, 'Triplet dynamics in fluorescent polymer light-emitting diodes', *Physical Review B*, vol. 85, p. 045209, 2012.
[29] C. Adachi, M. A. Baldo, M. E. Thompson, and S. R. Forrest, 'Nearly 100% internal phosphorescence efficiency in an organic light-emitting device', *Journal of Applied Physics*, vol. 90, pp. 5048–5051, 2001.
[30] K. Goushi, K. Yoshida, K. Sato, and C. Adachi, 'Organic light-emitting diodes employing efficient reverse intersystem crossing for triplet-to-singlet state conversion', *Nature Photon*, vol. 6, pp. 253–258, 2012.
[31] N. C. Greenham, R. H. Friend, and D. D. C. Bradley, 'Angular dependence of the emission from a conjugated polymer light-emitting diode: Implications for efficiency calculations', *Advanced Materials*, vol. 6, pp. 491–494, 1994.

[32] B. E. A. Saleh and M. C. Teich, *Fundamentals of Photonics*. New York: Wiley, 1991.
[33] J.-S. Kim, P. K. H. Ho, N. C. Greenham, and R. H. Friend, 'Electroluminescence emission pattern of organic light-emitting diodes: Implications for device efficiency calculations', *Journal of Applied Physics*, vol. 88, pp. 1073–1081, 2000.
[34] J. Kido, K. Hongawa, K. Okuyama, and K. Nagai, 'White light-emitting organic electroluminescent devices using the poly(N-vinylcarbazole) emitter layer doped with three fluorescent dyes', *Applied Physics Letters*, vol. 64, pp. 815–817, 1994.
[35] J. Kido, M. Kimura, and K. Nagai, 'Multilayer white light-emitting organic electroluminescent device', *Science*, vol. 267, pp. 1332–1334, 1995.
[36] Y. Sun, N. C. Giebink, H. Kanno, B. Ma, M. E. Thompson, and S. R. Forrest, 'Management of singlet and triplet excitons for efficient white organic light-emitting devices', *Nature*, vol. 440, pp. 908–912, 2006.
[37] F. So, J. Kido, and P. Burrows, 'Organic light-emitting devices for solid-state lighting', *Materials Research Society Bulletin*, vol. 33, pp. 663–669, 2008.
[38] D. Moses, 'High quantum efficiency luminescence from a conducting polymer in solution: A novel polymer laser dye', *Applied Physics Letters*, vol. 60, pp. 3215–3216, 1992.
[39] A. J. Heeger, 'Nobel Lecture: Semiconducting and metallic polymers: The fourth generation of polymeric materials', *Reviews of Modern Physics*, vol. 73, pp. 681–700, 2001.
[40] U. Scherf, S. Riechel, U. Lemmer, and R. F. Mahrt, 'Conjugated polymers: Lasing and stimulated emission', *Current Opinion in Solid State and Materials Science*, vol. 5, pp. 143–154, 2001.
[41] Y. Chen, J. Herrnsdorf, B. Guilhabert, A. L. Kanibolotsky, A. R. Mackintosh, Y. Wang, et al., 'Laser action in a surface-structured free-standing membrane based on a π-conjugated polymer-composite', *Organic Electronics*, vol. 12, pp. 62–69, 2011.
[42] W. Holzer, A. Penzkofer, T. Pertsch, N. Danz, A. Brauer, and E. B. Kley, 'Corrugated neat thin-film conjugated polymer distributed-feedback lasers', *Applied Physics B*, vol. 74, pp. 333–342, 2002.
[43] M. Leclerc, 'Polyfluorenes: 20 years of progress', *Journal of Polymer Science, Part A: Polymer Chemistry*, vol. 39, pp. 2867–2873, 2001.
[44] G. Heliotis, R. Xia, D. D. C. Bradley, G. A. Turnbull, I. D. W. Samuel, P. Andrew, and W. L. Barnes, 'Blue surface-emitting distributed feedback polyfluorene laser', *Applied Physics Letters*, vol. 83, pp. 2118–2120, 2003.
[45] M. Lehnhardt, T. Riedl, T. Weimann, and W. Kowalsky, 'Impact of triplet absorption and triplet-singlet annihilation on the dynamics of optically pumped organic solid-state lasers', *Physical Review B*, vol. 81, p. 165206, 2010.
[46] S. Chénais and S. Forget, 'Recent advances in solid-state organic lasers', *Polymer International*, vol. 61, pp. 390–406, 2012.
[47] B. H. Wallikewitz, M. de la Rosa, J. H. W. M. Kremer, D. Hertel, and K. Meerholz, 'A lasing organic light-emitting diode', *Advanced Materials*, vol. 22, pp. 531–534, 2010.
[48] C. Gartner, C. Karnutsch, U. Lemmer, and C. Pflumm, 'The influence of annihilation processes on the threshold current density of organic laser diodes', *Journal of Applied Physics*, vol. 101, p. 023107, 2007.
[49] N. C. Giebink and S. R. Forrest, 'Temporal response of optically pumped organic semiconductor lasers and its implication for reaching threshold under electrical excitation', *Physical Review B*, vol. 79, p. 073302, 2009.
[50] M. M. Mandoc, L. J. A. Koster, and P. W. M. Blom, 'Optimum charge carrier mobility in organic solar cells', *Applied Physics Letters*, vol. 90, p. 133504, 2007.

[51] C. L. Braun, 'Electric field assisted dissociation of charge transfer states as a mechanism of photocarrier production', *Journal of Chemical Physics*, vol. 80, pp. 4157–4161, 1984.

[52] M. M. Mandoc, W. Veurman, L. J. A. Koster, B. De Boer, and P. W. M. Blom, 'Origin of the reduced fill factor and photocurrent in MDMO-PPV:PCNEPV all-polymer solar cells', *Advanced Functional Materials*, vol. 17, pp. 2167–2173, 2007.

[53] A. C. Morteani, P. Sreearunothai, L. M. Herz, R. H. Friend, and C. Silva, 'Exciton regeneration at polymeric semiconductor heterojunctions', *Physical Review Letters*, vol. 92, p. 247402, 2004.

[54] H. Ohkita, S. Cook, Y. Astuti, W. Duffy, S. Tierney, W. Zhang, et al., 'Charge carrier formation in polythiophene/fullerene blend films studied by transient absorption spectroscopy', *Journal of the American Chemical Society*, vol. 130, pp. 3030–3042, 2008.

[55] H. Kallmann and M. Pope, 'Photovoltaic effect in organic crystals', *Journal of Chemical Physics*, vol. 30, pp. 585–586, 1959.

[56] C. W. Tang, 'Two-layer organic photovoltaic cell', *Applied Physics Letters*, vol. 48, pp. 183–185, 1986.

[57] N. S. Sariciftci, L. Smilowitz, A. J. Heeger, and F. Wudl, 'Semiconducting polymers (as donors) and buckminsterfullerene (as acceptor): Photoinduced electron transfer and heterojunction devices', *Synthetic Metals*, vol. 59, pp. 333–352, 1993.

[58] N. S. Sariciftci, D. Braun, C. Zhang, V. I. Srdanov, A. J. Heeger, G. Stucky, and F. Wudl, 'Semiconducting polymer-buckminsterfullerene heterojunctions: Diodes, photodiodes, and photovoltaic cells', *Applied Physics Letters*, vol. 62, pp. 585–587, 1993

[59] G. Yu, J. Gao, J. C. Hummelen, F. Wudl, and A. J. Heeger, 'Polymer photovoltaic cells: Enhanced efficiencies via a network of internal donor-acceptor heterojunctions', *Science*, vol. 270, pp. 1789–1791, 1995.

[60] S. E. Shaheen, C. J. Brabec, N. S. Sariciftci, F. Padinger, T. Fromherz, and J. C. Hummelen, '2.5% efficient organic plastic solar cells', *Applied Physics Letters*, vol. 78, pp. 841–843, 2001.

[61] W. Li, A. Furlan, K. H. Hendriks, M. M. Wienk, and R. A. J. Janssen, 'Efficient tandem and triple-junction polymer solar cells', *Journal of the American Chemical Society*, vol. 135, pp. 5529–5532, 2013.

[62] J. Y. Kim, K. Lee, N. E. Coates, D. Moses, T.-Q. Nguyen, M. Dante, and A. J. Heeger, 'Efficient tandem polymer solar cells fabricated by all-solution processing', *Science*, vol. 317, pp. 222–225, 2007.

[63] J. You, C.-C. Chen, Z. Hong, K. Yoshimura, K. Ohya, R. Xu, et al., '10.2% Power conversion efficiency polymer tandem solar cells consisting of two identical sub-cells', *Advanced Materials*, vol. 25, pp. 3973–3978, 2013.

[64] J. You, L. Dou, K. Yoshimura, T. Kato, K. Ohya, T. Moriarty, et al., 'A polymer tandem solar cell with 10.6% power conversion efficiency', *Nature Communications*, vol. 4, pp. 1446/1–1446/10, 2013, DOI: 10.1038/ncomms2411.

[65] T. Xu and L. Yu, 'How to design low bandgap polymers for highly efficient organic solar cells', *Materials Today*, vol. 17, pp. 11–15, 2014.

[66] J. Peet, J. Y. Kim, N. E. Coates, W. L. Ma, D. Moses, A. J. Heeger, and G. C. Bazan, 'Efficiency enhancement in low-bandgap polymer solar cells by processing with alkane dithiols', *Nature Materials*, vol. 6, pp. 497–500, 2007.

[67] H.-Y. Chen, J. Hou, S. Zhang, Y. Liang, G. Yang, L. Yu, et al., 'Polymer solar cells with enhanced open-circuit voltage and efficiency', *Nature Photonics*, vol. 3, pp. 649–653, 2009.

[68] Y. Liang, Z. Xu, J. Xia, S.-T. Tsai, Y. Wu, G. Li, et al., 'For the bright future–bulk heterojunction polymer solar cells with power conversion efficiency of 7.4%', *Advanced Materials*, vol. 22, pp. E135–E138, 2010.

[69] H.-J. Jhuo, P.-N. Yeh, S.-H. Liao, Y.-L. Li, Y.-S. Cheng, and S.-A. Chen, 'Review on the recent progress in low band gap conjugated polymers for bulk hetero-junction polymer solar cells', *Journal of the Chinese Chemical Society*, vol. 61, pp. 115–126, 2014.

[70] A. Tsumura, H. Koezuka, and T. Ando, 'Macromolecular electronic device: Field-effect transistor with a polythiophene thin film', *Applied Physics Letters*, vol. 49, pp. 1210–1212, 1986.

[71] J. H. Burroughes, C. A. Jones, and R. H. Friend, 'New semiconductor device physics in polymer diodes and transistors', *Nature*, vol. 335, pp. 137–141, 1988.

[72] F. Garnier, R. Hajlaoui, A. Yassar, and P. Srivastava, 'All-polymer field-effect transistor realized by printing techniques', *Science*, vol. 265, pp. 1684–1686, 1994.

[73] J. S. Ha, K. H. Kim, and D. H. Choi, '5-Bis(2-octyldodecyl)pyrrolo[3,4-c]pyrrole-1,4-(2H,5H)-dione-based donor-acceptor alternating copolymer bearing 5,5'-di(thiophen-2-yl)-2,2'-biselenophene exhibiting 1.5 $cm^2\ V^{-1}\ s^{-1}$ hole mobility in thin-film transistors', *Journal of the American Chemical Society*, vol. 133, pp. 10364–10367, 2011.

[74] J. Li, Y. Zhao, H. S. Tan, Y. Guo, C.-A. Di, G. Yu, et al., 'A stable solution-processed polymer semiconductor with record high-mobility for printed transistors', *Scientific Reports*, vol. 2, p. 754.

[75] X. Zhang, H. Bronstein, A. J. Kronemeijer, J. Smith, Y. Kim, R. J. Kline, et al., 'Molecular origin of high field effect mobility in an indacenodithiophene-benzothiadiazole copolymer', *Nature Communications*, vol. 4, p. 2238, 2013, DOI: 10.1038/ncomms3238.

[76] R. Noriega, J. Rivnay, K. Vandewal, F. P. V. Koch, N. Stingelin, P. Smith, et al., 'A general relationship between disorder, aggregation and charge transport in conjugated polymers'. *Nature Materials*, vol. 12, pp. 1038–1044, 2013.

[77] R. A. Street, 'Unravelling charge transport in conjugated polymers', *Science*, vol. 341, pp. 1072–1073, 2013.

[78] T. Someya, T. Sekitani, S. Iba, Y. Kato, H. Kawaguchi, and T. Sakurai, 'A large-area, flexible pressure sensor matrix with organic field-effect transistors for artificial skin applications', *Proceedings of the National Academy of Sciences of the United States of America*, vol. 101, pp. 9966–9970, 2004.

[79] T. Someya, Y. Kato, T. Sekitani, S. Iba, Y. Noguchi, Y. Murase, et al., 'Conformable, flexible, large-area networks of pressure and thermal sensors with organic transistor active matrixes', *Proceedings of the National Academy of Sciences of the United States of America*, vol. 102, pp. 12321–12325, 2005.

[80] D. Darwis, D. Elkington, S. Ulum, G. Bryant, W. Belcher, P. Dastoor, and X. Zhou, 'Novel low voltage and solution processible organic thin film transistors based on water dispersed polymer semiconductor nanoparticulates', *Journal of Colloid and Interface Science*, vol. 401, pp. 65–69, 2013.

[81] D. Elkington, N. Cooling, W. Belcher, P. C. Dastoor, and X. Zhou, 'OTFT-based sensors', *Electronics*, vol. 3, pp. 234–254, 2014.

[82] D. Elkington, W. J. Belcher, P. C. Dastoor, and X. J. Zhou, 'Detection of saliva-range glucose concentrations using organic thin-film transistors', *Applied Physics Letters*, vol. 105, p. 043303, 2014.

[83] D. Elkington, M. Wasson, W. Belcher, P. C. Dastoor, and X. J. Zhou, 'Printable organic thin film transistors for glucose detection incorporating inkjet-printing of the enzyme recognition element', *Applied Physics Letters*, vol. 106, p. 263301, 2015.

[84] T. Ito, H. Shirakawa, and S. Ikeda, 'Simultaneous polymerization and formation of polyacetylene film on the surface of concentrated Ziegler-type catalyst solution', *Journal of Polymer Science, Polymer Chemistry*, vol. 12, pp. 11–20, 1974.

[85] H. Shirakawa, 'The discovery of polyacetylene film: The dawning of an era of conducting polymers', *Angewandte Chemie International Edition*, vol. 40, pp. 2574–2580, 2001.

[86] A. G. MacDiarmid, '"Synthetic metals": A novel role for organic polymers', *Angewandte Chemie International Edition*, vol. 40, pp. 2581–2590, 2001.

[87] A. J. Heeger, 'Semiconducting and metallic polymers: The fourth generation of polymeric materials', *Angewandte Chemie International Edition*, vol. 40, pp. 2591–2611, 2001.

[88] J. H. Edwards and W. J. Feast, 'A new synthesis of poly(acetylene)', *Polymer*, vol. 21, pp. 595–596, 1980.

[89] W. J. Feast and J. N. Winter, 'An improved synthesis of polyacetylene', *Journal of the Chemical Society, Chemical Communications*, no. 4, pp. 202–203, 1985.

[90] A. J. Heeger, S. Kivelson, J. R. Schrieffer, and W.-P. Su, 'Solitons in conducting polymers', *Reviews of Modern Physics*, vol. 60, pp. 781–850, 1988.

[91] G. Wegner, 'Polymers with metal-like conductivity: A review of their synthesis, structure and properties', *Angewandte Chemie International Edition*, vol. 20, pp. 361–381, 1981.

[92] R. G. Sun, Y. Wang, X. Zou, M. Fahlam, Q. Zheng, T. Kobayashi, et al., 'Electroluminescence of carbazole-substituted polyacetylenes', *Proceedings of the Society of Photo-Optical Instrumentation Engineers*, vol. 3476, pp. 332–337, 1998.

[93] K. Tada, H. Sawada, J. Kyokane, and K. Yoshino, 'Optical properties of perfluoroalkylated poly(diphenylacetylene)', *Japanese Journal of Applied Physics*, vol. 34, no. 8B, pp. L1083–L1085, 1995.

[94] J. W. Y. Lam and B. Z. Tang, 'Liquid-crystalline and light-emitting polyacetylenes', *Journal of Polymer Science (A), Polymer Chemistry*, vol. 41, pp. 2607–2629, 2003.

[95] J. W. Y. Lam, H. Peng, M. Häussler, R. Zheng, and B. Tang, 'Conjugated polymers with linear and hyperbranched structures and advanced materials properties', *Molecular Crystals and Liquid Crystals*, vol. 415, pp. 43–60, 2004.

[96] J. W. Y. Lam and B. Z. Tang, 'Functional polyacetylenes', *Accounts of Chemical Research*, vol. 38, pp. 745–754, 2005.

[97] T. Masuda, 'Substituted polyacetylenes', *Journal of Polymer Science A, Polymer Chemistry*, vol. 45, pp. 165–180, 2007.

[98] J. Liu, J. W. Y. Lam, and B. Z. Tang, 'Acetylenic polymers: Synthesis, structures and functions', *Chemical Reviews*, vol. 109, pp. 5799–5867, 2009.

[99] K. Tada, R. Hidayat, M. Hirohata, M. Teraguchi, T. Masuda, and K. Yoshino, 'Optical properties and blue and green electroluminescence in soluble disubstituted acetylene polymers', *Japanese Journal of Applied Physics*, vol. 35, pp. L1138–L1141, 1996.

[100] R. Sun, T. Masuda, and T. Kobayashi, 'Green electroluminescent emission from substituted polyacetylenes', *Japanese Journal of Applied Physics*, vol. 35, pp. L1434–L1437, 1996.

[101] R. Hidayat, M. Hirohata, K. Tada, M. Teraguchi, T. Masuda, and K. Yoshino, 'Effect of molecular structure of substituents on green electroluminescence in disubstituted acetylene polymers', *Japanese Journal of Applied Physics*, vol. 36, pp. 3740–3743, 1997.

[102] K. Yoshino, M. Hirohata, R. Hidayat, K. Tada, T. Sada, M. Teraguchi, et al., 'Optical properties and electroluminescence characteristics of polyacetylene derivatives dependent on substituent and layer structure', *Synthetic Metals*, vol. 91, pp. 283–287, 1997.

[103] R. Sun, T. Masuda, and T. Kobayashi, 'Visible electroluminescence of polyacetylene derivatives', *Synthetic Metals*, vol. 91, pp. 301–303, 1997.
[104] K. Yoshino, M. Hirohata, R. Hidayat, D. W. Kim, K. Tada, M. Ozaki, et al., 'Photoluminescence and electroluminescence in polyacetylene derivatives', *Synthetic Metals*, vol. 102, p. 1159, 1999.
[105] R. G. Sun, Y. Z. Wang, D. K. Wang, Q. B. Zheng, and A. J. Epstein, 'Hole transport in substituted polydiphenylacetylenes', *Synthetic Metals*, vol. 111–112, pp. 403–408, 2000.
[106] R. Sun, T. Masuda, and T. Kobayashi, 'Blue electroluminescence of substituted polyacetylenes', *Japanese Journal of Applied Physics*, vol. 35, pp. L1673–L1676, 1997.
[107] S. M. Abdul Karim, R. Nomura, F. Sanda, S. Seki, M. Watanabe, and T. Masuda, 'Synthesis and properties of ion-conducting poly(anthrylacetylene)s', *Macromolecules*, vol. 36, pp. 4786–4789, 2003.
[108] P. Mastrorilli, C. F. Nobile, R. Grisorio, A. Rizzuti, G. P. Suranna, D. Acierno, et al., 'Polyacetylenes bearing chiral-substituted fluorene and terfluorene pendant groups: Synthesis and properties', *Macromolecules*, vol. 37, pp. 4488–4495, 2004.
[109] H. Xu, Q. Sun, P. P.-S. Lee, H. S. Kwok, and B. Z. Tang, 'Synthesis and light-emission of C60-containing poly(phenylbutynes)', *Thin and Solid Films*, vol. 363, pp. 143–145, 2000.
[110] B. Z. Tang, H. Xu, J. W. Y. Lam, P. P. S. Lee, K. Xu, Q. Sun, and K. K. L. Cheuk, 'C60-containing poly(1-phenyl-1-alkynes): Synthesis, light emission, and optical limiting', *Chemistry of Materials*, vol. 12, pp. 1446–1455, 2000.
[111] J. W. Y. Lam, C. K. Law, Y. Dong, J.-L. Wang, W. Ge, and B. Z. Tang, 'Mesomorphic and luminescent properties of disubstituted polyacetylenes bearing biphenyl pendants', *Optical Materials*, vol. 21, pp. 321–324, 2002.
[112] Z. Xie, J. W. Y. Lam, Y. Dong, C. Qiu, H.-S. Kwok, and B. Z. Tang, 'Blue luminescence of poly(phenyl-napthoxy-pentyne)', *Optical Materials*, vol. 21, pp. 231–234, 2002.
[113] Y. Dong, J. W. Y. Lam, H. Peng, K. K. L. Cheuk, H. S. Kwok, and B. Z. Tang, 'Synthesis and mesomorphic and luminescent properties of disubstituted polyacetylenes bearing biphenyl pendants', *Macromolecules*, vol. 37, pp. 6408–6417, 2004.
[114] J. W. Y. Lam, Y. Dong, H. S. Kwok, and B. Z. Tang, 'light-emitting polyacetylenes: Synthesis and electrooptical properties of poly(1-phenyl-1-alkyne)s bearing naphthyl pendants', *Macromolecules*, vol. 39, pp. 6997–7003, 2006.
[115] J. W. Y. Lam, J. Luo, Y. Dong, K. K. L. Cheuk, and B. Z. Tang, 'Functional polyacetylenes: Synthesis and light emission of polypropiolates', *Macromolecules*, vol. 35, pp. 8288–8299, 2002.
[116] J. Chen, Z. Xie, J. W. Y. Lam, C. C. W. Law, and B. Z. Tang, 'Silole-containing polyacetylenes. Synthesis, thermal stability, light emission, nanodimensional aggregation, and restricted intramolecular rotation', *Macromolecules*, vol. 36, pp. 1108–1117, 2003.
[117] J. W. Y. Lam, J. Chen, C. C. W. Law, H. Peng, Z. Xie, K. K. L. Cheuk, et al., 'Silole-containing linear and hyperbranched polymers', *Macromolecular Symposia*, vol. 196, pp. 289–300, 2003.
[118] J. Chen, H. S. Kwok, and B. Z. Tang, 'Silole-containing poly(diphenylacetylene): Snthesis, characterization, and light emission', *Journal of Polymer Science (A), Polymer Chemistry*, vol. 44, pp. 2487–2498, 2006.
[119] G. Yu, Y. Liu, X. Zhan, H. Li, M. Yang, and D. Zhu, 'Thermally stable substituted polyacetylenes', *Thin and Solid Films*, vol. 363, pp. 126–129, 2000.

[120] J.-W. Park, J.-W. Lee, Y.-S. Gal, J.-M. Ko, and J.-H. Chen, 'Poly(2-ethynyl pyrdinium tosylate)', *Polymer Preprints*, vol. 40(2), pp. 795–796, 1999.

[121] R. Hidayat, M. Hirohata, K. Tada, M. Teraguchi, T. Masuda, and K. Yoshino, 'Green electroluminescence from polydiphenylacetylene blends', *Japanese Journal of Applied Physics*, vol. 37, pp. L180–L183, 1998.

[122] Y. M. Huang, J. W. Y. Lam, K. K. L. Cheuk, W. Ge, and B. Z. Tang, 'Strong luminescence from poly(1-alkynes)', *Macromolecules*, vol. 32, pp. 5976–5978, 1999.

[123] Y. M. Huang, J. W. Y. Lam, K. K. L. Cheuk, W. Ge, and B. Z. Tang, 'Poly (alkylacetylene)s: A new class of luminescent polyacetylenes', *Thin and Solid Films*, vol. 363, pp. 146–148, 2000.

[124] P. P.-S. Lee, Y. Geng, H. S. Kwok, and B. Z. Tang, 'Synthesis and light-emission of poly(carbazolyl acetylene)s', *Thin and Solid Films*, vol. 363, pp. 149–151, 2000.

[125] R. A. Wessling and R. G. Zimmerman, 'U.S.Patent US 3,401,152', *Chemical Abstracts*, vol. 69, p. 87735q, 1968.

[126] R. A. Wessling, 'The polymerisation of xylylidene bisalkylsulfonium salts', *Journal of Polymer Science, Polymer Symposia*, vol. 72, pp. 55–66, 1985.

[127] R. O. Garay and R. W. Lenz, 'Effect of the reaction state on the polymerization of p-xylenesulfonium salts', *Journal of Polymer Science: Part A: Polymer Chemistry*, vol. 30, pp. 977–982, 1992.

[128] P. L. Burn, D. D. C. Bradley, R. H. Friend, D. A. Halliday, A. B. Holmes, R. W. Jackson, and A. Kraft, 'Precursor route chemistry and electronic properties of poly (p-phenylenevinylene), poly[(2,5-dimethyl-p-phenylene)vinylene] and poly[(2,5-dimethoxy-p-phenylene)vinylene]', *Journal of the Chemical Society, Perkin Transactions 1*, pp. 3225–3231, 1992.

[129] R. O. Garay, U. Baier, C. Bubeck, and K. Müllen, 'Low-temperature synthesis of poly(p-phenylenevinylene) by the sulfonium salt route', *Advanced Materials*, vol. 5, pp. 561–564, 1993.

[130] D. A. Halliday, P. L. Burn, D. D. C. Bradley, R. H. Friend, O. M. Gelsen, A. B. Holmes, et al., 'Large changes in optical response through chemical pre-ordering of poly(p-phenylenevinylene)', *Advanced Materials*, vol. 5, pp. 40–43, 1993.

[131] D. A. Halliday, P. L. Burn, R. H. Friend, D. D. C. Bradley, A. B. Holmes, and A. Kraft, 'Extended π-conjugation in poly(p-phenylenevinylene) from a chemically modified precursor polymer', *Synthetic Metals*, vol. 55, pp. 954–959, 1993.

[132] P. L. Burn, A. B. Holmes, A. Kraft, D. D. C. Bradley, A. R. Brown, and R. H. Friend, 'Synthesis of a segmented conjugated polymer chain giving a blue-shifted electroluminescence and improved efficiency', *Journal of the Chemical Society Chemical Communications*, pp. 32–34, 1992.

[133] A. R. Brown, P. L. Burn, D. D. C. Bradley, R. H. Friend, A. Kraft, and A. B. Holmes, 'Blue-shifted electroluminescence from a stable precursor to poly(P-phenylene vinylene)', *Molecular Crystals and Liquid Crystals*, vol. 216, pp. 111–116, 1992.

[134] J. C. Carter, I. Grizzi, S. K. Heeks, D. J. Lacey, S. G. Latham, P. G. May, et al., 'Operating stability of light-emitting polymer diodes based on poly(p-phenylene vinylene)', *Applied Physics Letters*, vol. 71, pp. 34–36, 1997.

[135] N. Tessler, N. T. Harrison, and R. H. Friend, 'High peak brightness polymer light-emitting diodes', *Advanced Materials*, vol. 10, pp. 64–68, 1998.

[136] G. Padmanaban and S. Ramakrishnan, 'Conjugation length control in MEHPPV', *Journal of the American Chemical Society*, vol. 122, pp. 2244–2251, 2000.

[137] G. Padmanaban and S. Ramakrishnan, 'Improved method for conjugation control in MEHPPV', *Synthetic Metals*, vol. 119, pp. 533–534, 2001.

[138] P. L. Burn, A. B. Holmes, A. Kraft, D. D. C. Bradley, A. R. Brown, R. H. Friend, and R. W. Gymer, 'Chemical tuning of electroluminescent copolymers to improve emission efficiencies and allow patterning', *Nature*, vol. 356, pp. 47–49, 1992.

[139] J. Gelan, D. Vanderzande, and F. Louwet, 'Eur. Patent Appl. EP 644,217', *Chemical Abstracts*, vol. 123, p. 144940m, 1995.

[140] F. Louwet, D. Vanderzande, J. Gelan, and J. Mullens, 'A new synthetic route to a soluble high molecular weight precursor to poly(*p*-phenylene vinylene)', *Macromolecules*, vol. 28, pp. 1330–1331, 1995.

[141] A. J. J. M. van Breemen, D. J. M. Vanderzande, P. J. Adriaensens, and J. M. J. V. Gelan, 'Highly selective route for producing unsymmetrically substituted monomers towards synthesis of conjugated polymers derived from poly(*p*-phenylene vinylene)', *Journal of Organic Chemistry*, vol. 64, pp. 3106–3112, 1999.

[142] A. J. Mozer, P. Denk, M. C. Scharber, H. Neugebauer, N. S. Sariciftci, P. Wagner, et al., 'Novel regiospecific MDMO-PPV copolymer with improved charge transport for bulk heterojunction solar cells', *The Journal of Physical Chemistry B*, vol. 108, pp. 5235–5242, 2004.

[143] H. G. Gilch and W. L. Wheelwright, 'Polymerisation of α-halogenated *p*-xylenes with base', *Journal of Polymer Science A-1*, vol. 4, pp. 1337–1349, 1966.

[144] F. Wudl and G. Srdanov, 'US Patent 5,189,136', *Chemical Abstracts*, vol. 123, p. 255575, 1993.

[145] G. J. Sarnecki, R. H. Friend, A. B. Holmes, and S. C. Moratti, 'The synthesis of a new regioregular conjugated polymer. Poly(2-bromo-5-dodecyloxy-1,4-phenylene-vinylene)', *Synthetic Metals*, vol. 69, pp. 545–546, 1995.

[146] R. M. Gurge, A. Sarker, P. M. Lahti, B. Hu, and F. E. Karasz, 'Red light emitting "push–pull" disubstituted poly(1,4-phenylenevinylenes)', *Macromolecules*, vol. 29, pp. 4287–4292, 1996.

[147] H. F. M. Schoo, R. C. J. E. Demandt, J. J. M. Vleggaar, and C. T. H. Liedenbaum, 'Organic polymer LEDs with mobile and immobile ions', *Macromolecular Symposia*, vol. 125, pp. 165–170, 1997.

[148] G. R. Webster and P. L. Burn, 'Controlling the conjugation length in poly[5-n-butyl-2-(2-ethylhexyl)-1,4-phenylene vinylene]: Exploring the scope of hydrogen radical substitution of leaving groups on precursor polymers', *Synthetic Metals*, vol. 145, pp. 159–169, 2004.

[149] R. O. Garay, M. N. Sarimalis, R. S. Montani, and S. A. Hernandez, 'Synthesis of conjugated polymers. Polymerizability studies of bis-sulfonium salts', *Designed Monomers and Polymers*, vol. 3, pp. 231–244, 2000.

[150] A. C. Grimsdale, K. L. Chan, R. E. Martin, P. G. Jokisz, and A. B. Holmes, 'Synthesis of light-emitting conjugated polymers for applications in electroluminescent devices', *Chemical Reviews*, vol. 109, pp. 987–1091, 2009.

[151] F. R. Denton, P. M. Lahti, and F. E. Karasz, 'The effect of radical trapping upon formation of poly(α-tetrahydrothiophenio paraxylene) polyelectrolytes by the Wessling soluble precursor method', *Journal of Polymer Science: Part A: Polymer Chemistry*, vol. 30, pp. 2223–2231, 1992.

[152] A. Issaris, D. Vanderzande, and J. Gelan, 'Polymerisationof a *p*-quinodimethane derivative to a precursor of poly(*p*-phenylene vinylene): Indications for a free radical mechanism', *Polymer*, vol. 38, pp. 2571–2574, 1997.

[153] B. R. Cho, 'Precursor polymers to poly(p-phenylene vinylene) and its heteroaromatic derivatives. Polymerization mechanism', *Progress in Polymer Science*, vol. 27, pp. 307–355, 2002.

[154] C. Yin and C.-Z. Yang, 'Mechanism analysis of the PEG-participated Gilch synthesis for soluble PPV derivatives', *Journal of Applied Polymer Science*, vol. 82, pp. 263–268, 2001.
[155] C. J. Neef and J. P. Ferraris, 'MEHPPV: Improved synthetic procedure and molecular weight control', *Macromolecules*, vol. 33, pp. 2311–2314, 2000.
[156] B. R. Hsieh, Y. Yu, A. C. Van Laeken, and H. Lee, 'General methodology towards soluble poly(*p*-phenylene vinylene) derivatives', *Macromolecules*, vol. 30, pp. 8094–8095, 1997.
[157] L. Hontis, V. Vrindts, L. Lutsen, D. Vanderzande, and J. Gelan, 'The Gilch polymerisation towards C1C10PPV: Indications for a radical mechanism', *Polymer*, vol. 42, pp. 5793–5799, 2001.
[158] T. Schwalm, J. Wiesecke, S. Immel, and M. Rehahn, 'The Gilch synthesis of poly(p-phenylene vinylenes): Mechanistic knowledge in the service of advanced materials', *Macromolecular Rapid Communications*, vol. 30, pp. 1295–1322, 2009.
[159] H. Becker, H. Spreitzer, K. Ibrom, and W. Kreuder, 'New insights into the microstructure of Gilch-polymerized PPVs', *Macromolecules*, vol. 32, pp. 4925–4932, 1999.
[160] H. Becker, O. Gelsen, E. Kluge, W. Kreuder, H. Schenk, and H. Spreitzer, 'Development of high-performance PPVs: Implications of the polymer microstructure', *Synthetic Metals*, vol. 111–112, pp. 145–149, 2000.
[161] H. Becker, H. Spreitzer, W. Kreuder, E. Kluge, H. Vestweber, H. Schenk, and K. Treacher, 'Advances in polymers for PLEDs: From a polymerization mechanism to industrial manufacturing', *Synthetic Metals*, vol. 122, pp. 105–110, 2001.
[162] A. Fleissner, K. Stegmaier, C. Melzer, H. von Seggern, T. Schwalm, and M. Rehahn, 'Residual halide groups in Gilch-polymerized PPV and their impact on performance and lifetime of OLEDs', *Chemistry of Materials*, vol. 21, pp. 4288–4298, 2009.
[163] H. Becker, H. Spreitzer, W. Kreuder, E. Kluge, H. Schenk, I. Parker, and Y. Cao, 'Soluble PPVs with enhanced performance: A mechanistic approach', *Advanced Materials*, vol. 12, pp. 42–48, 2000.
[164] A. Delmotte, M. Biesemans, B. V. Mele, M. Gielen, M. M. Bouman, and E. W. Meijer, 'Selective elimination in dialkoxy–PPV precursors yielding polymers with isolated tetraalkoxy-stilbene units', *Synthetic Metals*, vol. 68, pp. 269–273, 1995.
[165] Y.-L. Fan and K.-F. Lin, 'Dependence of the luminescence properties and chain length of poly[2-methoxy-5-(2'-ethylhexyloxy)-1,4-phenylene vinylene] on the formation of *cis*-vinylene bonds during the Gilch polymerization', *Journal of Polymer Science (A), Polymer Chemistry*, vol. 43, pp. 2520–2526, 2005.
[166] A. Greiner and W. Heitz, 'New synthetic approach to PPV and its derivatives by Pd catalyzed arylation of ethylene', *Macromolecular Chemistry, Rapid Communications*, vol. 9, p. 581, 1988.
[167] W. Heitz, W. Brügging, L. Freund, M. Gailberger, A. Greiner, H. Jung, et al., 'Synthesis of monomers and polymers by the Heck reaction', *Makromolekulare Chemie*, vol. 189, pp. 119–127, 1988.
[168] W. Heitz, 'Metal-catalyzed polycondensation reactions', *Pure and Applied Chemistry*, vol. 67, pp. 1951–1964, 1995.
[169] Y. Liu, P. M. Lahti, and F. La, 'Synthesis of a regiospecific, soluble poly(2-alkoxy-1,4-phenylenevinylene)', *Polymer*, vol. 39, pp. 5241–5244, 1998.
[170] S. Pfeiffer and H.-H. Hörhold, 'Synthesis of soluble MEHPPV and MEHPPB by Horner', *Synthetic Metals*, vol. 101, pp. 109–110, 1999.

[171] S. Pfeiffer and H.-H. Hörhold, 'Investigation of poly(arylenevinylene)s. Part 41. Synthesis of soluble dialkoxysubstituted poly(phenylene alkenylidenes) by applying the Horner reaction by condensation polymerization', *Macromolecular Chemistry and Physics*, vol. 200, pp. 1870–1878, 1999.
[172] H.-H. Hörhold, H. Tillmann, C. Bader, R. Stockmann, J. Nowotny, E. Klemm, et al., 'MEH-PPV and dialkoxy phenylene vinylene copolymers: Synthesis and lasing', *Synthetic Metals*, vol. 119, pp. 199–200, 2001.
[173] A. Drury, S. Maier, N. Rüther, and W. J. Blau, 'Investigation of different synthesis routes to and structure-property relationships of a poly(*m*-phenylenevinylene-*co*-2,5-dioctyloxy-*p*-phenyklene vinylene)', *Journal of Materials Chemistry*, vol. 13, pp. 485–490, 2003.
[174] L. Liao, Y. Pang, L. Ding, and F. E. Karasz, 'Effect of iodine-catalyzed isomerization on the optical properties of poly[(1,3-phenylenevinylene-*alt*-(2,5-dihexyloxy-1,4-phenylenevinylene)]s', *Macromolecules*, vol. 35, pp. 6055–6059, 2002.
[175] M. Rehahn and A.-D. Schlüter, 'Soluble poly(*p*-phenylenevinylene)s from 2,5-dihexylterephthaldehyde using the improved McMurry reagent', *Makromolecular Chemie Rapid Communications*, vol. 11, pp. 375–379, 1990.
[176] W. J. Feast and I. S. Millichamp, 'The synthesis of poly(4,4'-diphenylene diphenylvinylene) via condensation polymerization of 4,4'-dibenzoylbiphenyl', *Polymer Communications*, vol. 24, pp. 102–103, 1983.
[177] F. Cacialli, R. Daik, W. J. Feast, R. H. Friend, and C. Lartigau, 'Synthesis and properties of poly(arylene vinylene)s with controlled structures', *Optical Materials*, vol. 12, pp. 315–319, 1999.
[178] N. C. Greenham, S. C. Moratti, D. D. C. Bradley, R. H. Friend, and A. B. Holmes, 'Efficient light-emitting diodes based on polymers with high electron affinity', *Nature*, vol. 365, pp. 628–630, 1993.
[179] M. J. Marsella, D.-K. Fu, and T. M. Swager, 'Synthesis of regioregular poly (methylpyridinium vinylene). An isoelectronic analog to poly(paraphenylenevinylene)', *Advanced Materials*, vol. 7, pp. 145–147, 1995.
[180] Z. Bao, W. K. Chan, and L. Yu, 'Exploration of the Stille coupling reaction for the synthesis of functional polymers', *Journal of the American Chemical Society*, vol. 117, pp. 12426–12435, 1995.
[181] F. Babudri, S. R. Cicco, G. M. Farinola, F. Naso, A. Bolognesi, and W. Porzio, 'Synthesis, characterization, and properties of a soluble polymer with a poly(phenylenevinylene) structure', *Macromolecular Rapid Communications*, vol. 17, pp. 905–911, 1996.
[182] L. C. Lopez, P. Strohriegl, and T. Stübinger, 'Synthesis of poly(fluorenylene-co-phenylene vinylene) by Suzuki coupling', *Macromolecular Chemistry and Physics*, vol. 203, pp. 1926–1930, 2002.
[183] R. Grisorio, P. Mastrorilli, C. F. Nobile, G. Romanazzi, and E. P. Suranna, 'A novel synthetic protocol for poly(fluorenylenevinylene)s: A cascade Suzuki-Heck reaction', *Tetrahedron Letters*, vol. 46, pp. 2555–2558, 2005.
[184] E. Thorn-Csányi and P. Kraxner, 'All *trans*-oligomers of dialkyl-1,4-phenylenevinylenes: Metathesis preparation and characterization', *Macromolecular Chemistry and Physics*, vol. 198, pp. 3827–3843, 1997.
[185] R. Reetz, O. Norwark, O. Herzog, S. Brocke, and E. Thorn-Csányi, 'Substituted PPV oligomers by metathesis', *Synthetic Metals*, vol. 119, pp. 539–540, 2001.
[186] G. W. Oakley and K. B. Wagener, 'Solid-state olefin metathesis: ADMET of rigid-rod polymers', *Macromolecular Chemistry and Physics*, vol. 206, pp. 15–24, 2005.

[187] M. W. Wagaman and R. H. Grubbs, 'Synthesis of PNV homo- and copolymers by a ROMP precursor route', *Synthetic Metals*, vol. 84, pp. 327–328, 1997.

[188] T. Zyung, J.-J. Kim, W.-Y. Hwang, D. H. Hwang, and H. K. Shim, 'Electroluminescence from poly(*p*-phenylenevinylene) with monoalkoxy substituent on the aromatic ring', *Synthetic Metals*, vol. 71, pp. 2167–2169, 1995.

[189] D. Braun and A. J. Heeger, 'Visible light emission from semiconducting polymer diodes', *Applied Physics Letters*, vol. 58, pp. 1982–1984, 1991.

[190] S. Doi, M. Kuwabara, T. Noguchi, and T. Ohnishi, 'Organic electroluminescent devices having poly(dialkoxy-p-phenylene vinylenes) as a light emitting material', *Synthetic Metals*, vol. 57, pp. 4174–4179, 1993.

[191] M. R. Andersson, G. Yu, and A. J. Heeger, 'PL and EL of films from soluble PPV polymers', *Synthetic Metals*, vol. 85, pp. 1275–1276, 1997.

[192] D.-H. Hwang, S. T. Kim, H.-K. Shim, A. B. Holmes, S. C. Moratti, and R. H. Friend, 'Green light-emitting diodes from poly(2-dimethyloctylsilyl-1,4-phenylene-vinylene)', *Chemical Communications*, pp. 2241–2242, 1996.

[193] H. Vestweber, A. Greiner, U. Lemmer, R. F. Mahrt, R. Richert, W. Heitz, and H. Bässler, 'Progress towards processible materials for light-emitting devices using poly(*p*-phenylphenylenevinylene)', *Advanced Materials*, vol. 4, pp. 661–662, 1992.

[194] Y. Jin, J. Kim, S. H. Park, H. Kim, K. Lee, and H. Suh, 'Synthesis and color tuning of PPV containing terphenyl units for LEDs', *Bulletin of the Korean Chemical Society*, vol. 26, pp. 1807–1818, 2005.

[195] R. E. Martin, F. Generste, R. Riehn, B. S. Chuah, F. Cacialli, R. H. Friend, and A. B. Holmes, 'Efficient blue-green emitting poly(1,4-phenylene vinylene) copolymers', *Chemical Communications*, pp. 291–292, 2000.

[196] R. E. Martin, F. Geneste, B. S. Chuah, C. Fischmeister, Y. Ma, A. B. Holmes, et al., 'Versatile synthesis of conjugated aromatic homo- and copolymers', *Synthetic Metals*, vol. 122, pp. 1–6, 2001.

[197] B. R. Hsieh, W. C. Wan, Y. Yu, Y. Gao, T. E. Goodwin, S. A. Gonzalez, and W. A. Feld, 'Synthesis of highly phenylated poly(p-phenylenevinylenes) via a chlorine precursor route', *Macromolecules*, vol. 31, pp. 631–636, 1998.

[198] B. R. Hsieh, Y. Yu, E. W. Forsythe, G. M. Schaaf, and W. A. Feld, 'A new family of highly emissive soluble poly(p-phenylene vinylene) derivatives. A step toward fully conjugated blue-emitting poly(p-phenylene vinylenes)', *Journal of the American Chemical Society*, vol. 120, pp. 231–232, 1998.

[199] T. Ahn, M. S. Jang, H.-K. Shim, D.-H. Hwang, and T. Zyung, 'Blue electroluminescent polymers: Control of conjugation by kink linkages and substituents in the poly(*p*-phenylenevinylene)-related copolymers', *Macromolecules*, vol. 32, pp. 3279–3285, 1999.

[200] H. Spreitzer, H. Becker, E. Kluge, W. Kreuder, H. Schenk, R. Demandt, and H. Schoo, 'Soluble phenyl-substituted PPVs. New materials for highly efficient polymer LEDs', *Advanced Materials*, vol. 10, pp. 1340–1343, 1998.

[201] M. Van Der Borght, D. Vanderzande, and J. Gelan, 'Synthesis of high molecular weight poly(4,4'-biphenylene vinylene) and poly(2-6-naphthalene vinylene) via a non-ionic precursor route', *Polymer*, vol. 39, pp. 4171–4174, 1998.

[202] S.-H. Jin, H.-J. Park, J. Y. Kim, K. Lee, S.-P. Lee, D.-K. Moon, et al., 'Poly(fluorene vinylene) by Gilch for light emitting diode applications', *Macromolecules*, vol. 35, pp. 7532–7534, 2002.

[203] M. Remmers, D. Neher, J. Grüner, R. H. Friend, G. H. Gelinck, J. M. Warman, et al., 'The optical, electronic, and electroluminescent properties of novel poly(*p*-phenylene) related polymers', *Macromolecules*, vol. 29, pp. 7432–7445, 1996.

[204] J. D. Stenger-Smith, T. Sauer, G. Wegner, and R. W. Lenz, 'Preparation, spectroscopic and cyclic voltammetric studies of poly (1,4-naphthalene vinvlene) prepared from a cycloalkylene sulfonium precursor polymer', *Polymer*, vol. 31, pp. 1632–1636, 1990.

[205] Y.-J. Miao and G. C. Bazan, 'Paracyclophane route to poly(*p*-phenylene vinylene)', *Journal of the American Chemical Society*, vol. 116, pp. 9379–9380, 1994.

[206] D.-H. Hwang, K.-H. Choi, L.-M. Do, T. Ahn, H.-K. Shim, and T. Zyung, 'Two-colour emission from PPV-PTV copolymer and Alq3 heterostructure EL device', *Synthetic Metals*, vol. 102, pp. 1218–1219, 1999.

[207] D. R. Baigent, R. N. Marks, N. C. Greenham, R. H. Friend, S. C. Moratti, and A. B. Holmes, 'Surface-emitting polymer light emitting diodes', *Synthetic Metals*, vol. 71, pp. 2177–2178, 1995.

[208] S. C. Moratti, R. Cervini, A. B. Holmes, D. R. Baigent, R. H. Friend, N. C. Greenham, et al., 'High electron affinity polymers for LEDs', *Synthetic Metals*, vol. 71, pp. 2117–2120, 1995.

[209] Z. An, C. Q. Wu, and X. Sun, 'Dynamics of photogenerated polarons in conjugated polymers', *Physical Review Letters*, vol. 93, p. 216407, 2004.

[210] S. J. Blundell, F. L. Pratt, I. M. Marshall, C. Steer, W. Hayes, A. Husmann, et al., 'Muon-spin relaxation study of anisotropic charge carrier motion in polyphenylene vinylene-based polymers', *Journal of Physics: Condensed Matter*, vol. 14, pp. 9987–9995, 2002.

[211] A. J. Campbell, D. D. C. Bradley, and D. G. Lidzey, 'Space-charge limited conduction with traps in poly(phenylene vinylene) light emitting diodes', *Journal of Applied Physics*, vol. 82, pp. 6326–6342, 1997.

[212] D. Moses, H. Okumoto, C. H. Lee, A. J. Heeger, T. Ohnishi, and T. Noguchi, 'Mechanism of carrier generation in poly(phenylene vinylene): Transient photoconductivity and photoluminescence at high electric fields', *Physical Review B*, vol. 54, pp. 4748–4754, 1996.

[213] S. Barth and H. Bässler, 'Intrinsic photoconduction in PPV-type conjugated polymers', *Physical Review Letters*, vol. 79, pp. 4445–4448, 1997.

[214] J. Gmeiner, S. Karg, M. Meier, W. Riess, P. Strohriegl, and M. Schwoerer, 'Synthesis, electrical conductivity and electroluminescence of poly(*p*-phenylene vinylene) prepared by the precursor route', *Acta Polymerica*, vol. 44, pp. 201–205, 1993.

[215] D. R. Gagnon, J. D. Capistran, F. E. Karasz, R. W. Lenz, and S. Antoun, 'Synthesis, doping, and electrical conductivity of high molecular weight PPV', *Polymer*, vol. 28, p. 567, 1987.

[216] Y. Ohmori, K. Muro, M. Onada, and K. Yoshino, 'Fabrication and characterization of Schottky gated field-effect transistors utilizing poly(1,4-naphthalene vinylene) and poly(*p*-phenylene vinylene)', *Japanese Journal of Applied Physics*, vol. 31, p. L646, 1992.

[217] I. D. Parker, R. W. Gymer, M. G. Harrison, R. H. Friend, and H. Ahmed, 'Fabrication of a novel electro-optical intensity modulator from the conjugated polymer poly(2,5-dimethoxy-PPV)', *Applied Physics Letters*, vol. 62, pp. 1519–1521, 1993.

[218] R. M. Gurge, A. M. Sarker, P. M. Lahti, B. Hu, and F. E. Karasz, 'Light-emitting properties of fluorine-substituted poly(1,4-phenylene vinylenes)', *Macromolecules*, vol. 30, pp. 8286–8292, 1997.

[219] I.-N. Kang, H.-K. Shim, and T. Zyung, 'Yellow-light emitting fluorine-substituted PPV', *Chemistry of Materials*, vol. 9, pp. 746–749, 1997.

[220] N. C. Greenham, F. Cacialli, D. D. C. Bradley, R. H. Friend, S. C. Moratti, and A. B. Holmes, 'Cyano-derivatives of poly(*p*-phenylene vinylene) for use in thin-film

light-emitting diodes', *Materials Research Society Symposia Proceedings*, vol. 328, pp. 351–360, 1994.

[221] S.-H. Jin, M.-Y. Kim, J. Y. Kim, K. Lee, and Y.-S. Gal, 'High efficiency poly(*p*-phenylene vinylene)-based copolymers containing an oxadiazole pendant group for light-emitting diodes', *Journal of the American Chemical Society*, vol. 126, pp. 2474–2480, 2004.

[222] M. Grell and D. D. C. Bradley, 'Polarized luminescence from oriented molecular materials', *Advanced Materials*, vol. 11, pp. 895–905, 1999.

[223] M. O'Neill and S. M. Kelly, 'Liquid crystals for charge transport, luminescence and photonics', *Advanced Materials*, vol. 15, pp. 1135–1146, 2003.

[224] M. Hamaguchi and K. Yoshino, 'Lyotropic behavior of poly(2,5-dinonlyoxy-*p*-phenylene)', *Japanese Journal of Applied Physics*, vol. 33, pp. L1478–L1481, 1994.

[225] M. Hamaguchi and K. Yoshino, 'Rubbing-induced molecular orientation and polarized electroluminescence in conjugated polymer', *Japanese Journal of Applied Physics*, vol. 34, pp. L712–L715, 1995.

[226] M. Onoda, K. Tada, M. Ozaki, and K. Yoshino, 'An electroluminescent diode using liquid-crytalline conducting polymer', *Thin and Solid Films*, vol. 363, pp. 9–12, 2000.

[227] J. H. Park, C. H. Lee, K. Akagi, H. Shirakawa, and Y. W. Park, 'PL and EL of LC aromatic conjugated polymers', *Synthetic Metals*, vol. 119, pp. 633–634, 2001.

[228] Y. Nishikata, M. Kakimoto, and Y. Imai, 'Preparation and properties of poly(*p*-phenylene vinylene) Langmuir-Blodgett film', *Thin and Solid Films*, vol. 179, pp. 191–197, 1989.

[229] A. Wu, M. Jikei, M. Kakimoto, Y. Imai, S. Ukishima, and Y. Takahashi, 'Fabrication of polymeric light-emitting diodes based on poly(p-phenylene vinylene) LB films', *Chemistry Letters*, vol. 23, pp. 2319–2322, 1994.

[230] J. H. Kim, Y. K. Kim, B. C. Sohn, D.-Y. Kang, J.-I. Jin, C.-H. Kim, and C.-H. Pyun, 'Preparation and characterization of poly(*p*-phenylene vinylene) Langmuir-Blodgett film formed via precursor method', *Synthetic Metals*, vol. 71, pp. 2023–2024, 1995.

[231] E. Peeters, M. P. T. Christiaans, R. A. J. Janssen, H. F. M. Schoo, H. P. J. M. Dekkers, and E. W. Meijer, 'Circularly polarized electroluminescence from a polymer light-emitting diode', *Journal of the American Chemical Society*, vol. 119, pp. 9909–9910, 1997.

[232] L. Pu, 'Binaphthyl dimers, oligomers and polymers', *Chemical Reviews*, vol. 98, pp. 2405–2494, 1998.

[233] O. Hofmann, X. Wang, J. C. deMello, D. D. C. Bradley, and A. J. deMello, 'Towards microalbuminuria determination on a disposable diagnostic microchip with integrated fluorescence detection based on thin-film organic light emitting diodes', *Lab Chip*, vol. 5, pp. 863–868, 2005.

[234] S. Karg, W. Reiss, M. Meier, and M. Schwoerer, 'Characterization of LEDs and solar cells based on PPV', *Synthetic Metals*, vol. 57, pp. 4186–4191, 1993.

[235] A. C. Arias, M. Granstrom, D. S. Thomas, K. Petritsch, and R. H. Friend, 'Doped conducting-polymer-semiconducting-polymer interfaces: Their use in organic photovoltaic devices', *Physical Review B*, vol. 60, pp. 1854–1860, 1999.

[236] Y. Hayashi, I. Yamada, S. Takagi, A. Takasu, T. Soga, and T. Jimbo, 'Influence of structure and C_{60} composition on properties of blends and bilayers of organic donor-acceptor polymer/C_{60} photovoltaic devices', *Japanese Journal of Applied Physics*, vol. 44, pp. 1296–1300, 2005.

[237] T. Umeda, Y. Hashimoto, H. Mizukami, A. Fujii, and K. Yoshino, 'Fabrication of interpenetrating semilayered structure of semiconducting polymer and fullerene by

solvent corrosion method and its photovoltaic properties', *Japanese Journal of Applied Physics*, vol. 44, pp. 4155–4160, 2005.
[238] F. Zhang, M. Johansson, M. R. Andersson, J. C. Hummelen, and O. Inganas, 'Polymer photovoltaic cells with conducting polymer anodes', *Advanced Materials*, vol. 14, pp. 662–665, 2002.
[239] P. Lin, C.-J. Liang, Z.-B. Deng, D.-P. Xiong, L. Wang, Z.-F. Zhang, and X.-Q. Zhang, '[Photovoltaic character of organic EL devices MEH-PPV/Alq3]', *Guangpuxue Yu Guangpu Fenxi*, vol. 25, pp. 23–25, 2005; *Chemical Abstracts*, vol. 143, 88243.
[240] C. H. Lee, G. Yu, D. Moses, K. Pakbaz, C. Zhang, N. S. Sariciftci, et al., 'Sensitization of the photoconductivity of conducting polymers by C60: photoinduced electron transfer', *Physical Review B*, vol. 48, pp. 15425–15433, 1993.
[241] K. Yoshino, X. H. Yin, K. Muro, S. Kiyomatsu, S. Morita, A. A. Zakhidov, et al., 'Marked enhancement of photoconductivity and quenching of luminescence in poly (2,5-dialkoxy-p-phenylene vinylene) upon C_{60} doping', *Japanese Journal of Applied Physics*, vol. 32, p. L357, 1993.
[242] S. Morita, S. Kiyomatsu, X. H. Yin, A. A. Zakhidov, T. Noguchi, T. Ohnishi, and K. Yoshino, 'Doping effect of buckminsterfullerene in poly(2,5-dialkoxy-p-phenylene vinylene)', *Journal of Applied Physics*, vol. 74, pp. 2860–2865, 1993.
[243] L. Damodare, T. Soga, and T. Mieno, 'Studies on dark and photo-conductivities of poly[2-methoxy,5-(2'-ethylhexyloxy)-*p*-phenylene vinylene]:C_{60} thin films', *Japanese Journal of Applied Physics*, vol. 42, pp. 2498–2502, 2003.
[244] S. Günes, H. Neugebauer, and N. S. Sariciftci, 'Conjugated polymer-based organic solar cells', *Chemical Reviews*, vol. 107, pp. 1324–1338, 2007.
[245] A. W. Hains and T. J. Marks, 'High-efficiency hole extraction/electron-blocking layer to replace poly(3,4-ethylenedioxythiophene):poly(styrene sulfonate) in bulk-heterojunction polymer solar cells', *Applied Physics Letters*, vol. 92, p. 023504, 2008.
[246] Y. Sahin, S. Alem, R. de Bettignies, and J.-M. Nunzi, 'Development of air stable polymer solar cells using an inverted gold on top anode structure', *Thin and Solid Films*, vol. 476, pp. 340–343, 2005.
[247] C. J. Brabec, S. Gowrisanker, J. J. M. Halls, D. Laird, S. Jia, and S. P. Williams, 'Polymer–fullerene bulk-heterojunction solar cells', *Advanced Materials*, vol. 22, pp. 3839–3856, 2010.
[248] H. Hoppe and N. S. Sariciftci, 'Organic solar cells: An overview', *Journal of Materials Research*, vol. 19, p. 1924, 2004.
[249] Y. F. Huang, A. R. Inigo, C.-C. Chang, K.-C. Li, C.-F. Liang, C.-W. Chang, et al., 'Nanostructure-dependent vertical charge transport in MEH-PPV films', *Advanced Functional Materials*, vol. 17, pp. 2902–2910, 2007.
[250] Y.-F. Huang, C.-W. Chang, D.-M. Smilgies, U.-S. Jeng, A. R. Inigo, J. D. White, et al., 'Correlating nanomorphology with charge-transport anisotropy in conjugated-polymer thin films', *Advanced Materials*, vol. 21, pp. 2988–2992, 2009.
[251] J.-F. Nierengarten, 'Fullerene-(π-conjugated oligomer) dyads as active photovoltaic materials', *Solar Energy Materials and Solar Cells*, vol. 83, pp. 187–199, 2004.
[252] E. Peeters, P. van Hal, J. Knol, C. J. Brabec, N. S. Sariciftci, J. C. Hummelen, and R. Janssen, 'Synthesis, photphysical properties and photovoltaic devices of oligo(*p*-phenylene vinylene)-fullerene dyads', *The Journal of Physical Chemistry B*, vol. 104, pp. 10174–10190, 2000.
[253] U. Stalmach, B. de Boer, C. Videtot, P. F. van Hutten, and G. Hadziioannou, 'Semiconducting diblock copolymers synthesized by means of controlled radical

polymerization techniques', *Journal of the American Chemical Society*, vol. 122, pp. 5464–5472, 2000.
[254] J. Y. Park, S. B. Lee, Y. S. Park, Y. W. Park, C. H. Lee, J. I. Lee, and H. K. Shim, 'Doping effect of viologen on photoconductive device made of poly(*p*-phenylene vinylene)', *Applied Physics Letters*, vol. 72, pp. 2871–2873, 1998.
[255] M. A. Angadi, D. Gosztola, and M. R. Wasielewski, 'Characterization of photovoltaic cells using poly(phenylenevinylene) doped with perylenediimide electron acceptors', *Journal of Applied Physics*, vol. 83, pp. 6187–6189, 1998.
[256] K. Takahashi, K. Seto, T. Yamaguchi, J. Nakamura, C. Yokoe, and K. Murata, 'Performance enhancement by blending an electron acceptor in TiO_2/polyphenylenevinylene/Au solid-state solar cells', *Chemistry Letters*, vol. 33, pp. 1042–1043, 2004.
[257] N. Tanigaki, H. Mochizuki, X. Mo, T. Mizokuro, T. Hiraga, T. Taima, and K. Yase, 'Dye doping of poly(*p*-phenylenevinylene)s by vapor transportation for photovoltaic application', *Japanese Journal of Applied Physics*, vol. 44, pp. 630–632, 2005.
[258] E. Itoh, I. Suzuki, and K. Miyairi, 'Field emission from carbon-nanotube-dispersed conducting polymer thin film and its application to photovoltaic devices', *Japanese Journal of Applied Physics*, vol. 44, pp. 636–640, 2005.
[259] N. C. Greenham, X. Peng, and A. Alivisatos, 'A CdSe nanocrystal/MEHPPV polymer composite photovoltaic', *AIP Conference Proceedings*, vol. 404, pp. 295–301, 1997.
[260] W. J. E. Beek, M. M. Wienk, M. Kemerink, X. Yang, and R. A. J. Janssen, 'Hybrid zinc oxide conjugated polymer bulk heterojunction solar cells', *Journal of Physical Chemistry B*, vol. 109, pp. 9505–9516, 2005.
[261] A. J. Breeze, Z. Schlesinger, S. A. Carter, and P. J. Brock, 'Charge transport in TiO_2/MEHPPV polymer photovoltaics', *Physical Review B*, vol. 64, p. 125205, 2001.
[262] P. M. Sirimanne, T. Shirata, L. Damodare, Y. Hayashi, T. Soga, and T. Jimbo, 'An approach for utilization of organic polymer as a sensitizer in solid-state cells', *Solar Energy Materials and Solar Cells*, vol. 77, pp. 15–24, 2003.
[263] P. Ravirajan, S. A. Haque, J. R. Durrant, D. D. C. Bradley, and J. Nelson, 'The effect of polymer optoelectronic properties on the performance of multilayer hybrid polymer/TiO_2 solar cells', *Advanced Functional Materials*, vol. 15, pp. 609–618, 2005.
[264] A. Hagfeldt and M. Grätzel, 'Molecular photovoltaics', *Accounts of Chemical Research*, vol. 33, pp. 269–277, 2000.
[265] J. J. M. Halls, C. A. Walsh, N. C. Greenham, E. A. Marseglia, R. H. Friend, S. C. Moratti, and A. B. Holmes, 'Efficient photodiodes from interpenetrating polymer networks', *Nature*, vol. 376, pp. 498–500, 1995.
[266] G. Yu and A. J. Heeger, 'Charge separation and photovoltaic conversion in polymer composites with internal donor/acceptor heterojunctions', *Journal of Applied Physics*, vol. 78, pp. 4510–4515, 1995.
[267] J. Gao, G. Yu, and A. J. Heeger, 'Polymer p-i-n junction photovoltaic cells', *Advanced Materials*, vol. 10, pp. 692–695, 1998.
[268] S. C. Veenstra, W. J. H. Verhees, J. M. Kroon, M. M. Koetse, J. Sweelssen, J. J. A. M. Bastiaansen, et al., 'Photovoltaic properties of a conjugated polymer blend of MDMO-PPV and PCNEPV', *Chemistry of Materials*, vol. 16, pp. 2503–2508, 2004.
[269] A. J. Breeze, Z. Schlesinger, S. A. Carter, H. Tillmann, and H.-H. Hörhold, 'Improving power efficiencies in polymer-polymer blend photovoltaics', *Solar Energy Materials and Solar Cells*, vol. 83, pp. 263–271, 2004.
[270] J. Loos, X. Yang, M. M. Koetse, J. Sweelssen, H. F. M. Schoo, S. C. Veenstra, et al., 'Morphology determination of functional poly[2-methoxy-5-(3,7-dimethyloctyloxy)-

1,4-phenylenevinylene]/poly[oxa-1,1-phenylene-1,2-(1-cyanovinylene)-2-methoxy-5-(3,7-dimethyloctyloxy)-1,4-phenylene-1,2-(1-cyanovinylene)-1,4-phenylene] blends as used for all-polymer solar cells', *Journal of Applied Polymer Science*, vol. 97, pp. 1001–1007, 2005.

[271] K. Tada, K. Hosoda, M. Hirohata, R. Hidayat, T. Kawai, M. Onoda, et al., 'Donor polymer (PAT6)-acceptor polymer (CNPPV) fractal polymer networks', *Synthetic Metals*, vol. 85, pp. 1305–1306, 1997.

[272] Y. Greenwald, X. Xu, M. Fourmigué, G. Srdanov, C. Koss, F. Wudl, and A. J. Heeger, 'Polymer–polymer rectifying heterojunction based on poly(3,4-dicyanothiophene) and MEH-PPV', *Journal of Polymer Science Part A, Polymer Chemistry*, vol. 36, pp. 3115–3120, 1998.

[273] T. W. Holcombe, C. H. Woo, D. F. J. Kavaluk, B. C. Thompson, and J. M. J. Frechet, 'All-polymer photovoltaic devices of poly(3-(4-n-octyl)phenylthiophene) from Grignard metathesis (GRIM) polmerization', *Journal of the American Chemical Society*, vol. 131, pp. 14160–14161, 2009.

[274] H. Fuchigami, A. Tsumura, and H. Koezuka, 'Polythienylenevinylene thin-film transistor with high charge carrier mobility', *Applied Physics Letters*, vol. 63, pp. 1372–1374, 1993.

[275] Y. Qin and M. A. Hillmyer, 'Poly(3-hexyl-2,5-thienylene vinylene) by ADMET polymerization of a dipropenyl monomer', *Macromolecules*, vol. 42, pp. 6429–6432, 2009.

[276] R. S. Loewe and R. D. McCullogh, 'Effects of structural regularity on the properties of poly(3-alkylthienylenevinylenes)', *Chemistry of Materials*, vol. 12, pp. 3214–3221, 2000.

[277] J. Y. Kim, Y. Qin, D. M. Stevens, O. Ugurlu, V. Kalihari, M. A. Hillmyer, and C. D. Frisbie, 'Low band gap poly(thienylene vinylene)/fullerene bulk heterojunction photovoltaic cells', *Journal of Physical Chemistry C*, vol. 113, pp. 10790–10797, 2009.

[278] L. Huo, T. L. Chen, Y. Zhou, J. Hou, H.-Y. Chen, Y. Yang, and Y. Li, 'Improvement of photoluminescent and photovoltaic properties of poly(thienylene vinylene) by carboxylate substitution', *Macromolecules*, vol. 42, pp. 4377–4380, 2009.

[279] M. Al-Hashimi, M. A. Baklar, F. Colleaux, S. E. Watkins, T. D. Anthopoulos, N. Stingelin, and M. Heeney, 'Synthesis, characterization, and field-effect transistor properties of regioregular poly(3-alkyl-2,5-selenylenevinylene)', *Macromolecules*, vol. 44, pp. 5194–5199, 2011.

[280] U. H. F. Bunz, 'Poly(arylene ethynylene)s: Synthesis, properties, structures, and applications', *Chemical Reviews*, vol. 100, pp. 1605–1644, 2000.

[281] H. Häger and W. Heitz, 'Synthesis of poly(phenyleneethynylene) without diyne defects', *Macromolecular Chemistry and Physics*, vol. 199, pp. 1821–1826, 1998.

[282] W. Zhang and J. S. Moore, 'Synthesis of poly(2,5-thienyleneethynylene)s by alkyne metathesis', *Macromolecules*, vol. 37, pp. 3973–3975, 2004.

[283] A. Montali, P. Smith, and C. Weder, 'PPE-based light-emitting devices', *Synthetic Metals*, vol. 97, pp. 123–126, 1998.

[284] C. Schmitz, P. Pösch, M. Thelakkat, H.-W. Schmidt, A. Montali, K. Feldman, et al., 'Polymeric light-emitting diodes based on PPE, poly(triphenyldiamine) and spiro-quinoxaline', *Advanced Functional Materials*, vol. 11, pp. 41–46, 2001.

[285] C. E. Halkyard, M. E. Rampey, L. Kloppenburg, S. L. Studer-Martinez, and U. H. F. Bunz, 'Evidence of aggregate formation for 2,5-dialkylpoly(p-phenyleneethynylene) s in solution and films', *Macromolecules*, vol. 31, pp. 8655–8659, 1998.

[286] Z. Zhu and T. M. Swager, 'Conjugated polymers containing iptycene blocks', *Organic Letters*, vol. 3, pp. 3471–3474, 2001.
[287] C. Weder, C. Sarwa, C. Bastiaansen, and P. Smith, 'Highly polarised luminescence from oriented conjugated polymer/polyethylene blend films', *Advanced Materials*, vol. 9, pp. 1035–1039, 1997.
[288] A. Montali, C. Bastiaansen, P. Smith, and C. Weder, 'Polarising energy transfer in photoluminescent materials for display applications', *Nature*, vol. 392, pp. 261–264, 1998.
[289] C. Weder, C. Sarwa, A. Montali, C. Bastiaansen, and P. Smith, 'Incorporation of photoluminescent polarizers into liquid crystal displays', *Science*, vol. 279, pp. 835–837, 1998.
[290] J. Zheng and T. M. Swager, 'Poly(arylene ethynylene)s in chemosensing and biosensing', *Advances in Polymer Science*, vol. 177, pp. 151–179, 2005.
[291] S. W. I. Thomas, G. D. Joly, and T. M. Swager, 'Chemical sensors based on amplifying fluorescent conjugated polymers', *Chemical Reviews*, vol. 107, pp. 1339–1386, 2007.
[292] Q. Zhou and T. M. Swager, 'Fluorescent chemosensors based on energy migration in conjugated polymers: The molecular wire approach to increased sensitivity', *Journal of the American Chemical Society*, vol. 117, pp. 12593–12602, 1995.
[293] B. Erdogan, J. N. Wilson, and U. H. F. Bunz, 'Synthesis and mesoscopic order of a sugar-coated poly(*p*-phenyleneneethynylene)', *Macromolecules*, vol. 35, p. 7863, 2002.
[294] I.-B. Kim, B. Erdogan, J. N. Wilson, and U. H. F. Bunz, 'Sugar-poly(*para*-phenyleneethynylene) conjugates as sensory materials. Efficient quenching by Hg^{2+} and Pb^{2+} ions', *Chemistry – A European Journal*, vol. 10, pp. 6247–6254, 2004.
[295] J. N. Wilson, Y. Wang, J. J. Lavigne, and U. H. F. Bunz, 'A biosensing model system: Selective interaction of biotinylated PPEs with streptavidin-coated polystyrene microspheres', *Chemical Communications*, no. 14, pp. 1626–1627, 2003.
[296] J. Zheng and T. M. Swager, 'Biotinylated poly(p-phenyleenethynylene): Unexpected energy transfer results in the detection of biological analytes', *Chemical Communications*, no. 24, pp. 2798–2799, 2004.
[297] I. B. Kim, J. N. Wilson, and U. H. F. Bunz, 'Mannose-substituted PPEs detect lectins: A model for Ricin sensing', *Chemical Communications*, no. 10, pp. 1273–1275, 2005.
[298] S. A. Kushon, K. Bradford, V. Marin, C. Suhrada, B. A. Armitage, D. McBranch, and D. Whitten, 'Detection of single nucleotide mismatches via fluorescent polymer superquenching', *Langmuir*, vol. 19, pp. 6456–6464, 2003.
[299] J.-S. Yang and T. M. Swager, 'Porous shape persistent fluorescent polymer films: An approach to TNT sensory materials', *Journal of the American Chemical Society*, vol. 120, pp. 5321–5322, 1998.
[300] J.-S. Yang and T. M. Swager, 'Fluorescent porous polymer films as TNT chemosensors: Electronic and structural effects', *Journal of the American Chemical Society*, vol. 120, pp. 11864–11873, 1998.
[301] V. E. Williams and T. M. Swager, 'Iptycene-containing poly(aryleneethynylene)s', *Macromolecules*, vol. 33, pp. 4069–4073, 2000.
[302] M. J. la Grone, C. Cumming, M. E. Fisher, D. Reust, and R. Taylor, 'Land mine detection by chemical signature: Detection of vapors of nitroaromatic compounds by fluorescence quenching of novel polymer materials', *Proceedings of the Society of Photo-Optical Instrumentation Engineers*, vol. 3710, pp. 409–420, 2001.
[303] H. Dong, S. Jiang, L. Jiang, Y. Liu, H. Li, W. Hu, et al., 'Nanowire crystals of a rigid rod conjugated polymer', *Journal of the American Chemical Society*, vol. 131, pp. 17315–17320, 2009.

[304] T. Dallos, D. Beckmann, G. Brunklas, and M. Baumgarten, 'Thiaidazoloquinoxaline-acetylene containing polymers as semiconductors in ambupolar field effect transistors', *Journal of the American Chemical Society*, vol. 133, pp. 13898–13901, 2011.

[305] D. A. M. Egbe, H. Neugebauer, and N. S. Sariciftci, 'Alkoxy-substituted poly(arylelene-ethynylene)-*alt*-poly(arylene-vinylene)s: Synthesis, electroluminescence and photovoltaic applications', *Journal of Materials Chemistry*, vol. 21, pp. 1338–1349, 2011.

[306] D. A. M. Egbe, B. Carbonnier, L. Ding, D. Mühlbacher, E. Birckner, T. Pakula, et al., 'Supramolecular ordering, thermal behavior and photophysical, electrochemical, and electroluminescent properties of alkoxy-substituted yne-containing poly (phenylene-vinylene)s', *Macromolecules*, vol. 37, pp. 7451–7463, 2004.

[307] P. A. Troshin, O. A. Mukhacheva, Ö. Usluer, A. E. Goryachev, A. V. Akkuratov, D. K. Susarova, et al., 'Improved photovoltaic performance of PPV-based copolymers using optimized fullerene-based counterparts', *Advanced Energy Materials*, vol. 3, pp. 161–166, 2013.

[308] C. Kästner, C. Ulbricht, D. A. M. Egbe, and H. Hoppe, 'Polymer BHJ solar cell performance tuning by C60 fullerene derivative alkyl side-chain length', *Journal of Polymer Science: Part B: Polymer Physics*, vol. 50, pp. 1562–1566, 2012.

[309] O. Usluer, S. Boudiba, D. A. M. Egbe, L. Hirsch, and M. Abbas, 'Control of carrier mobilities for performance enhancement of anthracene-based polymer solar cells', *RSC Advances*, vol. 5, pp. 50668–50672, 2015.

[310] A. M. Ramos, M. T. Rispens, J. K. J. van Duren, J. C. Hummelen, and R. A. J. Janssen, 'Photoinduced electron transfer and photovoltaic devices of a conjugated polymer with pendant fullerenes', *Journal of the American Chemical Society*, vol. 123, pp. 6714–6715, 2001.

[311] D. A. M. Egbe, T. Kietzke, B. Carbonnier, D. Mühlbacher, H.-H. Hörhold, D. Neher, and T. Pakula, 'Synthesis, characterization, and photophysical, electrochemical, electroluminescent and hotovoltaic properties of yne-containg CN-PPVs', *Macromolecules*, vol. 37, pp. 8863–8873, 2004.

[312] P. Kovacic and M. B. Jones, 'Oxdiative coupling of arylenes', *Chemical Reviews*, vol. 87, pp. 357–379, 1987.

[313] T. Yamamoto, 'Electrically conducting and thermally stable π-conjugated poly (arylene)s prepared by organometallic processes', *Progress in Polymer Science*, vol. 17, pp. 1153–1205, 1992.

[314] G. Grem, V. Martin, F. Meghdadi, C. Paar, J. Stampfl, J. Sturm, et al., 'Stable poly (para-phenylene)s and their application in organic light emitting devices', *Synthetic Metals*, vol. 71, pp. 2193–2194, 1995.

[315] D. G. H. Ballard, A. Courtis, I. M. Shirley, and S. C. Taylor, 'A biotech route to polyphenylene', *Journal of the Chemical Society, Chemical Communications*, no. 17, pp. 954–955, 1983.

[316] D. G. H. Ballard, A. Courtis, I. M. Shirley, and S. C. Taylor, 'Synthesis of polyphenylene from a cis-dihydrocatechol, a biologically produced monomer', *Macromolecules*, vol. 21, pp. 294–304, 1988.

[317] D. L. Gin, V. P. Conticello, and R. H. Grubbs, 'Transition-metal-catalyzed polymerization of heteroatom-functionalized cyclohexadienes: Stereoregular precursors to poly(p-phenylene)', *Journal of the American Chemical Society*, vol. 114, pp. 3167–3169, 1992.

[318] D. L. Gin, V. P. Conticello, and R. H. Grubbs, 'Stereoregular precursors to poly(p-phenylene) via transition-metal-catalyzed polymerization. 1. Precursor design and synthesis', *Journal of the American Chemical Society*, vol. 116, pp. 10507–10519, 1994.

[319] D. L. Gin, V. P. Conticello, and R. H. Grubbs, 'Stereoregular precursors to poly(p-phenylene) via transition-metal-catalyzed polymerization. 2. The effects of polymer stereochemistry and acid catalysts on precursor aromatization: A characterization study', *Journal of the American Chemical Society*, vol. 116, pp. 10934–10947, 1994.

[320] D. L. Gin, J. K. Avlyanov, and A. G. MacDiarmid, 'Synthesis and processing of poly(p-phenylene) via the phosphoric acid-catalyzed pyrolysis of a stereoregular precursor polymer: a characterization study', *Synthetic Metals*, vol. 66, pp. 169–175, 1994.

[321] G. Grem, G. Leditzky, B. Ullrich, and G. Leising, 'Realization of a blue-light-emitting device using poly(p-phenylene)', *Advanced Materials*, vol. 4, pp. 36–37, 1992.

[322] U. Scherf, 'Oligo-and polyarylenes, oligo- and polyarylenevinylenes', *Topics in Current Chemistry*, vol. 201, pp. 163–222, 1999.

[323] A. C. Grimsdale and K. Müllen, 'Polyphenylene-type emissive materials: Polyphenylenes, polyfluorenes and ladder polymers', *Advances in Polymer Science*, vol. 199, pp. 1–82, 2006.

[324] A. D. Schlüter, 'The tenth anniversary of Suzuki polycondensation', *Journal of Polymer Science A: Polymer Chemistry*, vol. 39, pp. 1533–1556, 2001.

[325] J. Sakamoto, M. Rehahn, G. Wegner, and A. D. Schlüter, 'Suzuki polycondensation: Polyarylenes a la carte', *Macromolecular Rapid Communications*, vol. 30, pp. 653–687, 2009.

[326] N. Tanigaki, H. Masuda, and K. Kaeriyama, 'Substituted polyphenylenes from diheptylbenzene and aryldiboronates', *Polymer*, vol. 38, pp. 1221–1226, 1997.

[327] B. S. Nehls, U. Asawapirom, S. Füldner, E. Preis, T. Farrell, and U. Scherf, 'Semiconducting polymers via microwave-assisted Suzuki and Stille cross-coupling reactions', *Advanced Functional Materials*, vol. 14, pp. 352–356, 2004.

[328] Y. Yang, Q. Pei, and A. J. Heeger, 'Efficient blue polymer light-emitting diodes from a series of soluble poly(paraphenylene)s', *Journal of Applied Physics*, vol. 79, pp. 934–939, 1996.

[329] V. Cimrová, M. Remmers, D. Neher, and G. Wegner, 'Polarized light emission from LEDs prepared by the Langmuir-Blodgett technique', *Advanced Materials*, vol. 8, pp. 146–149, 1996.

[330] V. Cimrová, W. Schmidt, R. Rulkens, M. Schulze, W. Meyer, and D. Neher, 'Efficient blue light emitting devices based on rigid-rod polyelectrolytes', *Advanced Materials*, vol. 8, pp. 585–588, 1996.

[331] M. Remmers, B. Müller, K. Martin, H.-J. Räder, and W. Köhler, 'Synthesis, optical properties and analysis of PPPs', *Macromolecules*, vol. 32, pp. 1073–1079, 1999.

[332] J. L. Reddinger and J. R. Reynolds, 'UV-emitting poly(m-phenylene)s', *Macromolecules*, vol. 30, pp. 479–481, 1997.

[333] V. Cimrová, D. Výprachtický, J. Pecka, and R. Kotva, 'OLEDs based on novel blends', *Proceedings of the Society of Photo-Optical Instrumentation Engineers*, vol. 3939, pp. 164–171, 2000.

[334] U. Scherf, 'Ladder-type materials', *Journal of Materials Chemistry*, vol. 9, pp. 1853–1864, 1999.

[335] J. Grimme, M. Kreyenschmidt, F. Uckert, K. Müllen, and U. Scherf, 'On the conjugation length in poly(para-phenylene)-type polymers', *Advanced Materials*, vol. 7, pp. 292–295, 1995.

[336] P. Lu, H. Zhang, F. Shen, B. Yang, D. Li, Y. Ma, et al., 'A wide bandgap semiconducting polymer for ultraviolet and blue light-emitting diodes', *Macromolecular Chemistry and Physics*, vol. 204, pp. 2274–2280, 2003.

[337] U. Scherf and K. Müllen, 'Poly(arylene)s and poly(arylenevinylene)s. 7. A soluble ladder polymer via bridging of functionalised poly(p-phenylene)-precursors', *Makromolecular Chemie Rapid Communications*, vol. 12, pp. 489–497, 1991.
[338] U. Scherf and K. Müllen, 'Poly(arylene)s and poly(arylenevinylene)s. 11. A modified two-step route to soluble phenylene-type ladder polymers', *Macromolecules*, vol. 25, pp. 3546–3548, 1992.
[339] U. Scherf and K. Müllen, 'The synthesis of ladder polymers', *Advances in Polymer Science*, vol. 123, pp. 1–40, 1995.
[340] P. Hickl, M. Ballauff, U. Scherf, K. Müllen, and P. Lindner, 'Characterization of a ladder polymer by small-angle X-ray and neutron scattering', *Macromolecules*, vol. 30, pp. 273–279, 1997.
[341] G. Petekidis, G. Fytas, U. Scherf, K. Müllen, and G. Fleischer, 'Dynamics of polyphenylene ladder polymers in solution', *Journal of Polymer Science (A), Polymer Chemistry*, vol. 37, pp. 2211–2220, 1999.
[342] E. Somma, B. Loppinet, G. Fytas, S. Setayesh, J. Jacob, A. C. Grimsdale, and K. Müllen, 'Collective orientation dynamics in semi-rigid polymers', *Colloid and Polymer Science*, vol. 282, pp. 867–873, 2004.
[343] F. Schindler, J. Jacob, A. C. Grimsdale, U. Scherf, K. Müllen, J. M. Lupton, and J. Feldmann, 'Counting chromophores in conjugated polymers', *Angewandte Chemie, International Edition*, vol. 44, pp. 1520–1525, 2005.
[344] W. Graupner, G. Grem, F. Meghdadi, C. Paar, G. Leising, U. Scherf, et al., 'Electroluminesence with conjugated polymers and oligomers', *Molecular Crystals and Liquid Crystals*, vol. 256, pp. 549–554, 1994.
[345] J. Stampfl, W. Graupner, G. Leising, and U. Scherf, 'Photoluminescence and UV-VIS absorption study of poly(para-phenylene)-type ladder polymers', *Journal of Luminescence*, vol. 63, pp. 117–123, 1995.
[346] G. Grem and G. Leising, 'Electroluminescence of "wide-bandgap" chemically tunable cyclic conjugated polymers', *Synthetic Metals*, vol. 55–57, pp. 4105–4110, 1993.
[347] J. Hüber, K. Müllen, J. Salbeck, H. Schenk, U. Scherf, T. Stehlin, and R. Stern, 'Blue light-emitting diodes based on ladder polymers of the PPP type', *Acta Polymerica*, vol. 45, pp. 244–247, 1994.
[348] J. Grüner, H. F. Wittmann, P. J. Hamer, R. H. Friend, J. Huber, U. Scherf, et al., 'Electroluminescence and photoluminescence investigations of the yellow emission of devices based on ladder-type oligo(para-phenylene)s', *Synthetic Metals*, vol. 67, pp. 181–185, 1994.
[349] U. Lemmer, S. Heun, R. F. Mahrt, U. Scherf, M. Hopmeier, U. Siegner, et al., 'Aggregate fluorescence in conjugated polymers', *Chemical Physics Letters*, vol. 240, pp. 373–380, 1995.
[350] A. Köhler, J. Grüner, R. H. Friend, K. Müllen, and U. Scherf, 'Photocurrent measurements on aggregates in ladder-type polyphenylene', *Chemical Physics Letters*, vol. 243, pp. 456–461, 1995.
[351] X. Yang, Y. Hou, Z. Wang, X. Chen, Z. Xu, and X. Xu, 'Blue polymer light-emitting diode based on ladder-type polyphenylene', *Thin and Solid Films*, vol. 363, pp. 211–213, 2000.
[352] U. Scherf, A. Bohnen, and K. Müllen, 'The oxidized states of a (1,4-phenylene) ladder polymer', *Makromolekulare Chemie*, vol. 193, pp. 1127–1133, 1992.
[353] S. Tasch, A. Niko, G. Leising, and U. Scherf, 'Highly efficient electroluminescence of new wide band gap ladder-type poly(para-phenylenes)', *Applied Physics Letters*, vol. 68, pp. 1090–1092, 1996.

[354] A. Haugeneder, U. Lemmer, and U. Scherf, 'Exciton diffussion dynamics in a conjugated polymer containing aggregate states', *Chemical Physics Letters*, vol. 351, pp. 354–358, 2002.
[355] J. M. Lupton, 'On-chain defect emission in conjugated polymers', *Chemical Physics Letters*, vol. 365, pp. 366–368, 2002.
[356] J. M. Lupton, A. Pogantsch, T. Piok, E. W. J. List, S. Patil, and U. Scherf, 'Intrinsic room-temperature electrophosphorescence from a pi-conjugated polymer', *Physical Review Letters*, vol. 89, pp. 7401–7404, 2002.
[357] S. Qiu, P. Lu, X. Liu, F. Shen, L. Liu, Y. Ma, and J. Shen, 'New ladder-type poly(p-phenylene)s containing fluorene units exhibiting highly efficient electroluminescence', *Macromolecules*, vol. 36, pp. 9823–9829, 2003.
[358] G. Grem, C. Paar, J. Stampfl, G. Leising, J. Huber, and U. Scherf, 'Soluble segmented stepladder poly(p-phenylenes) for blue-light-emitting diodes', *Chemistry of Materials*, vol. 7, pp. 2–4, 1995.
[359] J. Grüner, P. J. Hamer, R. H. Friend, H.-J. Huber, U. Scherf, and A. B. Holmes, 'A high efficiency blue-light-emitting diode based on novel ladder poly(p-phenylene)s', *Advanced Materials*, vol. 6, pp. 748–752, 1994.
[360] J. Grüner, R. H. Friend, J. Huber, and U. Scherf, 'A blue-luminescent ladder-type poly(para-phenylene) copolymer containing oxadiazole groups', *Chemical Physics Letters*, vol. 251, pp. 204–210, 1996.
[361] M. Forster and U. Scherf, 'Strongly fluorescent ethylene-bridged poly(para-phenylene) ladder polymers', *Macromolecular Rapid Communications*, vol. 21, pp. 810–813, 2000.
[362] K. Chmil and U. Scherf, 'A simple two-step syntheis of a novel, fully aromatic ladder-type polymer', *Makromolekulare Chemie, Raid Communications*, vol. 14, pp. 217–222, 1993.
[363] K. Chmil and U. Scherf, 'Conjugated all-carbon ladder polymers: Improved solubility and molecular weights', *Acta Polymerica*, vol. 48, pp. 208–211, 1997.
[364] S. Kirstein, G. Cohen, D. Davidov, U. Scherf, M. Klapper, K. Chmil, and K. Müllen, 'Transient and DC electroluminescence of some new conjugated polymers', *Synthetic Metals*, vol. 69, pp. 415–418, 1995.
[365] M. E. Kleybolte, S. I. Vagin, and B. Rieger, 'A polymer lost in the shuffle: The perspective of poly(para)phenylenes', *Macromolecular Chemistry and Physics*, vol. 324, p. 2200441, 2023.
[366] C. Li, M. S. Liu, N. G. Pschirer, M. Baumgarten, and K. Müllen, 'Polyphenylene-based materials for organic photovoltaics', *Chemical Reviews*, vol. 110, no. 11, pp. 6817–6855, 2010.
[367] A. Nejim and G. Carter, 'Ion implantation induced conductivity changes in polyphenylene', *Nuclear Instruments and Methods in Physics Research, B*, vol. 61, no. 4, pp. 502–508, 1991
[368] H. D. Sehgal, Y. Pratap, and S. Kabra, 'Detection of poisonous gases using JL FinFET with conducting gate polymer,' in *2022 6th International Conference on Devices, Circuits and Systems (ICDCS)*, Coimbatore, India: IEEE, pp. 440–444, 2022, DOI: 10.1109/ICDCS54290.2022.9780853.
[369] F. Mehrdad and Z. Ahangari, 'Design and simulation of a gas sensitive junctionless FinFET based on conducting polymer as the gate material', *Physica Scripta*, vol. 97, no. 7, p. 075805, 2022.
[370] C. Rothe, C.-J. Chiang, V. Jankus, K. Abdullah, X. Zeng, R. Jitchati, et al., 'Ionic iridium (III) complexes with bulky side groups for use in light emitting cells: Reduction of concentration quenching', *Advanced Functional Materials*, vol. 19, no. 13, pp. 2038–2044, 2009.

[371] D. Neher, 'Polyfluorene homopolymers', *Macromolecular Rapid Communications*, vol. 22, pp. 1365–1385, 2001.

[372] U. Scherf and E. W. J. List, 'Polyfluorenes: Towards reliable structure–activity relationships', *Advanced Materials*, vol. 14, pp. 477–487, 2002.

[373] A. C. Grimsdale and K. Müllen, 'Oligomers and polymers based on bridged phenylenes as electronic materials'. *Macromolecular Rapid Communications*, vol. 28, pp. 1676–1702, 2007.

[374] A. C. Grimsdale and K. Müllen, 'Bridged polyphenylenes: From polyfluorenes to ladder polymers'. *Advances in Polymer Science*, vol. 212, pp. 1–48, 2008.

[375] M. Fukuda, K. Sawada, and K. Yoshino, 'Fusible conducting poly(9-alkylfluorene) and poly(9,9-dialkylfluorene) and their characteristics', *Japanese Journal of Applied Physics*, vol. 28, pp. L1433–L1435, 1989.

[376] M. Fukuda, K. Sawada, and K. Yoshino, 'Synthesis of fusible and soluble conducting polyfluorene derivatives and their characteristics'. *Journal of Polymer Science, Polymer Chemistry*, vol. 31, pp. 2465–2471, 1993.

[377] Y. Ohmori, K. Uchida, K. Muro, and K. Yoshino, 'Blue electroluminescent diodes utilizing poly(alkylfluorene)', *Japanese Journal of Applied Physics*, vol. 30, pp. L1941–L1943, 1991.

[378] Y. Ohmori, M. Uchida, C. Morishima, A. Fujii, and K. Yoshino, 'Enhancement of emission efficiency in electroluminescent diode utilizing vapor-deposited poly(alkylfluorene)', *Japanese Journal of Applied Physics*, vol. 32, pp. L1663–L1666, 1993.

[379] M. Uchida, Y. Ohmori, C. Morishima, and K. Yoshino, 'Visible and blue electroluminescent diodes utilizing poly(3-alkylthiophene)s and poly(alkylfluorene)s', *Synthetic Metals*, vol. 57, pp. 4168–4173, 1993.

[380] E. P. Woo, M. Inbasekaran, W. Shiang, and G. R. Roof, 'Polydialkylfluorenes by Suzuki coupling', *Chemical Abstracts*, 126, 225700y, p. WO97/05184, 1997.

[381] M. Inbasekaran, W. Wu, and E. P. Woo, 'Polydialkylfluorene copolymers by Suzuki', *Chemical Abstracts*, 129, 123010t, p. 5777070, 1998.

[382] M. Inbasekaran, E. Woo, M. Bernius, and L. Wujkowski, 'Fluorene homopolymers and copolymers', *Synthetic Metals*, vol. 111–112, pp. 397–401, 2000.

[383] M. Bernius, M. Inbasekaran, J. O'Brien, and W. Wu, 'Progress with light-emitting polymers', *Advanced Materials*, vol. 12, pp. 1737–1750, 2000.

[384] M. Bernius, M. Inbasekaran, E. Woo, W. Wu, and L. Wujkowski, 'Fluorene-based polymers: Synthesis and applications', *Journal of Materials Science; Materials for Electronics*, vol. 11, pp. 111–116, 2000.

[385] M. Bernius, M. Inbasekaran, E. Woo, W. Wu, and L. Wujkowski, 'LEDs based on fluorene polymers', *Thin and Solid Films*, vol. 363, pp. 55–57, 2000.

[386] A. W. Grice, D. D. C. Bradley, M. T. Bernius, M. Inbasekaran, W. W. Wu, and E. P. Woo, 'High brightness and efficiency blue polymer light-emitting diodes', *Applied Physics Letters*, vol. 73, pp. 629–631, 1998.

[387] M. Grell, et al., 'Blue polarized electroluminescence from liquid crystalline polyfluorene', *Advanced Materials*, vol. 11, pp. 671–675, 1999.

[388] G. Klaerner and R. D. Miller, 'Well-defined oligomers of poly(9,9-dihexylfluorene)', *Macromolecules*, vol. 31, pp. 2007–2009, 1998.

[389] Q. Wang, Y. Qu, H. Tian, Y. Geng, and F. Wang, 'Iterative binomial synthesis of monodisperse polyfluorenes up to 64-mers and trheir chain-length-dependent properties', *Macromolecules*, vol. 44, pp. 1256–1260, 2011.

[390] M. Grell, D. D. C. Bradley, M. Inbasekaran, and E. P. Woo, 'Liquid crystalline polymer for polarized electroluminescence applications', *Advanced Materials*, vol. 9, pp. 798–802, 1997.

[391] J. Teetsov and M. A. Fox, 'Photophysical characterisation of polydialkylfluorene films and solutions', *Journal of Materials Chemistry*, vol. 9, pp. 2117–2122, 1999.

[392] H.-G. Nothofer, A. Meisel, T. Miteva, D. Neher, M. Forster, M. Oda, et al., 'Liquid crystalline polyfluorenes for blue polarised electroluminescence', *Macromolecular Symposia*, vol. 154, pp. 139–146, 2000.

[393] K. S. Whitehead, M. Grell, D. D. C. Bradley, M. Inbasekaran, and E. P. Woo, 'Polarized emission from liquid crystalline polymers', *Synthetic Metals*, vol. 111–112, pp. 181–185, 2000.

[394] K. S. Whitehead, M. Grell, D. D. C. Bradley, M. Jandke, and P. Strohriegl, 'Highly polarized blue electroluminescence from aligned polydialkylfluorene films', *Applied Physics Letters*, vol. 76, pp. 2946–2948, 2000.

[395] M. Oda, S. C. J. Meskers, H. G. Nothofer, U. Scherf, and D. Neher, 'Chiroptical properties of chiral-substituted polyfluorenes', *Synthetic Metals*, vol. 111–112, pp. 575–577, 2000.

[396] M. Oda, H.-G. Nothofer, G. Lieser, U. Scherf, S. C. J. Meskers, and D. Neher, 'Circularly polarized electroluminescence from liquid crystalline chiral polyfluorenes', *Advanced Materials*, vol. 12, no. 5, pp. 362–365, 2000.

[397] A. C. Grimsdale, 'In search of stable blue emission from phenylene-based conjugated polymers', *Current Organic Chemistry*, vol. 14, pp. 2196–2217, 2010.

[398] J. M. Lupton, M. R. Craig, and E. W. Meijer, 'On-chain defect emission in electroluminescent polyfluorenes', *Applied Physics Letters*, vol. 80, pp. 4489–4491, 2002.

[399] E. W. J. List, R. Guentner, P. Scanducci de Freitas, and U. Scherf, 'Effect of keto defect sites on emission of polyfluorenes', *Advanced Materials*, vol. 14, pp. 374–378, 2002.

[400] L. Ilharco, A. R. Garcia, J. L. da Silva, M. J. Lemos, and L. F. Vieira Ferreira, 'Ultraviolet-visible and Fourier transform infrared diffuse reflectance studies of benzophenone and fluorenone adsorped onto microcrystalline cellulose', *Langmuir*, vol. 13, pp. 3787–3793, 1997.

[401] J. Pei, J. Ni, X.-H. Zhou, X.-Y. Cao, and Y.-H. Lai, 'Head-to-tail regioregular oligothiophene functionalized spirobifluorene derivatives', *Journal of Organic Chemistry*, vol. 67, pp. 4924–4936, 2002.

[402] M. Gaal, E. J. W. List, and U. Scherf, 'Excimers or emissive on-chain defects', *Macromolecules*, vol. 36, pp. 4236–4238, 2003.

[403] G. Zeng, W.-L. Yu, S.-J. Chua, and W. Huang, 'Spectral and thermal stability study for fluorene-based conjugated polymers', *Macromolecules*, vol. 35, pp. 6907–6914, 2002.

[404] X. Gong, P. K. Iyer, D. Moses, G. C. Bazan, A. J. Heeger, and S. S. Xiao, 'Stabilized blue emission from polyfluorene-based LEDs: Elimination of fluorenone defects', *Advanced Functional Materials*, vol. 13, pp. 325–330, 2003.

[405] K. L. Chan, M. Sims, S. I. Pascu, M. Ariu, A. B. Holmes, and D. D. C. Bradley, 'Understanding the nature of the states responsible for green emission in oxidised PDAFs', *Advanced Functional Materials*, vol. 19, pp. 2147–2154, 2009.

[406] K.-H. Weinfurtner, H. Fujikawa, S. Tokito, and Y. Taga, 'Influence of molecular weight distribution on aggregation tendency of polydialkylfluorenes', *Applied Physics Letters*, vol. 76, pp. 2502–2504, 2000.

[407] D. Sainova, T. Miteva, H. G. Nothofer, U. Scherf, I. Glowacki, J. Ulanski, et al., 'Control of colour and efficiency of light-emitting diodes using polydialkylfluorene and triphenylamine blend', *Applied Physics Letters*, vol. 76, pp. 1810–1812, 2000.

[408] M. R. Sirtonski, S. L. McFarlane, and J. G. C. Veinot, 'Stabilizing the optical properties of PFO through addition of a low molecular weight aromatic ether', *Journal of Materials Chemistry*, vol. 20, pp. 8147–8152, 2010.

[409] C. Ego, et al., 'Attaching perylene dyes to polyfluorene: Three simple efficient methods for facile colour tuning of light-emitting polymers', *Journal of the American Chemical Society*, vol. 125, pp. 437–443, 2003.
[410] G. Klärner, M. H. Davey, W.-D. Chen, J. C. Scott, and R. D. Miller, 'Colourfast blue light-emitting dihexylfluorene copolymers', *Advanced Materials*, vol. 10, pp. 993–997, 1998.
[411] G. Klärner, J.-I. Lee, M. H. Davey, and R. D. Miller, 'Exciton migration and trapping in polydialkylfluorene copolymers', *Advanced Materials*, vol. 11, pp. 115–118, 1999.
[412] R. D. Miller, et al., 'Thermally stable polyfluorene copolymers for blue light emission', *Molecular Crystals and Liquid Crystals, Part B, Nonlinear Optics*, vol. 20, pp. 269–295, 1999.
[413] X. Guo, Y. Cheng, Z. Xie, Y. Geng, L. Wang, X. Jing, and F. Wang, 'Fluorene-based copolymers containing dinaphto-s-indacene as new building blocks for high-efficiency and colour-stable blue LEDs', *Macromolecular Rapid Communications*, vol. 30, pp. 816–825, 2009.
[414] P. Lu, H. Q. Zhang, Y. Zheng, Y. G. Ma, G. Zhang, X. F. Chen, and J. C. Shen, 'New ultraviolet-emissive wide bandgap semiconductive polymers', *Synthetic Metals*, vol. 135–136, pp. 205–206, 2003.
[415] S. F. Lim, R. H. Friend, I. D. Rees, J. Li, Y. G. Ma, K. Robinson, et al., 'Suppression of green emission in a new class of blue-emitting PF copolymers with twisted biphenyl moieties', *Advanced Functional Materials*, vol. 15, pp. 981–988, 2005.
[416] H. Zhang, E. Guo, Y. Fang, P. Ren, and W. Yang, 'Synthesis and optoelectronic properties of alternating benzofuran/terfluorene copolymer with stable blue emission', *Journal of Polymer Science (A), Polymer Chemistry*, vol. 47, pp. 5488–5497, 2008.
[417] C. Xia and R. C. Advincula, 'Decreased aggregation in polyfluorenes by introducing carbazole copolymer units', *Macromolecules*, vol. 34, pp. 5854–5859, 2001.
[418] J. S. Park, M. Song, S.-H. Jin, J. W. Lee, C. W. Lee, and J.-S. Gal, 'Poly(carbazole)-based copolymers containing deep-blue-chromophores for polymer light-emitting diode applications', *Macromolecular Chemistry and Physics*, vol. 210, pp. 1572–1578, 2009.
[419] G. Klärner, J.-I. Lee, V. Y. Lee, E. Chan, J.-P. Chen, A. Nelson, et al., 'Cross-linkable polymers based on polydialkylfluorenes', *Chemistry of Materials*, vol. 11, pp. 1800–1805, 1999.
[420] S. Setayesh, A. C. Grimsdale, T. Weil, V. Enkelmann, K. Müllen, F. Meghdadi, et al., 'Polyfluorenes with polyphenylene dendron side chains: Toward non-aggregating, light-emitting polymers', *Journal of the American Chemical Society*, vol. 123, pp. 946–953, 2001.
[421] D. Marsitzky, R. Vestberg, P. Blainey, B. T. Tang, C. J. Hawker, and K. R. Carter, 'Self-encapsulation of polyfluorenes in a dendrimer matrix', *Journal of the American Chemical Society*, vol. 123, pp. 6965–6972, 2001.
[422] J. Lee, H.-J. Cho, B.-J. Jung, N. S. Cho, and H.-K. Shim, 'Stabilized blue luminescent polyfluorenes: Introducing polyhedral oligomeric silsesquioxane', *Macromolecules*, vol. 37, pp. 8523–8529, 2004.
[423] K. Takagi, S. Kunii, and Y. Yuki, 'Synthesis and photophysical properties of polyfluorenes bearing Si-based functional groups', *Journal of Polymer Science (A), Polymer Chemistry*, vol. 43, pp. 2119–2127, 2005.
[424] M. R. Craig, M. M. de Kok, J. W. Hofstraat, A. P. H. J. Schenning, and E. W. Meijer, 'Improving color purity and stability in a blue emitting polyfluorene by monomer purification', *Journal of Materials Chemistry*, vol. 13, pp. 2861–2862, 2003.

[425] S. Y. Cho, A. C. Grimsdale, D. J. Jones, S. E. Watkins, and A. B. Holmes, 'Polyfluorenes without monoalkylfluorene defects', *Journal of the American Chemical Society*, vol. 129, pp. 11910–11911, 2007.

[426] R. Abbel, M. Wolffs, R. A. A. Bovee, J. L. J. van Dongen, X. Lou, O. Henze, et al., 'Side-chain degradation of ultrapure π-conjugted oligomers: Implications for organic electronics', *Advanced Materials*, vol. 21, pp. 597–602, 2009.

[427] J. Jacob, L. Oldridge, J. Zhang, M. Gaal, E. J. W. List, A. C. Grimsdale, and K. Müllen, 'Progress towards stable blue light-emitting polymers', *Current Applied Physics*, vol. 3, pp. 339–342, 2004.

[428] A. F. Pogantsch, F. P. Wenzl, E. W. J. List, G. Leising, A. C. Grimsdale, and K. Müllen, 'Polyfluorenes with dendron sidechains as the active materials for polymer light-emitting diodes', *Advanced Materials*, vol. 14, pp. 1061–1064, 2002.

[429] E. J. W. List-Kratochvil, A. Pogantsch, F. P. Wenzl, C.-H. Kim, J. Shinar, M. A. Loi, et al., 'A comparative study of the photophysics in polyfluorenes with alkyl and dendron sidechains', *Materials Research Society Symposia Proceedings*, vol. 665, pp. C5.47.1–C5.47.6, 2001.

[430] J. M. Lupton, P. Schouwink, P. E. Keivanidis, A. C. Grimsdale, and K. Müllen, 'Influence of dendronisation on spectral diffusion and aggregation in conjugated polymers', *Advanced Functional Materials*, vol. 13, pp. 154–158, 2003.

[431] A. Pogantsch, C. Gadermaier, G. Cerullo, G. Lanzani, U. Scherf, A. C. Grimsdale, et al., 'Photophysics of poly(fluorene)s with dendronic side chains', *Synthetic Metals*, vol. 139, pp. 847–849, 2003.

[432] A. Pogantsch, F. P. Wenzl, U. Scherf, A. C. Grimsdale, K. Müllen, and E. J. W. List, 'Long lived photoexcitation dynamics in a dendronically substituted poly(fluorene)', *Journal of Chemical Physics*, vol. 119, pp. 6904–6910, 2003.

[433] C. Ego, A. C. Grimsdale, F. Uckert, G. Yu, G. Srdanov, and K. Müllen, 'Triphenylamine-substituted polyfluorene: A stable blue-emitter with improved charge injection for light-emitting diodes', *Advanced Materials*, vol. 14, pp. 809–811, 2002.

[434] J.-H. Lee and D.-H. Hwang, 'Alkoxyphenyl-substituted polyfluorene: A stable blue-light emitting polymer with good solution processability', *Chemical Communications*, no. 22, pp. 2836–2837, 2003.

[435] M. Surin, E. Hennebicq, C. Ego, D. Marsitzky, A. C. Grimsdale, K. Müllen, et al., 'Correlation between the microscopic morphology and the solid-state photoluminescence properties in fluorene-based polymers and copolymers', *Chemistry of Materials*, vol. 16, pp. 994–1001, 2004.

[436] F.-I. Wu, S. Reddy, C.-F. Shu, M. S. Liu, and A. K.-Y. Jen, 'Novel oxadiazole containing polyfluorene with efficient blue electroluminescence', *Chemistry of Materials*, vol. 15, pp. 269–274, 2003.

[437] C.-F. Shu, R. Dodda, F.-I. Wu, M. S. Liu, and A. K.-Y. Jen, 'Highly efficient blue light-emitting diodes from polyfluorene contining bipolar pendant groups', *Macromolecules*, vol. 36, pp. 6698–6703, 2003.

[438] J. Salbeck, 'Electroluminescence with organic compounds', *Berichte der Bunsengesellschaft Physical Chemistry*, vol. 100, pp. 1666–1677, 1996.

[439] Y. Wu, J. Li, Y. Fu, and Z. Bo, 'Synthesis of extremely stable blue light emitting poly(spirobifluorene)s by Suzuki polycondensation', *Organic Letters*, vol. 6, pp. 3485–3487, 2004.

[440] W.-L. Yu, J. Pei, W. Huang, and A. J. Heeger, 'Spiro-functionalised polyfluorenes as light emitting materials', *Advanced Materials*, vol. 12, pp. 828–831, 2000.

[441] R. Grisorio, P. Mastrorilli, C. F. Nobile, G. Romanazzi, G. P. Suranna, D. Acierno, and E. Amendola, 'New spiro-functionalized polyfluorenes: Synthesis and properties', *Macromolecular Chemistry and Physics*, vol. 206, pp. 448–455, 2005.
[442] D. Vak, C. Chun, C. L. Lee, J.-J. Kim, and D.-Y. Kim, 'Novel spiro-functionalized polyfluorene derivative with solubilizing sidechains', *Journal of Materials Chemistry*, vol. 14, pp. 1342–1346, 2004.
[443] C. D. Müller, A. Falcou, N. Reckefuss, M. Rojahn, V. Wiederhirn, P. Rudati, et al., 'Multi-colour organic light-emitting diodes by solution processing', *Nature*, vol. 421, pp. 829–833, 2003.
[444] M. Redecker, D. D. C. Bradley, M. Inbasekaran, and E. P. Woo, 'Nondispersive hole transport in an electroluminescent polyfluorene', *Applied Physics Letters*, vol. 73, pp. 1565–1567, 1998.
[445] P. Sonar, A. C. Grimsdale, M. Heeney, M. Shkunov, I. McCulloch, and K. Müllen, 'A study of the effects metal residues in poly(dioctylfluorene) have on field-effect transistor device characteristics', *Synthetic Metals*, vol. 157, pp. 872–875, 2007.
[446] H. Sirringhaus, R. J. Wilson, R. H. Friend, M. Inbasekaran, W. Wu, E. P. Woo, et al., 'Mobility enhancement in conjugated polymer field-effect transistors through chain alignment in a liquid-crystalline phase', *Applied Physics Letters*, vol. 77, pp. 406–408, 2000.
[447] H. Sirringhaus, T. Kawase, R. H. Friend, T. Shimoda, M. Inbasekaran, W. Wu, and E. P. Woo, 'High-resolution inkjet printing of all-polymer transistor circuits', *Science*, vol. 290, pp. 2123–2126, 2000.
[448] M. Surin, P. Sonar, A. C. Grimsdale, K. Müllen, R. Lazzaroni, and P. Leclere, 'Supramolecular organization in fluorene/indenofluorene-oligothiophene alternating conjugated coplymers', *Advanced Functional Materials*, vol. 15, pp. 1426–1434, 2005.
[449] J. J. M. Halls, A. C. Arias, J. D. MacKenzie, W. Wu, M. Inbasekaran, E. P. Woo, and R. H. Friend, 'Photodiodes based on polyfluorene composites: Influence of morphology', *Advanced Materials*, vol. 12, pp. 498–502, 2000.
[450] M. A. Stevens, C. Silva, D. M. Russell, and R. H. Friend, 'Exciton dissociation mechanisms in the polymeric semiconductors poly(9,9-dioctylfluorene) and poly (9,9-dioctylfluorene-co-benzothiadiazole)', *Physical Review B*, vol. 63, p. 165213, 2001.
[451] A. C. Arias, J. D. MacKenzie, R. Stevenson, J. J. M. Halls, M. Inbasekaran, E. P. Woo, et al., 'Photovoltaic performance and morphology of polyfluorene blends: A combined microscopic and photovoltaic investigation', *Macromolecules*, vol. 34, pp. 6005–6013, 2001.
[452] C. R. McNeill, H. Frohne, J. L. Holdsworth, and P. C. Dastoor, 'Near-field scanning photocurrent measurements of polyfluorene blend devices: Directly correlating morphology with current generation', *Nano Letters*, vol. 4, pp. 2503–2507, 2004.
[453] C. R. McNeill, B. Watts, L. Thomsen, W. J. Belcher, N. C. Greenham, and P. C. Dastoor, 'Nano-scale quantitative chemical mapping of conjugated polymer blends', *Nano Letters*, vol. 6, pp. 1202–1206, 2006.
[454] C. R. McNeill, B. Watts, L. Thomsen, H. Ade, N. C. Greenham, and P. C. Dastoor, 'X-ray microscopy of photovoltaic polyfluorene blends:relating nanomorphology to device performance', *Macromolecules*, vol. 40, pp. 3263–3270, 2007.
[455] C. R. McNeill, B. Watts, S. Swaraj, H. Ade, L. Thomsen, W. Belcher, and P. C. Dastoor, 'Evolution of the nanomorphology of photovoltaic polyfluorene blends: Sub-100 nm resolution with X-ray spectromicroscopy', *Nanotechnology*, vol. 19, p. 424015, 2008.

[456] C. R. McNeill, B. Watts, L. Thomsen, W. J. Belcher, N. C. Greenham, P. C. Dastoor, and H. Ade, 'Evolution of laterally phase-separated polyfluorene blend morphology studied by X-ray spectromicroscopy', *Macromolecules*, vol. 42, pp. 3347–3352, 2009.
[457] B. Watts and C. R. McNeill, 'Simultaneous surface and bulk imaging of polymer blends with x-ray spectromicroscopy', *Macromolecular Rapid Communications*, vol. 31, pp. 1706–1712, 2010.
[458] M. Abdulla, C. Renero-Lecuna, J. S. Kim, and R. H. Friend, 'Morphological study of F8BT:PFB thin film blends', *Organic Electronics*, vol. 23, pp. 87–98, 2015.
[459] T. Michinobu, H. Osako, and K. Shigehara, 'Synthesis of poly(1,8-carbazole) derivatives', *Macromolecular Rapid Communications*, vol. 29, pp. 111–116, 2008.
[460] J. V. Grazulevicius, P. Strohriegl, J. Pielichowski, and K. Pielichowski, 'Carbazole-containing polymers: synthesis, properties, and applications', *Progress in Polymer Science*, vol. 28, pp. 1297–1353, 2003.
[461] J.-F. Morin, M. Leclerc, D. Ades, and A. Siove, 'Polycarbazoles: 25 years of progress', *Macromolecular Rapid Communications*, vol. 26, pp. 761–778, 2005.
[462] J. Li and A. C. Grimsdale, 'Carbazole-based copolymers for organic photovoltaic devices', *Chemical Society Reviews*, vol. 39, pp. 2399–2410, 2010.
[463] J.-F. Morin and M. Leclerc, 'Synthesis of conjugated polymers derived from N-alkyl-2,7-carbazoles', *Macromolecules*, vol. 34, pp. 4680–4682, 2001.
[464] F. Dierschke, A. C. Grimsdale, and K. Müllen, 'Efficient synthesis of 2,7-dibromocarbazoles as components for electroactive materials', *Synthesis*, pp. 2470–2472, 2003.
[465] J. I. G. Cadogan, M. Cameron-Wood, R. K. Mackie, and R. J. G. Searle, 'Reduction of nitro-compounds by triethyl phosphite. Convenient route to carbazoles, indoles, indazoles, triazoles and related compounds', *Journal of the Chemical Society*, pp. 4831–4837, 1965.
[466] A. W. Freeman, M. Urvoy, and M. E. Criswell, 'Triphenylphosphine-mediated reductive cyclisation of 2-nitrobiphenyls: a practical and convenient synthesis of carbazoles', *Journal of Organic Chemistry*, vol. 70, pp. 5014–5019, 2005.
[467] J.-F. Morin and M. Leclerc, '2,7-Carbazole-based conjugated polymers for blue, green and red light emission', *Macromolecules*, vol. 35, pp. 8413–8417, 2002.
[468] A. Iraqi and I. Wataru, 'Preparation and properties of 2,7-linked N-alkyl-9H-carbazole main-chain polymers', *Chemistry of Materials*, vol. 16, pp. 442–448, 2004.
[469] G. Zotti, G. Schiavon, S. Zecchin, J.-F. Morin, and M. Leclerc, 'Electrochemical, conductive and magnetic properties of 2,7-cvarbazole-based conjugated polymers', *Macromolecules*, vol. 35, pp. 2122–2128, 2002.
[470] J. Li, F. Dierschke, J. Wu, A. C. Grimsdale, and K. Müllen, 'Poly(2,7-carbazole) and PDI: A promising donor/acceptor pair for polymer solar cells', *Journal of Materials Chemistry*, vol. 16, pp. 96–100, 2006.
[471] K. Lmimouni, C. Legrand, and A. Chapoton, 'Poly(N-alkylcarbazole) LEDs', *Synthetic Metals*, vol. 97, pp. 151–155, 1998.
[472] J.-F. Morin, S. Beaupre, M. Leclerc, I. Levesque, and M. D'Iorio, 'Blue LEDs from new conjugated poly(N-substituted-2,7-carbazole) derivatives', *Applied Physics Letters*, vol. 80, pp. 341–343, 2002.
[473] S. A. Patil, U. Scherf, and A. Kadashchuk, 'New conjugated ladder polymer containing carbazole moieties', *Advanced Functional Materials*, vol. 13, pp. 609–614, 2003.
[474] F. Dierschke, A. C. Grimsdale, and K. Müllen, 'Novel carbazole-based ladder-type polymers for electronic applications', *Macromolecular Chemistry and Physics*, vol. 205, pp. 1147–1153, 2004.

[475] A. K. Mishra, M. Graf, F. Grasse, J. Jacob, E. J. W. List, and K. Müllen, 'Blue-emitting carbon and nitrogen-bridged poly(ladder-type tetraphenylene)s', *Chemistry of Materials*, vol. 17, pp. 2879–2885, 2006.

[476] K. L. Chan, M. J. McKiernan, C. R. Towns, and A. B. Holmes, 'Poly(2,7-dibenzosilole): A blue light emitting polymer', *Journal of the American Chemical Society*, vol. 127, pp. 7662–7663, 2005.

[477] Z. E. Ooi, T. L. Tam, R. Y. C. Shin, Z. K. Chen, T. Kietzke, A. Sellinger, et al., 'Solution processable bulk heterojunction solar cells using a small molecule acceptor', *Journal of Materials Chemistry*, vol. 18, pp. 4619–4622, 2008.

[478] C. Yang, H. Scheiber, E. J. W. List, J. Jacob, and K. Müllen, 'Poly(2,7-phenanthrylene)s and poly(3,6-phenanthrylene)s as PPP and PPV analogues', *Macromolecules*, vol. 39, pp. 5213–5221, 2006.

[479] S. Setayesh, D. Marsitzky, and K. Müllen, 'Polyindenofluorenes: Bridging the gap between LPPP and PF', *Macromolecules*, vol. 33, pp. 2016–2020, 2000.

[480] A. C. Grimsdale, P. Leclère, R. Lazzaroni, J. D. MacKenzie, C. Murphy, S. Setayesh, et al., 'Correlation between molecular structure, microscopic morphology and optical properties of poly(tetraalkylindenofluorene)s', *Advanced Functional Materials*, vol. 12, pp. 729–733, 2002.

[481] P. E. Keivanidis, J. Jacob, L. Oldridge, P. Sonar, B. Carbonnier, S. Baluschev, et al., 'Photophysical characterization of light-emitting polyindenofluorenes', *ChemPhysChem*, vol. 6, pp. 1650–1660, 2005.

[482] D. Marsitzky, C. Scott, J.-P. Chen, V. Y. Lee, R. D. Miller, S. Setayesh, and K. Müllen, 'Poly-2,8-(indenofluorene-co-anthracene): A colorfast blue light-emitting random copolymer', *Advanced Materials*, vol. 13, pp. 1096–1099, 2001.

[483] J. Jacob, J. Zhang, A. C. Grimsdale, K. Müllen, M. Gaal, and E. J. W. List, 'Poly (tetraarylindenofluorene)s: New stable blue-emitting polymers', *Macromolecules*, vol. 36, pp. 8240–8245, 2003.

[484] D. Vak, B. Lim, S.-H. Lee, and D.-Y. Kim, 'Synthesis of a double spiro-indenofluorene with a stable blue emission', *Organic Letters*, vol. 7, pp. 4229–4232, 2005.

[485] F. Laquai, A. K. Mishra, M. R. Ribas, A. Petrozza, J. Jacob, L. Akcelrud, et al., 'Photophysical properties of a series of poly(ladder-type phenylene)s', *Advanced Functional Materials*, vol. 17, pp. 3231–3240, 2007.

[486] J. Jacob, S. Sax, T. Piok, E. J. W. List, A. C. Grimsdale, and K. Müllen, 'Ladder-type pentaphenylenes and their polymers: Efficient blue-light emitters and electron-acepting materials via a common intermediate', *Journal of the American Chemical Society*, vol. 126, pp. 6987–6995, 2004.

[487] J. Jacob, A. C. Grimsdale, K. Müllen, S. Sax, M. Gaal, and E. J. W. List, 'Fully aryl-substituted poly(ladder-type pentaphenylene): A remarkably stable blue light-emitting polymer', *Macromolecules*, pp. 9933–9938, 2005.

[488] C. E. Finlayson, J.-S. Kim, M. J. Liddell, R. H. Friend, S.-H. Jung, A. C. Grimsdale, and K. Müllen, 'Exciplex emission from electroluminescent ladder-type pentaphenylene oligomers bearing both electron and hole-accepting substituents', *Journal of Chemical Physics*, vol. 128, p. 044703, 2008.

[489] F. Uckert, S. Setayesh, and K. Müllen, 'Precursor route to polyfluorenone', *Macromolecules*, vol. 32, pp. 4519–4524, 1999.

[490] F. Uckert, Y.-H. Tak, K. Müllen, and H. Bässler, 'Polyfluorenone: A trap-free electron injection material', *Advanced Materials*, vol. 12, pp. 905–908, 2000.

[491] A. P. Kulkarni, X. Kong, and S. A. Jenekhe, 'Fluorenone containing PFs and oligomers: Photophysics, origin of the green emission and efficient green EL', *Journal of Physical Chemistry B*, vol. 108, pp. 8689–8701, 2004.

[492] H. Usta, A. Facchetti, and T. J. Marks, 'Air-stable, solution processable n-channel and ambipolar semiconductors for thin-film transistors based on the indenofloure-nebis(dicyanovinylene) core', *Journal of the American Chemical Society*, vol. 130, pp. 8580–8581, 2008.

[493] H. Usta, A. Facchetti, and T. J. Marks, 'Synthesis and characterization of electron-deficient and highly soluble (bis)indenofluorene building blocks for n-type semiconducting polymers', *Organic Letters*, vol. 10, pp. 1385–1388, 2008.

[494] S. Wakim and M. Leclerc, 'Towards the synthesis of ladder oligo(p-aniline)s', *Synlett*, no. 8, pp. 1223–1234, 2005.

[495] Y. Wu, Y. Li, S. Gardner, and B. S. Ong, 'Indolo[3,2-b]carbazole-based thin-film transistors with high mobility and stability', *Journal of the American Chemical Society*, vol. 127, pp. 614–618, 2005.

[496] Y. Li, Y. Wu, S. Gardner, and B. S. Ong, 'Novel peripherally substituted indolo[3,2-b]carbazoles for high mobility organic thin-film transistors', *Advanced Materials*, vol. 17, pp. 849–853, 2005.

[497] Y. Li, Y. Wu, and B. S. Ong, 'Poly[3,2-b]carbazoles: A new class of p-channel semiconductor polmers for organic thin-film transistors', *Macromolecules*, vol. 39, pp. 6521–6527, 2006.

[498] N. Blouin, A. Michaud, S. Wakim, P.-L. T. Boudreault, M. Leclerc, B. Vercelli, et al., 'Optical, electrochemical, magnetic and conductive properties of new poly-indolocarbazoles and polydiindolocarbazoles', *Macromolecular Chemistry and Physics*, vol. 207, pp. 166–174, 2006.

[499] W. Pisula, A. K. Mishra, J. Li, M. Baumgarten, and K. Müllen, 'Carbazole-based conjugated polymers as donor-material for photovoltaic devices,' in *Organic Photovoltaics: Materials, Device Physics, and Manufacturing Technologies*, C. J. Brabec, V. Dyakononv, and U. Scherf (eds). Weinheim: Wiley-VCH, 2008, ch. 3, pp. 93–127.

[500] K. Kaneto, Y. Kohno, K. Yoshino, and Y. Inuishi, 'Electrochemical preparation of a metallic polythiophene film', *Journal of the Chemical Society Chemical Communications*, no. 7, pp. 382–383, 1983.

[501] R. D. McCullough, 'The chemistry of conducting polythiophenes', *Advanced Materials*, vol. 10, pp. 93–116, 1998.

[502] M. Leclerc, F. Martinez Diaz, and G. Wegner, 'Structural analysis of poly(3-alkylthiophene)s', *Makromolekulare Chemie*, vol. 190, pp. 3105–3116, 1989.

[503] M. R. Andersson, D. Selse, M. Berggren, H. Jaervinen, T. Hjertberg, O. Inganaes, et al., 'Regioselective polymerization of 3-(4-octylphenyl)thiophene with $FeCl_3$', *Macromolecules*, vol. 27, pp. 6503–6506, 1994.

[504] R. E. Gill, G. E. Malliaras, J. Wildeman, and G. Hadziioannou, 'Tuning of photo- and electroluminescence in alkylated polythiophenes with well-defined regioregularity', *Advanced Materials*, vol. 6, pp. 132–135, 1994.

[505] R. D. McCullogh and R. D. Lowe, 'Enhanced electrical conductivity in regioselectively synthesized poly(3-alkylthiophenes)', *Journal of the Chemical Society, Chemical Comminucations*, no. 1, pp. 70–72, 1992.

[506] R. D. McCullough, R. D. Lowe, M. Jayaraman, and D. L. Anderson, 'Design, synthesis, and control of conducting polymer architectures: Structurally homogeneous poly(3-alkylthiophenes)', *Journal of Organic Chemistry*, vol. 58, pp. 904–912, 1993.

[507] T.-A. Chen and R. D. Rieke, 'The first regioregular head-to-tail poly(3-hexylthiophene-2,5-diyl) and a regiorandom isopolymer: Nickel versus palladium catalysis of 2(5)-bromo-5(2)-(bromozincio)-3-hexylthiophene polymerization', *Journal of the American Chemical Society*, vol. 114, pp. 10087–10088, 1992.

[508] T.-A. Chen, X. Wu, and R. D. Rieke, 'Regiocontrolled synthesis of poly(3-alkylthiophenes) mediated by Rieke zinc: Their characterization and solid-state properties', *Journal of the American Chemical Society*, vol. 117, pp. 233–244, 1995.
[509] A. Iraqi and G. W. Barker, 'Synthesis and characterisation of telechelic regioregular head-to-tail poly(3-alkylthiophene)s', *Journal of Materials Chemistry*, vol. 8, pp. 25–29, 1998.
[510] S. Guillerez and G. Bidan, 'New convenient synthesis of highly regioregular poly(3-octylthiophene) based on the Suzuki coupling reaction', *Synthetic Metals*, vol. 93, pp. 123–126, 1998.
[511] R. S. Loewe, S. M. Khersonsky, and R. D. McCullough, 'A simple method to prepare head-to-tail coupled, regioregular poly(3-alkylthiophene)s using Grignard metathesis', *Advanced Materials*, vol. 11, pp. 127–130, 1999.
[512] R. S. Loewe, P. C. Ewbank, J. Liu, L. Zhai, and R. D. McCullough, 'Regioregular, head-to-tail coupled poly(3-alkylthiophene)s made easy by the GRIM method: Investigation of the reaction and the origin of regioselectivity', *Macromolecules*, vol. 34, pp. 4324–4333, 2001.
[513] M. C. Iovu, E. E. Sheina, R. R. Gil, and R. D. McCullough, 'Experimental evidence for the quasi-"living" nature of the Grignard metathesis (GRIM) for the synthesis of regioregular poly(3-alkylthiophene)s', *Macromolecules*, vol. 38, pp. 8649–8656, 2005.
[514] M. Jeffries-El, G. Sauve, and R. D. McCullough, 'In-situ endgroup functionalization of regioregular poly(3-alkylthiophene)s using the Grignard metathesis polymerization method', *Advanced Materials*, vol. 16, pp. 1017–1019, 2004.
[515] M. Jeffries-El, G. Sauve, and R. D. McCullough, 'Facile synthesis of end-functionalized regioregular poly(3-alkylthiophene)s via modified Grignard metathesis reaction', *Macromolecules*, vol. 38, pp. 10346–10352, 2005.
[516] P. Kohn, S. Huettner, H. Komber, V. Senkovskyy, R. Tkachov, A. Kiriy, et al., 'On the role of single regiodefects and polydispersity in regioregular poly(3-hexylthiophene): Defect distribution, synthesis of defect-free chains, and a simple model for the determination of crystallinity', *Journal of the American Chemical Society*, vol. 134, pp. 4790–4805, 2012.
[517] T. Yokozawa and A. Yokoyama, 'Chain-growth condensation polymerization for the synthesis of well-defined condensation polymers and π-conjugated polymers', *Chemical Reviews*, vol. 109, pp. 5595–5619, 2009.
[518] H. Sirringhaus, P. J. Brown, R. H. Friend, M. M. Nielsen, K. Bechgaard, B. M. W. Langeveld-Voss, et al., 'Two-dimensional charge transport in self-organized, high-mobility conjugated polymers', *Nature*, vol. 401, pp. 685–688, 1999.
[519] Q. Wang, J. Swensen, D. Moses, and A. J. Heeger, 'Increased mobility from regioregular poly(3-hexylthiophene) field-effect transistors', *Journal of Applied Physics*, vol. 93, pp. 6137–6141, 2003.
[520] R. J. Kline, M. D. McGehee, E. N. Kadnikova, J. Liu, and J. M. J. Frechet, 'Controlling the field-effect mobility of regioregular polythiophene by changing the molecular weight', *Advanced Materials*, vol. 15, pp. 1519–1522, 2003.
[521] B. S. Ong, Y. Wu, P. Liu, and S. Gardner, 'High-performance semiconducting polythiophenes for organic thin-film transistors', *Journal of the American Chemical Society*, vol. 126, pp. 3378–3379, 2004.
[522] S. Ko, E. Verploegen, S. Hong, R. Mondal, E. T. Hoke, M. F. Toney, et al., '3,4-Disubstituted polyalkylthiophenes for high-performance thin-film transistors and photvoltaics', *Journal of the American Chemical Society*, vol. 133, pp. 16722–16725, 2011.

[523] J. Kim, B. Lim, K.-J. Baeg, Y.-Y. Noh, D. Khim, H.-G. Jeong, et al., 'Highly soluble poly(thienylenevinylene) derivatives with charge-carrier mobility exceeding 1 $cm^2V^{-1}s^{-1}$', *Chemistry of Materials*, vol. 23, pp. 4663–4665, 2011.
[524] S. Kirchmeyer and K. Reuter, 'Scientific importance, properties and growing applications of poly(3,4-ethylenedioxythiophene)', *Journal of Materials Chemistry*, vol. 15, pp. 2077–2088, 2005.
[525] U. Lang, E. Müller, N. Naujoks, and J. Dual, 'Microscopical investigations of PEDOT:PSS thin films', *Advanced Functional Materials*, vol. 19, pp. 1215–1220, 2009.
[526] Y. Wen and J. Xu, 'Scientific importance of water-processable PEDOT-PSS and preparation, challenge and new application in sensors of its film electrode: A review', *Journal of Polymer Science A: Polymer Chemistry*, vol. 55, pp. 1121–1150, 2017.
[527] N. Toshima, 'Recent progress of organic and hybrid thermoelectric materials', *Synthetic Metals*, vol. 225, pp. 3–21, 2017.
[528] J. D. Ryan, D. A. Mengistie, R. Gabrielsson, A. Lund, and C. Müller, 'Machine-washable PEDOT:PSS dyed silk yarns for electronic textiles', *ACS Applied Materials and Interfaces*, vol. 9, pp. 9045–9050, 2017.
[529] M. Nikolou and G. G. Malliaras, 'Applications of poly(3,4-ethylenedioxythiophene) doped with poly(styrene sulfonic acid) transistors in chemical and biological sensors', *The Chemical Record*, vol. 8, pp. 13–22, 2008.
[530] Z.-T. Zhu, J. T. Mabeck, C. Zhu, N. C. Cady, C. A. Batt, and G. G. Malliaras, 'A simple poly(3,4-ethylene dioxythiophene)/poly(styrene sulfonic acid) transistor for glucose sensing at neutral pH', *Chemical Communications*, vol. 40, pp. 1556–1557, 2004.
[531] D. A. Bernards, D. J. Macaya, M. Nikolou, J. A. DeFranco, S. Takamatsu, and G. G. Malliaras, 'Enzymatic sensing with organic electrochemical transistors', *Journal of Materials Chemistry*, vol. 18, pp. 116–120, 2007.
[532] J. Liu, M. Agarwal, and K. Varahramyan, 'Glucose sensor based on organic thin film transistor using glucose oxidase and conducting polymer', *Sensors and Actuators B: Chemical*, vol. 135, no. 1, pp. 195–199, 2008.
[533] A. Phongphut, C. Sriprachuabwong, A. Wisitsoraat, A. Tuantranont, S. Prichanont, and P. Sritongkham, 'A disposable amperometric biosensor based on inkjet-printed Au/PEDOT-PSS nanocomposite for triglyceride determination', *Sensors and Actuators B: Chemical*, vol. 178, pp. 501–507, 2013.
[534] M. R. Andersson, O. Thomas, W. Mamma, M. Svensson, M. Theander, and O. Inganäs, 'Substituted polythiophenes designed for optoelectronic devices and conductors', *Journal of Materials Chemistry*, vol. 9, pp. 1933–1940, 1999.
[535] R. D. McCullogh and S. P. Williams, 'A dramatic conformational transformation of a regioregular polythiophene via a chemoselective metal-ion assisted deconjugation', *Chemistry of Materials*, vol. 7, pp. 2001–2003, 1995.
[536] M. J. Marsella and T. M. Swager, 'Designing conducting polymer-based sensors: Selective ionochromic response in crown ether containing polythiophenes', *Journal of the American Chemical Society*, vol. 115, pp. 12214–12215, 1993.
[537] H. A. Ho and M. Leclerc, 'New colorimetric and fluorimetric chemosensor based on a cationic polythiophene derivative for iodide-specific detection', *Journal of the American Chemical Society*, vol. 125, pp. 4412–4413, 2003.
[538] H. A. Ho and M. Leclerc, 'Optical sensors based on hybrid aptamer/conjugated polymer complexes', *Journal of the American Chemical Society*, vol. 126, pp. 1384–1387, 2004.

[539] K. Doré, S. Dubus, H.-A. Ho, I. Lévesque, M. Brunette, G. Corbeil, et al., 'Fluorescent polymeric transducer for the rapid, simple, and specific detection of nucleic acids at the zeptomole level', *Journal of the American Chemical Society*, vol. 126, pp. 4240–4244, 2004.
[540] H. A. Ho, K. Doré, M. Boissinot, M. G. Bergeron, R. M. Tanguay, D. Boudreau, and M. Leclerc, 'Direct molecular detection of nucleic acids by fluorescence signal amplification', *Journal of the American Chemical Society*, vol. 127, pp. 12673–12676, 2005.
[541] B. S. Gaylord, A. J. Heeger, and G. C. Bazan, 'DNA detection using water-soluble conjugated polymers and peptide nucleic acid probes', *Proceedings of the National Academy of Sciences of the United States of America*, vol. 99, pp. 10954–10957, 2002.
[542] B. S. Gaylord, A. J. Heeger, and G. C. Bazan, 'DNA hybridization detection with water-soluble conjugated polymers and chromophore-labelled single-stranded DNA', *Journal of the American Chemical Society*, vol. 125, pp. 896–900, 2003.
[543] G. Dennler, M. C. Scharber, and C. J. Brabec, 'Polymer-fullerene bulk-heterojunction solar cells', *Advanced Materials*, vol. 21, pp. 1323–1338, 2009.
[544] M. T. Dang, L. Hirsch, and G. Wantz, 'P3HT:PCBM, best seller in polymer photovoltaic research', *Advanced Materials*, vol. 23, pp. 3597–3602, 2011.
[545] P. Schilinsky, U. Asawapirom, U. Scherf, M. Biele, and C. J. Brabec, 'Influence of the molecular weight of P3HT on the performance of BHJ solar cells', *Chemistry of Materials*, vol. 27, pp. 2175–2180, 2005.
[546] Y. Kim, et al., 'A strong regioregularity effect in self-organizing conjugated polymer films and high-efficiency polythiophene:fullerene solar cells', *Nature Materials*, vol. 5, no. 3, pp. 197–203, 2006.
[547] N. Chandrasekaran, E. Gann, N. Jain, A. Kumar, S. Gopinathan, A. Sadhanala, et al., 'Correlation between photovoltaic performance and interchain ordering induced delocalization of electronics states in conjugated polymer blends', *ACS Applied Materials & Interfaces*, vol. 8, pp. 20243–20250, 2016.
[548] K. Sivula, C. K. Luscombe, B. C. Thompson, and J. M. J. Frechet, 'Enhancing the thermal stability of polythiophene:fullerene solar cells by decreasing effective polymer regioregularity', *Journal of the American Chemical Society*, vol. 128, no. 43, pp. 13988–13989, 2006.
[549] B. A. Collins, E. Gann, L. Guignard, X. He, C. R. McNeill, and H. Ade, 'Molecular miscibility of polymer-fullerene blends', *Journal of Physical Chemistry Letters*, vol. 1, pp. 3160–3166, 2010.
[550] D. R. Kozub, K. Vakhshouri, L. M. Orme, C. Wang, A. Hexemer, and E. D. Gomez, 'Polymer crystallization of partially miscible polythiophene/fullerene mixtures controls morphology', *Macromolecules*, vol. 44, pp. 5722–5726, 2011.
[551] B. Watts, W. J. Belcher, L. Thomsen, H. Ade, and P. C. Dastoor, 'A quantitative study of PCBM diffusion during annealing of P3HT:PCBM blend films', *Macromolecules*, vol. 42, pp. 8392–8397, 2009.
[552] G. A. Berriman, N. P. Holmes, J. L. Holdsworth, X. Zhou, W. J. Belcher, and P. C. Dastoor, 'A new model for PCBM phase segregation in P3HT:PCBM blends', *Organic Electronics*, vol. 30, pp. 12–17, 2016.
[553] G. A. Berriman, N. P. Holmes, J. L. Holdsworth, X. Zhou, W. J. Belcher, and P. C. Dastoor, 'Molecular versus crystallite PCBM diffusion in P3HT:PCBM blends', *AIP Advances*, vol. 5, p. 097220, 2015.
[554] C.-Y. Chang, et al., 'Enhanced performance and stability of a polymer solar cell by incorporation of vertically aligned, cross-linked fullerene nanorods', *Angewandte Chemie, International Edition*, vol. 50, pp. 9386–9390, 2011.

[555] D. Wang, V. Noël, and B. Piro, 'Electrolytic gated organic field-effect transistors for application in biosensors: A review', *Electronics*, vol. 5, p. 9, 2017.
[556] A. Lv, Y. Pan, and L. Chi, 'Gas sensors based on polymer field-effect transistors', *Sensors*, vol. 17, p. 213, 2017.
[557] G. Ryu, J. Huang, O. Hofmann, C. A. Walshe, J. Y. Y. Sze, G. D. McClean, et al., 'Highly sensitive fluorescence detection system for microfluidic lab-on-a-chip', *Lab on a Chip*, vol. 11, pp. 1664–1670, 2011.
[558] M. Magliulo, A. Mallardi, R. Gristina, F. Ridl, L. Sabbatini, N. Cioffi, et al., 'Part per trillion label-free electronic bioanalytical detection', *Analytical Chemistry*, vol. 85, pp. 3849–3857, 2013.
[559] Y.-C. Chao, W.-J. Lai, C.-Y. Chen, H.-F. Meng, H.-W. Zan, and S.-F. Horng, 'Low voltage active pressure sensor based on polymer space-charge-limited transistor', *Applied Physics Letters*, vol. 95, p. 253306, 2009.
[560] B. Li and D. N. Lambeth, 'Chemical sensing using nanostructured polythiophene transistors', *Nano Letters*, vol. 8, pp. 3563–3567, 2008.
[561] D. Elkington, M. Wasson, W. J. Belcher, P. C. Dastoor, and X. J. Zhou, 'Printable organic thin film transistors for glucose detection incorporating inkjet-printing of the enzyme recognition element', *Applied Physics Letters*, vol. 106, p. 263301, 2015.
[562] Y. Liu, Y. Liu, and X. Zhan, 'High-mobility conjugated polymers based on fused-thiophene building blocks', *Macromolecular Chemistry and Physics*, vol. 212, pp. 428–433, 2011.
[563] I. McCulloch, M. Heeney, M. L. Chabinyc, D. DeLongchamp, R. J. Kline, M. Cölle, et al., 'Semiconducting thienothiophene copolymers; design, synthesis, morphology, and performance in thin-film organic transistors', *Advanced Materials*, vol. 21, pp. 1091–1109, 2009.
[564] I. McCulloch, M. Heeney, C. Bailey, K. Genevicius, I. MacDonald, M. Shkunov, et al., 'Liquid-crystalline semiconducting polymers with high charge-carrier mobility', *Nature Materials*, vol. 5, pp. 328–333, 2006.
[565] Y. Li, Y. Wu, P. Liu, M. Birau, H. Pan, and B. S. Ong, 'Poly(2,5-bis(2-thienyl)-3,6-dialkylthieno[3,2-b]thiophene)s: High mobility semiconductors for thin-film transistors', *Advanced Materials*, vol. 18, pp. 3029–3032, 2006.
[566] J. Li, F. Qin, C. M. Li, Q. Bao, M. B. Chan-Park, W. Zhang, et al., 'High-performance thin-film transistors from solution-processed dithienothiophene polymer semiconductor nanoparticles', *Chemistry of Materials*, vol. 20, pp. 2057–2059, 2008.
[567] H. H. Fong, V. A. Pozdin, A. Amassian, G. G. Malliaras, D.-M. Smilgies, M. He, et al., 'Tetrathienoacene copolymers as high mobility, soluble organic semiconductors', *Journal of the American Chemical Society*, vol. 130, pp. 13202–13203, 2009.
[568] H. Pan, Y. Li, Y. Wu, P. Liu, B. S. Ong, S. Zhu, and G. Xu, 'Low-temperature, solution-processed, high-mobility polymer semiconductors for thin-film transistors', *Journal of the American Chemical Society*, vol. 129, pp. 4112–4113, 2007.
[569] R. Rieger, D. Beckmann, W. Pisula, W. Steffen, M. Kastler, and K. Müllen, 'Rational optimization of benzo[2,1-b;3,4-b']dithiophene-containing polymers for organic field-effect transistors', *Advanced Materials*, vol. 22, pp. 83–86, 2010.
[570] Y.-J. Cheng, S.-H. Yang, and C.-S. Hsu, 'Synthesis of conjugated polymers for organic solar cell applications', *Chemical Reviews*, vol. 109, pp. 5868–5923, 2009.
[571] A. Facchetti, 'π-conjugated polymers for organic electronics and photovoltaic cell applications', *Chemistry of Materials*, vol. 23, pp. 733–758, 2011.
[572] A. C. Arias, J. D. MacKenzie, I. McCulloch, J. Rivnay, and A. Salleo, 'Materials and applications for large area electronics: Solution-based approaches', *Chemical Reviews*, vol. 110, pp. 3–24, 2010.

[573] M. J. Robb, S.-Y. Ku, and C. J. Hawker, 'No assembly required: Recent advances in fully conjugated block copolymers', *Advanced Materials*, vol. 25, pp. 5686–5700, 2013.

[574] R. C. Mulherin, S. Jung, S. Huettner, K. Johnson, P. Kohn, M. Sommer, et al., 'Ternary photovoltaic blends incorporating an all-conjugated donor-acceptor diblock copolymer', *Nano Letters*, vol. 11, pp. 4846–4851, 2011.

[575] C. R. McNeill, A. Abrusci, J. Zaumseil, R. Wilson, M. J. McKiernan, J. H. Burroughes, et al., 'Dual electron donor/electron acceptor character of a conjugated polymer in efficient photovoltaic diodes', *Applied Physics Letters*, vol. 90, p. 193506, 2007.

[576] C. Guo, Y-H. Lin, M. D. Witman, K. A. Smith, C. Wang, A. Hexemer, et al., 'Conjugated block copolymer photovoltaics with near 3% efficiency through microphase separation', *Nano Letters*, vol. 13, pp. 2957–2963, 2013.

[577] M. Fahlman, J. Birgersson, K. Kaeriyama, and W. R. Salaneck, 'Poly(2,5-diheptyl-1,4-phenylene-alt-2,5-thienylene): A new material for blue-light-emitting diodes', *Synthetic Metals*, vol. 75, pp. 223–228, 1995.

[578] J. P. Lère-Porte, J. J. E. Moreau, and C. Torreilles, 'Thienylene-dialkoxyphenylene copolymers', *Synthetic Metals*, vol. 101, p. 104, 1999.

[579] A. Donat-Bouillud, I. Lévesque, Y. Tao, M. D'Iorio, S. Beaupré, P. Blondin, et al., 'Light-emitting diodes from fluorene-based π-conjugated polymers', *Chemistry of Materials*, vol. 12, pp. 1931–1936, 2000.

[580] B. Liu, Y.-H. Niu, W.-L. Yu, Y. Cao, and W. Huang, 'Applications of alternating fluorene-thiophene copolymers in PLEDS', *Synthetic Metals*, vol. 129, pp. 129–134, 2002.

[581] M. Forster, K. O. Annan, and U. Scherf, 'Conjugated ladder polymers containing thiophene units', *Macromolecules*, vol. 32, pp. 3159–3162, 1999.

[582] P. K. H. Ho, D. S. Thomas, R. H. Friend, and N. Tessler, 'All-polymer optoelectronic devices', *Science*, vol. 285, pp. 233–237, 1999.

[583] Y. He, S. Gong, R. Hatori, and J. Kanicki, 'High performance light-emitting heterostructure devices', *Applied Physics Letters*, vol. 74, pp. 2265–2267, 1999.

[584] R. B. Fletcher, D. G. Lidzey, D. D. C. Bradley, S. Walker, M. Inbasekaran, and E. P. Woo, 'High brightness conjugated polymer LEDs', *Synthetic Metals*, vol. 111–112, pp. 151–153, 2000.

[585] F. Zhu, K. Zhang, E. Guenther, and C. S. Jin, 'Optimized ITO contact for OLED applications', *Thin and Solid Films*, vol. 363, pp. 314–317, 2000.

[586] P. Herguth, X. Jiang, M. S. Liu, and A. K.-Y. Jen, 'Highly efficient F-BT copolymers for PLEDs', *Macromolecules*, vol. 35, pp. 6094–6100, 2002.

[587] Q. Hou, Y. Xu, W. Yang, M. Yuan, J. Peng, and Y. Cao, 'Novel red-emitting fluorene-based copolymers', *Journal of Materials Chemistry*, vol. 12, pp. 2887–2892, 2002.

[588] Q. Hou, Y. S. Xu, W. Yang, R. Q. Yang, M. Yuan, J. B. Peng, and Y. Cao, 'Synthesis and electroluminescent properties of fluorene-based copolymers', *Synthetic Metals*, vol. 135–136, pp. 179–180, 2003.

[589] Q. Hou, Q. Zhou, Y. Zhang, W. Yang, R. Yang, and Y. Cao, 'Synthesis and EL properties of high-efficiency red emitter based on copolymers of fluorene and dithienylbenzothiadiazole', *Macromolecules*, vol. 37, pp. 6299–6305, 2004.

[590] J. Luo, J. Peng, Y. Cao, and Q. Hou, 'High-efficiency red LEDs based on PF copolymers with extremely low content of dithienylbenzothiadiazole: Comparative studies of intrachain and interchain interaction', *Applied Physics Letters*, vol. 87, p. 261103, 2005.

[591] Y. Xia, J. Luo, X. Deng, X. Li, D. Li, X. Zhu, et al., 'Novel random low-band-gap fluorene-based copolymers for deep red/near infrared light-emitting diodes and bulk heterojunction photovoltaic cells', *Macromolecular Chemistry and Physics*, vol. 207, pp. 511–520, 2006.

[592] M. Chen, E. Perzon, M. R. Andersson, S. Marcinkevicius, S. K. M. Jönsson, M. Fahlman, and M. Berggren, '1 micron wavelength photo- and electroluminescence from a conjugated polymer', *Applied Physics Letters*, vol. 84, pp. 3570–3572, 2004.

[593] M. X. Chen, E. Perzon, N. Robisson, S. K. M. Jönsson, M. R. Andersson, M. Fahlman, and M. Berggren, 'Low bandgap acceptor-donor-donor polymers for infrared electroluminescence and transistors', *Synthetic Metals*, vol. 146, pp. 233–236, 2004.

[594] E. Hennebicq, G. Pourtois, G. D. Scholes, L. M. Herz, D. M. Russell, C. Silva, et al., 'Exciton migration in rigid-rod conjugated polymers: An improved Förster model', *Journal of the American Chemical Society*, vol. 127, pp. 4744–4762, 2005.

[595] S. Baluschev, J. Jacob, Y. Avlasevich, P. E. Keivanidis, T. Miteva, A. Yasuda, et al., 'Enhanced operational stability of the up-conversion fluoresence in films of palladium-porphyrin end-capped poly(pentaphenylene)', *ChemPhysChem*, vol. 6, pp. 1250–1253, 2005.

[596] S. Baluschev, P. E. Keivanidis, G. Wegner, J. Jacob, A. C. Grimsdale, K. Mullen, et al., 'Upconversion photoluminescence in poly(ladder-type-pentaphenylene) doped with metal(II)octaethyl porphyrins', *Applied Physics Letters*, vol. 86, p. 061904, 2005.

[597] H.-R. Tseng, H. Phan, C. Luo, M. Wang, L. A. Perez, S. N. Patel, et al., 'High-mobility field-effect transistors fabricated with macroscopic aligned semiconductiung polymers', *Advanced Materials*, vol. 26, pp. 2993–2998, 2014.

[598] C. Luo, A. K. K. Kyaw, L. A. Perez, S. Patel, M. Wang, B. Grimm, et al., 'General strategy for self-assembly of highly oriented nanocrystalline semiconducting polymers with high mobility', *Nano Letters*, vol. 14, pp. 2764–2771, 2014.

[599] A. Saeki, Y. Koizumi, T. Aida, and S. Seki, 'Comprehensive approach to intrinsic charge carrier mobility in conjugated organic molecules, macromolecules, and supramolecular architectures', *Accounts of Chemical Research*, vol. 45, pp. 1193–1202, 2012.

[600] X. Zhang, H. Bronstein, A. J. Kronemeijer, J. Smith, Y. Kim, R. J. Kline, et al., 'Molecular origin of high field effect mobility in an indacenodithiophene-benzothiadiazole copolymer', *Nature Communications*, vol. 4, p. 2238, 2013.

[601] R. Noriega, J. Rivnay, K. Vandewal, F. P. V. Koch, N. Stingelin, P. Smith, et al., 'A general relationship between disorder, aggregation and charge transport in conjugated polymers', *Nature Materials*, vol. 12, pp. 1038–1044, 2013.

[602] U. Koldemir, S. R. Puniredd, M. Wagner, S. Tongay, T. D. McCarley, G. D. Kamenov, et al., 'End capping does matter: Enhanced order and charge transport in conjugated donor-acceptor polymers', *Macromolecules*, vol. 48, pp. 6369–6377, 2015.

[603] H. N. Tsao, D. M. Cho, I. Park, M. R. Hansen, A. Mavrinskiy, D. Y. Yoon, et al., 'Ultrahigh mobility in polymer field-effect transistors by design', *Journal of the American Chemical Society*, vol. 133, pp. 2605–2612, 2011.

[604] P. Sonar, L. Oldridge, A. C. Grimsdale, K. Müllen, M. Surin, R. Lazzaroni, et al., 'Synthesis, characterization and comparative OFET behaviour of indenofluorene-bithiophene and terthiophene alternating copolymers', *Synthetic Metals*, vol. 160, pp. 468–474, 2010.

[605] H. Bronstein, D. S. Leem, R. Hamilton, P. Woebkenberg, S. King, W. Zhang, et al., 'Indacenodithiophene-*co*-benzothiadiazole copolymers for high performance solar

cells or transistors via alkyl chain optimization', *Macromolecules*, vol. 44, pp. 6649–6652, 2011.

[606] D. Venkateshvaran, M. Nikolka, A. Sadhanala, V. Lemaur, M. Zelazny, M. Kepa, et al., 'Approaching disorder-free transport in high-mobility conjugated polymers', *Nature*, vol. 515, pp. 384–388, 2014.

[607] H. Bronstein, Z. Chen, R. S. Ashraf, W. Zhang, J. Du, J. R. Durrant, et al., 'Thieno[3,2-*b*]thiophene-diketopyrrolopyrrole-containing polymers for high performance organic field-effect transistors and organic photovoltaic devices', *Journal of the American Chemical Society*, vol. 133, pp. 3272–3275, 2011.

[608] Z. Chen, M. J. Lee, R. S. Ashraf, Y. Gu, S. Albert-Seifried, M. M. Nielsen, B. Schroeder, et al., 'High-performance ambipolar diketopyrrolopyrrole-thieno[3,2-*b*]-thiophene copolymer filed-effect transistors with balanced hole and electron mobilities', *Advanced Materials*, vol. 24, pp. 647–652, 2012.

[609] J. D. Yuen, J. Fan, J. Seifter, B. Lim, R. Hufschmid, A. J. Heeger, and F. Wudl, 'High performance weak-donor-acceptor polymers in thin film transistors: Effect of the acceptor on electronic properties, ambipolar conductivity, mobility and thermal stability', *Journal of the American Chemical Society*, vol. 133, pp. 20799–20807, 2011.

[610] J. Mei, D. H. Kim, A. L. Ayzner, M. F. Toney, and Z. Bao, 'Siloxane-terminated solubilizing side chains: bringing conjugated backbones closer and boosting hole mobilities in thin-film transistors', *Journal of the American Chemical Society*, vol. 133, pp. 20130–20133, 2011.

[611] J. Mei, H.-C. Wu, Y. Diao, A. Appleton, H. Wang, Y. Zhou, et al., 'Effect of spacer length of siloxane-terminated side chains on charge transport in isoindigo-based polymer semiconductor thin films', *Advanced Functional Materials*, vol. 25, pp. 3455–3462, 2015.

[612] J. Lee, A.-R. Han, H. Yu, T. J. Shin, C. Yang, and J. H. Oh, 'Boosting the ambipolar performance of solution-processable polymer semiconductors via hybrid side-chain engineering', *Journal of the American Chemical Society*, vol. 135, pp. 9540–9547, 2013.

[613] A.-R. Han, G. K. Dutta, J. Lee, H. R. Lee, S. M. Lee, H. Ahn, et al., 'ε-branched flexible side chain substituted diketopyrrolopyrrole-containing polymers designed for high hole and electron mobilities', *Advanced Functional Materials*, vol. 25, pp. 247–254, 2015.

[614] I.-N. Kang, H.-J. Yun, D. S. Chung, S.-K. Kwon, and Y.-H. Kim, 'Record high mobility in polymer semiconductors via side-chain engineering', *Journal of the American Chemical Society*, vol. 135, pp. 14896–14899, 2013.

[615] B. Meng, H. Song, X. Chen, Z. Xie, J. Liu, and L. Wang, 'Replacing alkyl with oligo(ethylene glycol) as side chains of conjugated polymers for close $\pi-\pi$ stacking', *Macromolecules*, vol. 48, pp. 4357–4363, 2015.

[616] S. Chen, B. Sun, W. Hong, H. Azia, Y. Meng, and Y. Li, 'Influence of side chain length and bifurcation point on the crystalline structure and charge transport of diketopyyrolopyrrole-quaterthiophene copolymers (PDQTs)', *Journal of Materials Chemistry C*, vol. 2, pp. 2183–2190, 2014.

[617] X. Zhao and X. Zhan, 'Electron transporting semiconducting polymers in organic electronics', *Chemical Society Reviews*, vol. 40, pp. 3728–3743, 2011.

[618] H. Yan, Z. Chen, Y. Zhang, C. Newman, J. R. Quinn, F. Dötz, et al., 'A high-mobility electron transporting polymer for printed transistors', *Nature*, vol. 457, pp. 679–687, 2009.

[619] V. Senkovskyy, R. Tkachiv, H. Komber, M. Sommer, M. Heuken, B. Voit, et al., 'Chain-growth polymerization of unusual anion-radical monomers based on

naphthalene diimide: A new route to well-defined n-type conjugated copolymers', *Journal of the American Chemical Society*, vol. 133, pp. 19966–19970, 2011.

[620] B. Sun, W. Hong, Z. Yan, H. Aziz, and Y. Li, 'Record high electron mobility of 6.3 cm2V-1s-1 achieved for, polymer semiconductors using a new building block', *Advanced Materials*, vol. 26, pp. 2636–2642, 2014.

[621] T. Lei, X. Xia, J.-Y. Wang, C.-J. Liu, and J. Pei, '"Conformation locked" strong electron deficient poly(*p*-phenylene vinylene) derivatives for ambient stable n-type field effect transistors: Synthesis, propoerties and effects of fluorine substitution position', *Journal of the American Chemical Society*, vol. 136, pp. 2135–2141, 2014.

[622] T. Lei, J.-H. Dou, X.-Y. Cao, J.-Y. Wang, and J. Pei, 'A BDOPV-based donor acceptor polymer for high performance n-type and oxygen-doped ambipolar field-effect transitors', *Advanced Materials*, vol. 25, pp. 6589–6593, 2013.

[623] J. Lee, S. Cho, and C. Yang, 'Highly reproducible organic field-effect transistor from pseudo 3-dimensional triphenylamine-based amorphous conjugated copolymer', *Journal of Materials Chemistry*, vol. 21, pp. 8528–8531, 2011.

[624] P.-T. Wu, F. S. Kim, and S. A. Jenekhe, 'New poly(arylene vinylene)s based on diketopyrrolopyrrole for ambipolar transistors', *Chemistry of Materials*, vol. 23, pp. 4618–4624, 2011.

[625] S. Cho, J. Lee, M. Tong, J. H. Seo, and C. Yang, 'Poly(diketopyrrolopyrrole-benzothiadiazole) with ambipolarity approaching 100% equivalency', *Advanced Functional Materials*, vol. 21, pp. 1910–1916, 2011.

[626] J. Lee, S. Cho, J. H. Seo, P. Anant, J. Jacob, and C. Yang, 'Swapping field-effect transistor characteristics in polymeric diketopyrrolopyrrole semiconductors: Debut of an electron dominant transporting polymer', *Journal of Materials Chemistry*, vol. 22, pp. 1504–1510, 2012.

[627] Y. Wang, H. Masunaga, T. Hikima, H. Matsumoto, T. Mori, and T. Michinobu, 'New semiconducting polymers based on benzobisthiadiazole analogues: Tuning of charge polarity in thin film transistors via heteroatom substitution', *Macromolecules*, vol. 48, pp. 4012–4023, 2015.

[628] C. Xiao, G. Zhao, A. Zhang, W. Jiang, R. A. J. Janssen, W. Li, et al., 'High performance polymer nanowire field-effect transistors with distinct molecular orientations', *Advanced Materials*, vol. 27, pp. 4963–4968, 2015.

[629] D. E. Schwartz, J. Rivnay, G. L. Whiting, P. Mei, Y. Zhang, B. Krusor, et al., 'Flexible hybrid electronic circuits and systems', *IEEE Journal on Emerging and Selected Topics in Circuits and Systems*, vol. 7, pp. 28–37, 2017.

[630] J. S. Chang, A. F. Facchetti, and R. A. Reuss, 'Circuits and systems perspective of organic/printed electronics: Review, challenges, and contemporary and emerging design approaches', *IEEE Journal on Emerging and Selected Topics in Circuits and Systems*, vol. 7, pp. 7–26, 2017.

[631] S. Casalini, C. A. Bortolotti, F. Leonardi, and F. Biscarini, 'Self-assembled monolayers in organic electronics', *Chemical Society Reviews*, vol. 46, pp. 4–71, 2017.

[632] S. Holliday, J. E. Donaghey, and I. McCulloch, 'Advances in charge carrier mobilities of semiconducting polymers used in organic transistors', *Chemistry of Materials*, vol. 26, pp. 647–663, 2014.

[633] J. Zhou, T. Ge, E. Ng, and J. S. Chang, 'Fully additive low-cost printed electronics with very low process variations', *IEEE Transactions on Electron Devices*, vol. 63, pp. 793–799, 2016.

[634] S. D. Dimitrov and J. R. Durrant, 'Materials design considerations for charge generation in organic solar cells', *Chemistry of Materials*, vol. 26, pp. 616–630, 2014.

[635] D. Baran, T. Kirchartz, S. Wheeler, S. Dimitrov, M. Abdelsamie, J. Gorman, et al., 'Reduced voltage losses yield 10% efficient fullerene free organic solar cells with >1 V open circuit voltages', *Energy and Environmental Science*, vol. 9, pp. 3783–3793, 2016.

[636] H. Zhou, L. Yang, S. Stoneking, and W. You, 'A weak donor-strong acceptor strategy to design ideal polymers for organic solar cells', *ACS Applied Materials and Interfaces*, vol. 2, pp. 1377–1383, 2010.

[637] H. Zhou, L. Yang, and W. You, 'Rational design of high performance conjugated polymers for organic solar cells', *Macromolecules*, vol. 45, pp. 607–632, 2012.

[638] Z.-G. Zhang and J. Wang, 'Structures and properties of conjugated donor-acceptor copolymers for solar cell applications', *Journal of Materials Chemistry*, vol. 22, pp. 4178–4187, 2012.

[639] O. O. Adegoke, I. H. Jung, M. Orr, L. Yu, and T. Goodson, 'Effect of acceptor strength on optical and electronic properties in conjugated polymers for solar applications', *Journal of the American Chemical Society*, vol. 137, pp. 5759–5769, 2015.

[640] E. Zhou, M. Nakamura, T. Nishizawa, Y. Zhang, Q. Wei, K. Tajima, et al., 'Synthesis and photovoltaic propetries of a novel low band gap polymer based on N-substituted dithieno[3,2-*b*:2',3'-*d*]pyrrole', *Macromolecules*, vol. 41, pp. 8302–8305, 2008.

[641] T. T. Steckler, X. Zhang, J. Hwang, R. Honeyager, S. Ohira, X.-H. Zhang, et al., 'A spray-processable, low bandgap, and ambipolar donor-acceptor conjugated polymer', *Journal of the American Chemical Society*, vol. 131, pp. 2824–2826, 2009.

[642] Z. Wu, W. Yao, A. E. London, J. D. Azoulay, and T. N. Ng, 'Temperature-dependent detectivity of near-infrared organic bulk heterojunction photodiodes', *ACS Applied Materials and Interfaces*, vol. 9, pp. 1654–1660, 2017.

[643] L. Dou, Y. Liu, Z. Hong, G. Li, and Y. Yang, 'Low-bandgap near-IR conjugated polymers/molecules for organic electronics', *Chemical Reviews*, vol. 115, pp. 12633–12665, 2015.

[644] H. Li, T. L. Tam, Y. M. Lam, S. G. Mhaisalkar, and A. C. Grimsdale, 'Synthesis of low bandgap [1,2,5]-thiadiazolo[3,4-*g*]quinoxaline and pyrazino[2,3-*g*]quinoxaline derivatives by selective reduction of benzo[1,2-*c*;4,5-*c*']bis[1,2,5]thiadiazole', *Organic Letters*, vol. 13, pp. 46–49, 2011.

[645] T. L. Tam, H. Li, Y. M. Lam, S. G. Mhaisalkar, and A. C. Grimsdale, 'Synthesis and characterization of [1,2,5]chalcogenazolo[3,4-*f*]benzotriazole and [1,2,3]triazolo [3,4-*g*]quinoxaline derivatives', *Organic Letters*, vol. 13, pp. 4612–4615, 2011.

[646] D. G. Patel, F. Feng, Y.-Y. Ohnishi, K. A. Abboud, K. S. Schanze, and J. R. Reynolds, 'It takes more than an imine: The role of the central atom on the electron-accepting ability of benzotriazole and benzothiadiazole oligomers', *Journal of the American Chemical Society*, vol. 134, pp. 2599–2612, 2012.

[647] T. L. Tam, F. Zhou, H. Li, J. C. Y. Pang, Y. M. Lam, S. G. Mhaisalkar, et al., 'Substituent effects on the electronic properties of pyrazino[2,3-*g*]quinoxaline molecules', *Journal of Materials Chemistry*, vol. 21, pp. 17798–17804, 2011.

[648] H. Meng and F. Wudl, 'A robust low band gap processable n-type conducting polymer based on poly(isothianaphthene)', *Macromolecules*, vol. 34, pp. 1810–1816, 2001.

[649] H. Li, S. Sun, T. Salim, S. Bomma, A. C. Grimsdale, and Y. M. Lam, 'Conjugated polymers based on dicarboxylic imide-substituted isothianaphthene and their applications in solar cells', *Journal of Polymer Science (A), Polymer Chemistry*, vol. 50, pp. 250–260, 2012.

[650] H. J. Son, F. He, B. Carsten, and L. Yu, 'Are we there yet? Design of better conjugated polymers for polymer solar cells', *Journal of Materials Chemistry*, vol. 21, pp. 18934–18945, 2011.
[651] N. Blouin, A. Michaud, and M. Leclerc, 'A low bandgap poly(2,7-carbazole) derivative for use in high-performance solar cells', *Advanced Materials*, vol. 19, pp. 2295–2300, 2007.
[652] S. H. Park, A. Roy, S. Beaupré, S. Cho, N. Coates, J. S. Moon, et al., 'Bulk heterojunction solar cells with internal quantum efficiency approaching 100%', *Nature Photonics*, vol. 3, pp. 297–303, 2009.
[653] T.-Y. Chu, S. Alem, S.-W. Tsang, S.-C. Tse, S. Wakim, J. Lu, et al., 'Morphology control in polycarbazole based bulk hetrojunction solar cells and its impact on device performance', *Applied Physics Letters*, vol. 98, p. 253301, 2011.
[654] S. Beaupre and M. Leclerc, 'PCDTBT: En route for low cost plastic solar cells', *Journal of Materials Chemistry A*, vol. 1, pp. 11097–11105, 2013.
[655] N. Blouin, A. Michaud, D. Gendron, S. Wakim, E. Blair, R. Neagu-Plesu, et al., 'Towards a rational design of poly(2,7-carbazole) derivatives for solar cells', *Journal of the American Chemical Society*, vol. 130, pp. 732–742, 2008.
[656] Z. He, C. Zhong, X. Huang, W.-Y. Wong, H. Wu, L. Chen, et al., 'Simultaneous enhancement of open-circuit violtage, short-circuit current density, and fill factor in polymer solar cells', *Advanced Materials*, vol. 23, pp. 4636–4643, 2011.
[657] S. Liu, K. Zhang, J. Lu, J. Zhang, H.-P. Yip, F. Huang, and Y. Cao, 'High-efficiency polymer solar cells via the incorporation of an amino-functionalized conjugated metallopolymer as a cathode interlayer', *Journal of the American Chemical Society*, vol. 135, pp. 15326–15329, 2013.
[658] J.-D. Chen, C. Cui, Y.-Q. Li, L. Zhou, Q.-D. Ou, C. Li, et al., 'Single-junction polymer solar cells exceeeding 10% power conversion efficiency', *Advanced Materials*, vol. 27, pp. 1035–1041, 2015.
[659] H. Zhou, L. Yang, A. C. Stuart, S. C. Price, S. Liu, and W. You, 'Development of fluorinated benzothiadiazole as a structural unit for a polymer solar cell of 7% efficiency', *Angewandte Chemie, International Edition*, vol. 50, pp. 2995–2998, 2011.
[660] Y. Huang, L. Huo, S. Zhang, X. Guo, C. C. Han, Y. Li, and J. Hou, 'Sulfonyl: A new application of electron-withdrawing substituent in highly efficient photovoltaic polymer', *Chemical Communications*, vol. 47, pp. 8904–8906, 2011.
[661] J. W. Jo, S. Bae, F. Liu, T. P. Russell, and W. H. Jo, 'Comparison of two D-A type polymers with each being fluorinated on D and A unit for high performance solar cells', *Advanced Functional Materials*, vol. 25, pp. 120–125, 2015.
[662] Y. Liu, J. Zhao, Z. Li, C. Mu, W. Ma, H. Hu, et al., 'Aggregation and morphology control enables multiple cases of high-efficiency polymer solar cells', *Nature Communications*, vol. 5, p. 5293, 2014.
[663] L. Huo, S. Zhang, X. Guo, F. Xu, F. Li, and J. Hou, 'Replacing alkoxy groups with alkylthienyl groups: A feasible approach to improve the properties of photovoltaic polymers', *Angewandte Chemie, International Edition*, vol. 50, pp. 9697–9702, 2011.
[664] J.-H. Kim, C. E. Song, B. J. Kim, I.-N. Kang, W. S. Shin, and D. H. Hwang, 'Thieno [3,2-b]thiophene-substituted benzo[1,2-*b*:4,5-*b*']dithiophene as apromising building block for low bandgap semiconducting polymers for high performance singale and tandem organic photovoltaic cells', *Chemistry of Materials*, vol. 26, pp. 1234–1242, 2014.
[665] S.-H. Liao, H.-J. Jhuo, P.-N. Yeh, Y.-S. Cheng, Y.-L. Li, Y.-H. Lee, et al., 'Single-junction inverted polymer solar cell reaching power conversion efficiency by

employing dual-doped zinc oxide nano-film as cathode interlayer', *Scientific Reports*, vol. 4, pp. 6813/1–6813/7, 2014.

[666] L. Nian, W. Zhang, N. Zhu, L. Liu, Z. Xie, H. Wu, et al., 'Photoconductive cathode interlayer for highly efficient inverted polymer solar cells', *Journal of the American Chemical Society*, vol. 137, pp. 6995–6998, 2015.

[667] H. Zhou, Y. Zhang, C.-K. Mai, S. D. Collins, G. C. Bazan, T.-Q. Nguyen, and A. J. Heeger, 'Polymer homo-tandem solar cells with best efficiency of 11.3%', *Advanced Materials*, vol. 27, pp. 1767–1773, 2015.

[668] W.-H. Chang, L. Meng, L. Dou, J. You, C.-C. Chen, Y. Yang, et al., 'A selenophene containign benzodithiophene-*alt*-thienothiophene polymer for additive-free high performance solar cell', *Macromolecules*, vol. 48, pp. 562–568, 2015.

[669] L. Huo, T. Liu, X. Sun, Y. Cai, A. J. Heeger, and Y. Sun, 'Single-junction organic solar cells based on a novel wde-bandgap polymer with efficiency of 9.7%', *Advanced Materials*, vol. 27, pp. 2938–2944, 2015.

[670] W. Zhao, S. Li, H. Yao, S. Zhang, Y. Zhang, B. Yang, and J. Hou, 'Molecular optimization enables over 13% efficiency in organic solar cells', *Journal of the American Chemical Society*, vol. 139, p. 7148–7151, 2017.

[671] Z. Fei, F. D. Eisner, X. Jiao, M. Azzouzi, J. A. Röhr, Y. Han, et al., 'An alkylated indacenodithieno[3,2-b]thiophene-based nonfullerene acceptor with high crystallinity exhibiting single junction solar cell efficiencies greater than 13% with low voltage losses', *Advanced Materials*, vol. 30, p. 1705209, 2018.

[672] Y.-Z. Cui, H. Yao, J. Zhang, T. Zhang, Y. Wang, L. Hong, et al., 'Over 16% efficiency organic photovoltaic cells enabled by a chlorinated acceptor with increased open-circuit voltages', *Nature Communications*, vol. 10, p. 2515, 2019.

[673] J. Wang, Z. Zheng, Y. Zu, Y. Wang, X. Liu, S. Zhang, et al., 'A tandem organic photovoltaic cell with 19.6% efficiency enabled by light distribution control', *Advanced Materials*, vol. 33, p. 2102787, 2021.

[674] Q. Liu, Y. Jiang, K. Jin, J. Qin, J. Xu, W. Li, et al., '18% Efficiency solar cells', *Science Bulletin*, vol. 65, pp. 272–275, 2020.

[675] J. You, L. Dou, K. Yoshimura, T. Kato, K. Ohya, T. Moriarty, et al., 'A polymer tandem solar cell with 10.6% power conversion efficiency', *Nature Communications*, vol. 4, p. 1446, 2013.

[676] J. Cao, L. Qian, F. Lu, J. Zhang, Y. Feng, X. Qiu, et al., 'A lactam building block for efficient polymer solar cells', *Chemical Communications*, vol. 51, pp. 11830–11833, 2015.

[677] C.-C. Chen, W.-H. Chang, K. Yoshimura, K. Ohya, J. You, J. Gao, et al., 'An efficient triple-junction polymer solar cell having a power efficiency exceeding 11%', *Advanced Materials*, vol. 26, pp. 5670–5677, 2014.

[678] A. R. bin Mohd Yusoff, D. H. Kim, H. P. Kim, F. K. Shneider, W. J. da Silva, and J. Jang, 'A high efficiency solution processed polymer inverted triple-junction solar cell exhibiting a power conversion efficiency of 11.83%', *Energy and Environmental Science*, vol. 8, pp. 303–316, 2015.

[679] L. Meng, Y. Zhang, X. Wan, C. Li, X. Zhang, Y. Wang, et al., 'Organic and solution-processed tandem solar cells with 17.3% efficiency', *Science*, vol. 361, pp. 1094–1098, 2019.

[680] C. J. Mulligan, M. Wilson, G. Bryant, B. Vaughan, X. Zhou, W. J. Belcher, and P. C. Dastoor, 'A projection of commercial-scale organic photovoltaic module costs', *Solar Energy Materials and Solar Cells*, vol. 120, pp. 9–17, 2014.

[681] C. J. Mulligan, C. Bilen, X. Zhou, W. J. Belcher, and P. C. Dastoor, 'Levelised cost of electricity for organic photovoltaics', *Solar Energy Materials and Solar Cells*, vol. 133, pp. 26–31, 2015.

[682] A. Gambhir, P. Sandwell, and J. Nelson, 'The future costs of OPV: A bottom-up model of material and manufacturing costs with uncertainty analysis', *Solar Energy Materials and Solar Cells*, vol. 156, pp. 49–58, 2016.

[683] N. A. Cooling, E. F. Barnes, F. Almyadi, K. Feron, M. F. Al-Mudhaffer, A. Al-Ahmad, et al., 'A low-cost mixed fullerene acceptor blend for printed electronics', *Journal of Materials Chemistry A*, vol. 4, pp. 10274–10281, 2016.

[684] R. Po, G. Bianchi, C. Carbonera, and A. Pellegrino, '"All that glisters is not gold": An analysis of the synthetic complexity of efficient polymer donors for polymer cells', *Macromolecules*, vol. 48, pp. 453–461, 2015.

[685] C.-Z. Li, H.-L. Yip, and A. K.-Y. Jen, 'Functional fullerenes for organic photovoltaics', *Journal of Materials Chemistry*, vol. 22, pp. 4161–4177, 2012.

[686] Y. Zhong, M. T. Trinh, R. Chen, G. E. Purdam, P. P. Khlyabich, M. Sezen, et al., 'Molecular helices as electron acceptors in high-performance bulk heterojunction solar cells', *Nature Communications*, vol. 6, pp. 8242/1–8242/8, 2015.

[687] D. Mori, H. Benten, I. Okada, H. Ohkita, and S. Ito, 'Low-bandgap donor/acceptor polymer blend solar cell with efficiency exceeding 4%', *Advanced Energy Materials*, vol. 4, p. 1301006, 2014.

[688] Y.-J. Hwang, B. A. E. Courtright, A. S. Ferreira, S. H. Tolbert, and S. A. Jenekhe, '7.7% Efficient all-polymer solar cells', *Advanced Materials*, vol. 27, pp. 4578–4584, 2015.

[689] B. Fan, L. Ying, P. Zhu, F. Pan, F. Liu, J. Chen, et al., 'All-polymer solar cells based on a conjugated polymer containing siloxane-functionalized side chains with efficiency over 10%', *Advanced Materials*, vol. 29, p. 1703906, 2017.

[690] Z. Genene, W. Mammo, E. Wang, and M. R. Andersson, 'Recent advances in n-type polymers for all-polymer solar cells', *Advanced Materials*, vol. 31, p. 18027275, 2019.

[691] M.-H. Chen, J. Hou, Z. Hong, G. Yan, L.-M. Chen, and Y. Yang, 'Efficient polymer solar cells with thin active layers based on alternating fluorene copolymer/fullerene bulk heterojunction', *Advanced Materials*, vol. 21, pp. 4238–4242, 2009.

[692] C. Du, C. Li, W. Li, X. Chen, Z. Bo, C. Veit, et al., '9-Alkylidene-*9H*-fluorene-containing polymer for high-efficiency polymer solar cells', *Macromolecules*, vol. 44, pp. 7617–7624, 2011.

[693] H. Li, S. Sun, S. G. Mhaisalkar, M. T. Zin, Y. M. Lam, and A. C. Grimsdale, 'A high voltage solar cell using a donor-acceptor conjugated polymer based on pyrrolo[3,4-*f*]-2,1,3-benzothiadiazole-5,7-dione', *Journal of Materials Chemistry A*, vol. 2, pp. 17925–17933, 2014.

[694] M. Manceau, E. Bundgaard, J. E. Carlé, O. Hagemann, M. Helgesen, R. Søndergaard, et al., 'Photochemical stability of π-conjugated polymers for polymer solar cells: A rule of thumb', *Journal of Materials Chemistry*, vol. 21, pp. 4132–4141, 2011.

[695] J. Song, C. Du, C. Li, and Z. Bo, 'Silole-containing polymers for high-efficiency polymer solar cells', *Journal of Polymer Science (A), Polymer Chemistry*, vol. 49, pp. 4267–4278, 2011.

[696] N. Shin, H.-J. Yun, Y. Yoon, H. J. Son, S.-Y. Ju, S.-K. Kwon, et al., 'Highly stable polymer solar cells based on poly(dithienobenzodithiophene-*co*-thienothiophene)', *Macromolecules*, vol. 48, pp. 3890–3899, 2015.

[697] U. Dettinger, H.-J. Egelhaaf, C. J. Brabec, F. Latteyer, H. Peisert, and T. Chasse, 'FTIR study of the impact of PC[60]BM on the photodegradation of the low band gap polymer PCPDTBT under O_2 environment', *Chemistry of Materials*, vol. 27, pp. 2299–2308, 2015.

[698] J. Kesters, P. Verstappen, J. Raymakers, W. Vanormelingen, J. Drijkoningen, J. D'Haen, et al., 'Enhanced organic solar cell stability by polymer (PCPDTBT) side chain functionalization', *Chemistry of Materials*, vol. 27, pp. 1332–1341, 2015.
[699] M. P. Nikiforov, B. Lai, W. Chen, S. Chen, R. D. Schaller, J. Strzalka, et al., 'Detection and role of trace impurities in high-performance organic solar cells', *Energy and Environmental Science*, vol. 6, pp. 1513–1520, 2013.
[700] T. Vangerven, P. Verstappen, J. Drijkoningen, W. Dierckx, S. Himmelberger, A. Salleo, et al., 'Molar mass versus polymer solar cell performance: Highlighting the role of homocouplings', *Chemistry of Materials*, vol. 27, pp. 3726–3732, 2015.
[701] L. Lu, T. Zheng, T. Xu, D. Zhao, and L. Yu, 'Mechanistic study of effect of dispersity on the photovoltaic performance of PTB7 polymer solar cells', *Chemistry of Materials*, vol. 27, pp. 537–543, 2015.
[702] J. K. Park, J. Jo, J. H. Seo, J. S. Moon, Y. D. Park, K. Lee, et al., 'End-capping effect of a narrow bandgap conjugated polymer on bulk heterojunction solar cells', *Advanced Materials*, vol. 23, pp. 2430–2435, 2011.
[703] W. Chang, D. N. Congreve, E. Hontz, M. E. Bahlke, D. P. McMahon, S. Reineke, et al., 'Spin-dependent charge transfer state design rules in organic photovoltaics', *Nature Communications*, vol. 6, pp. 6415/1–6415/6, 2015.
[704] M. Qian, R. Zhang, J. Hao, W. Zhang, Q. Zhang, J. Wang, et al., 'Dramatic enhancement of power conversion efficiency in polymer solar cells by conjugated very low ratio of triplet iridium complexes to PTB7', *Advanced Materials*, vol. 27, pp. 3546–3552, 2015.
[705] J. Kesters, P. Verstappen, J. Raymakers, W. Vanormelingen, J. Drijkoningen, J. D'Haen, et al., 'Enhanced organic solar cell stability by polymer (PBPDTBT) side chain functionalization', *Chemistry of Materials*, vol. 27, pp. 1332–1341, 2015.
[706] M. Jikei and M. Kakimoto, 'Hyperbranched polymers: A promising new class of materials', *Progress in Polymer Science*, vol. 26, pp. 1233–1285, 2001.
[707] T. Lin, Q. He, F. Bai, and L. Dai, 'Synthesis and photophysics of a hyperbranched conjugated polymer', *Thin and Solid Films*, vol. 363, pp. 122–125, 2000.
[708] S.-J. Lim, D. Y. Seok, B.-K. An, S. D. Jung, and S. Y. Park, 'Modified strategy for the synthesis of hyperbranched PPV: Achieving extended π-conjugation with growing molecular weight', *Macromolecules*, vol. 39, pp. 9–11, 2006.
[709] Z. Xie, S.-J. Yoon, and S. Y. Park, 'Synthesis of highly fluorescent and soluble 1,2,4-linking hyperbranched poly(arylenevinylene) featuring intramolecular energy funneling', *Advanced Functional Materials*, vol. 20, pp. 1638–1644, 2010.
[710] H. S. Mangold, T. V. Richter, S. Link, U. Würfel, and S. Ludwigs, 'Optoelectronic properties of hyperbranched polythiophenes', *Journal of Physical Chemistry B*, vol. 116, pp. 154–159, 2012.
[711] C.-Q. Ma, E. Mena-Osteritz, T. Debaerdemaeker, M. M. Wienk, R. A. J. Janssen, and P. Bäuerle, 'Functionalized 3D oligothiophene dendrons and dendrimers: Novel macromolecules for organic electronics', *Angewandte Chemie, International Edition*, vol. 46, pp. 1679–1683, 2007.
[712] C.-Q. Ma, E. Mena-Osteritz, M. Wunderlin, G. Schulz, and P. Bäuerle, '2,2':3',2''-Terthiophene-based all-thiophene dendrons and dendrimers: Synthesis, structuural characterization, and properties', *Chemistry: A European Journal*, vol. 18, pp. 12880–12901, 2012.
[713] H. Dong, J. W. Y. Lam, M. Haeussler, R. Zheng, H. Peng, C. C. W. Law, and B. Z. Tang, 'Constructions of conjugated macromolecules from acetylenic building blocks: Syntheses and properties of hyperbranched polyarylenes and polyynes', *Current Trends in Polymer Science*, vol. 9, pp. 15–31, 2004.

[714] H. Dong, Y. Dong, J. W. Y. Lam, M. Häussler, H. Peng, and B. Z. Tang, 'Synthesis of a hyperbranched polyarylene with phenylenevinylene chromophore', *Polymer Preprints*, vol. 45, no. 2, pp. 831–832, 2004.

[715] H. Peng, R. Zheng, H. Dong, D. Jia, and B. Z. Tang, 'Synthesis of hyperbranched conjugative poly(aryleneethynylene)s by alkyne polycyclotrimerization', *Chinese Journal of Polymer Science*, vol. 23, pp. 1–3, 2005.

[716] H. Peng, J. W. Y. Lam, and B. Z. Tang, 'Hyperbranched poly(aryleneethynylene)s: Synthesis, thermal stability and optical properties', *Macromolecular Rapid Communications*, vol. 26, pp. 673–677, 2005.

[717] H.-M. Shih, R.-C. Wu, P.-I. Shih, C.-L. Wang, and C.-S. Hsu, 'Synthesis of fluorene-based hyperbranched polymers for solution processable blue, green, red, and white light-emitting devices', *Journal of Polymer Science (A), Polymer Chemistry*, vol. 50, pp. 696–710, 2012.

[718] C. Coya, A. L. Alvarez, W. S. Yoon, and S. Y. Park, 'Influence of the 1,2,4-linking hyperbranched poly(arylenevinylene) structure on organic light emitting diode performance as compared to conventional 1,3,5-linking one', *Journal of Applied Physics*, vol. 109, p. 094507, 2011.

[719] X.-S. Ma, Y.-Z. Cui, Y.-Q. Ding, F.-R. Tao, B. Zheng, R.-H. Yu, and W. Huang, '2D hyperbranched conjugated polymer for detecting TNT with excellent exciton migration', *Sensors and Actuators B*, vol. 238, pp. 48–57, 2017.

[720] X. Wu, Z. Zhang, H. Hang, Y. Chen, Y. Xu, H. Tong, and L. Wang, 'Solution-processable hyperbranched conjugated polymer nanoparticles based on C3h-symmetric benzotrithiophene for polymer solar cells', *Macromolecular Rapid Communications*, vol. 38, p. 1700001, 2017.

[721] P. L. Burn, S.-C. Lo, and I. D. W. Samuel, 'The development of light-emitting dendrimers for displays', *Advanced Materials*, vol. 19, pp. 1675–1688, 2007.

[722] A. C. Grimsdale, T. Vosch, M. Lor, M. Cotlet, S. Habuchi, J. Hofkens, et al., 'Synthesis of and excited state processes in multichromophoric dendritic systems', *Journal of Luminescence*, vol. 111, pp. 239–253, 2005.

[723] A. C. Grimsdale and K. Müllen, 'The chemistry of organic nanomaterials', *Angewandte Chemie, International Edition*, vol. 44, pp. 5592–5629, 2005.

[724] J. N. G. Pillow, P. L. Burn, I. D. W. Samuel, and M. Halim, 'Synthetic routes to phenylene vinylene dendrimers', *Synthetic Metals*, vol. 102, pp. 1468–1469, 1999.

[725] J. N. G. Pillow, M. Halim, J. M. Lupton, P. L. Burn, and I. D. W. Samul, 'Facile iterative procedure for making luminescent dendrimers', *Macromolecules*, vol. 32, pp. 5985–5993, 1999.

[726] M. Halim, I. D. W. Samuel, J. N. G. Pillow, and P. L. Burn, 'Conjugated dendrimers for OLEDs: Control of colour', *Synthetic Metals*, vol. 102, pp. 1113–1114, 1999.

[727] M. Halim, J. N. G. Pillow, I. D. W. Samuel, and P. L. Burn, 'Conjugated dendrimers for LEDs: Effect of generation', *Advanced Materials*, vol. 11, pp. 371–374, 1999.

[728] R. Bauer, A. C. Grimsdale, and K. Müllen, 'Functionalised polyphenylene dendrimers and their applications', *Topics in Current Chemistry*, vol. 245, p. 253, 2004.

[729] T. Weil, E. Reuther, and K. Müllen, 'Polyphenylene dyads and triad for transduction of excitation energy', *Angewandte Chemie International Edition*, vol. 41, pp. 1900–1904, 2002.

[730] A. W. Freeman, S. C. Koene, P. R. L. Malefant, M. E. Thompson, and J. M. J. Fréchet, 'Dendrimer-containing LEDs', *Journal of the American Chemical Society*, vol. 122, pp. 12385–12386, 2000.

[731] T. W. Kwon, M. M. Alam, and S. A. Jenekhe, 'N-type conjugated dendrimers: Synthesis, electroluminescence and use as electron transporting materials for LEDs', *Chemistry of Materials*, vol. 16, pp. 4657–4666, 2004.
[732] H. Ren, J. Li, R. Wang, T. Zhang, Z. Gao, and D. Liu, 'Synthesis and luminescent properties of perylene bisimide-cored dendrimers witrh carbazole surface groups', *Polymer*, vol. 52, pp. 3639–3646, 2011.
[733] S.-C. Lo and P. L. Burn, 'Development of dendrimers: Macromolecules for use in organic light-emitting diodes and solar cells', *Chemical Reviews*, vol. 107, pp. 1097–1116, 2007.
[734] G. Tang, S. S. Y. Chen, P. E. Shaw, K. Hegedus, X. Wang, P. L. Burn, and P. Meredith, 'Fluorescent carbazole dendrimers for the detection of explosives', *Polymer Chemistry*, vol. 2, pp. 2360–2368, 2011.
[735] M. A. Ali, S. S. Y. Chen, H. Cavaye, A. R. G. Smith, P. L. Burn, I. R. Gentle, et al., 'Diffusion of nitroaromatic vapours into fluorescent dendrimer films for explosives detection', *Sensors and Actuators B*, vol. 210, pp. 550–557, 2015.
[736] S. Shoaee, S. S. Y. Chen, H. Cavaye, A. R. G. Smith, P. L. Burn, I. R. Gentle, et al., 'Assessing the sensing limits of fluorescent dendrimer thin films for the detection of explosive vapors', *Sensors and Actuators B*, vol. 239, pp. 727–733, 2017.
[737] N. Kopidakis, W. J. Mitchell, J. van de Lagemaat, D. S. Ginley, G. Rumbles, S. E. Shaheen, and W. L. Rance, 'Bulk heterojunction organic photovoltaic devices based on phenyl-cored thiophene dendrimers', *Applied Physics Letters*, vol. 89, p. 103524, 2006.
[738] C.-Q. Ma, M. Fonrodona, M. C. Schikora, M. M. Wienk, R. A. J. Janssen, and P. Bäuerle, 'Solution-processed bulk-heterojunction solar cells based on monodisperse dendritic oligothiophenes', *Advanced Functional Materials*, vol. 18, pp. 3323–3331, 2008.
[739] D. M. Stoltzfus, C.-Q. Ma, R. C. R. Nagiri, A. J. Clulow, P. Bäuerle, P. L. Burn, et al., 'Thiophene dendrimer-based low donor content solar cells', *Applied Physics Letters*, vol. 109, p. 103302, 2016.
[740] B. Li, X. Xu, M. Sun, Y. Fu, G. Yu, Y. Liu, and Z. Bo, 'Porphyrin-cored star polymers as efficient nondoped red light-emitting materials', *Macromolecules*, vol. 39, pp. 456–461, 2006.
[741] H. J. Kim, K. Y. Cho, S. S. Hwang, D. H. Choi, M. J. Ko, and K. Y. Baek, 'Controlled synthesis of multi-armed P3HT star polymers with gold nanoparticle core', *RSC Advances*, vol. 6, pp. 49206–49213, 2016.
[742] X.-C. Li, A. C. Grimsdale, R. Cervini, A. B. Holmes, S. C. Moratti, T. M. Yong, et al., 'Synthesis and properties of novel high-electron-affinity polymers for electroluminescent devices', *ACS Symposium Series*, vol. 672, pp. 322–344, 1997.
[743] M. Sommer, S. Huettner, and M. Thelekkat, 'Donor-acceptor block copolymers for photovoltaic applications', *Journal of Materials Chemistry*, vol. 20, pp. 10788–10797, 2010.
[744] J.-K. Lee, R. R. Schrock, D. R. Baigent, and R. H. Friend, 'A new type of blue-light-emitting electroluminescent polymer', *Macromolecules*, vol. 28, pp. 1966–1971, 1995.
[745] P. Hesemann, H. Vestweber, J. Pommerehne, R. F. Mahrt, and A. Greiner, 'A blue light emitting polymer with phenylenevinylene segments in the side-chains', *Advanced Materials*, vol. 7, pp. 388–390, 1995.
[746] J. Sanetra, P. Armatys, S. Niziol, D. Bogdal, M. Warzala, and J. Pielichowski, 'Light from electroluminescent diodes made from copolymers', *Synthetic Metals*, vol. 121, pp. 1733–1734, 2001.

[747] J. Sanetra, D. Bogdal, S. Niziol, P. Armatys, and J. Pielichowski, 'Electroluminescence of poly(N-vinylcarbazole) (PVK) and its blends with 3-(-2-methacrylate-ethaoxy)carbonyl dimethylocoumarine', *Synthetic Metals*, vol. 121, pp. 1731–1732, 2001.

[748] J. Vollbrecht, 'Excimers in organic electronics', *New Journal of Chemistry*, vol. 42, pp. 11249–11254, 2018.

[749] C. Li and H. Wonneberger, 'Perylene imides for organic photovoltaics: Yesterday, today, and tomorrow', *Advanced Materials*, vol. 24 pp. 613–636, 2012.

[750] A. L. Stevens, S. Novakovic, J. M. White, W. W. H. Wong, T. A. Smith, K. P. Ghiggino, et al., 'Exciton dynamics of photoexcited pendant porphyrin polymers in solution and in thin films', *Journal of Physical Chemistry A*, vol. 122, pp. 9605–9614, 2018.

[751] N. Leclerc, M.-C. Pasareanu, and A.-J. Attias, 'Synthesis and photophysical properties of polymers containing a novel class of light emitters', *Macromolecules*, vol. 38, pp. 1531–1534, 2005.

[752] Z. Yang, F. E. Karasz, and H. J. Geise, 'Intrinsically soluble copolymers with well-defined alternating substituted p-phenylenevinylene and ethylene oxide blocks', *Macromolecules*, vol. 26, pp. 6570–6575, 1993.

[753] Z. Yang, I. Sokolik, and F. E. Karasz, 'A soluble blue-light-emitting polymer', *Macromolecules*, vol. 26, pp. 1188–1190, 1993.

[754] S. T. Pasco, P. M. Lahti, and F. E. Karasz, 'Synthesis of substituted poly(p-phenylenevinylene) copolymers by the Heck method for luminescence studies', *Macromolecules*, vol. 32, pp. 6933–6937, 1999.

[755] N. Benfaremo, D. J. Sandman, S. Tripathy, J. Kumar, K. Yang, M. F. Rubner, and C. Lyons, 'Synthesis and characterization of luminescent polymers of distrylbenzenes with oligo(ethylene glycol) spacers', *Macromolecules*, vol. 31, pp. 3595–3599, 1998.

[756] I. D. Parker, Q. Pei, and M. Marrocco, 'Efficient blue electroluminescence from a fluorinated polyquinoline', *Applied Physics Letters*, vol. 65, pp. 1272–1274, 1994.

[757] V. B. Yasarapudi, L. Frazer, N. J. L. K. Davis, E. P. Booker, A. Macmillan, J. K. Gallaher, et al., 'Optimization of energy transfer in a polymer composite with perylene chromophores', *Journal of Materials Chemistry C*, vol. 6, pp. 7333–7342, 2018.

[758] Z. Yang, B. Hu, and F. E. Karsz, 'Contributions of nonconjugated spacers to EL properties of copolymers', *Journal of Macromolecular Science: Pure and Applied Chemistry*, vol. A35, pp. 233–247, 1998.

[759] F. Cacialli, R. H. Friend, W. J. Feast, and P. W. Lovenich, 'Poly(distyrylbenzene-block-sexi(ethylene oxide)), a highly luminescent processable derivative of PPV', *Chemical Communications*, pp. 1778–1779, 2001.

[760] L. Duan, Y. Qiu, and H. Wang, 'Blue electroluminescence from a processable derivative of PPV based copolymer with tri(ethylene oxide) segements in the backbone', *Synthetic Metals*, vol. 137, pp. 1133–1135, 2003.

[761] B. S. Aitken, P. M. Wieruszewski, K. R. Graham, J. R. Reynolds, and K. B. Wagener, 'Perfeectly regioregular electroactive polyolefin: Impact of interchromophore distance on PLED EQE', *Macromolecules*, vol. 45, pp. 705–712, 2012.

[762] M. Zheng, A. M. Sarker, E. E. Gürel, P. M. Lahti, and F. E. Karasz, 'Structure–property relationships in light-emitting polymers', *Macromolecules*, vol. 33, pp. 7426–7430, 2000.

[763] F. D. Konstandakopoulou and J. K. Kallitsis, 'Polyethers containing bisphenyl- or bisstyrylanthracenes', *Journal of Polymer Science: Part A: Polymer Chemistry*, vol. 37, pp. 3826–3837, 1999.

[764] L. S. Park, Y. S. Han, J. S. Hwang, and S. D. Kim, 'Copolymer of poly(anthracene vinylene) and MEHPPV', *Journal of Polymer Science: Part A: Polymer Chemistry*, vol. 38, pp. 3173–3180, 2000.
[765] M. C. Gather, S. Köber, S. Heun, and K. Meerholz, 'Improving the lifetime of white polymeric organic light-emitting diodes', *Journal of Applied Physics*, vol. 106, p. 024506, 2009.
[766] S.-Y. Liu, Z.-K. Chen, L.-H. Wang, E.-T. Kang, Y.-H. Lai, S.-J. Chua, and W. Huang, 'Novel blue photoluminescent copolymers containing bipyridine and organosilicon', *Synthetic Metals*, vol. 114, pp. 101–104, 2000.
[767] A. Wu, T. Akagi, M. Jikei, M.-A. Kakimoto, Y. Imai, S. Ukishima, and Y. Takahashi, 'New fluorescent polyimides for electroluminescent devices based on 2,5-distyrylpyrazine', *Thin Solid Films*, vol. 273, pp. 214–217, 1996.
[768] A. C. Grimsdale, R. Cervini, R. H. Friend, A. B. Holmes, S. T. Kim, and S. C. Moratti, 'Model compounds for novel high electron affinity polymers', *Synthetic Metals*, vol. 85, pp. 1257–1258, 1997.
[769] Z. Peng and M. E. Galvin, 'Novel polymers for light emitting diodes', *Acta Polymerica*, vol. 49, pp. 244–247, 1998.
[770] F. D. Konstandakopoulou, S. M. Iconomopoulou, K. G. Gravalos, and J. K. Kallitsis, 'Rigid-flexible copolymers containing blue and yellow emitting units', *Chemistry of Materials*, vol. 12, pp. 2957–2963, 2000.
[771] S.-C. Yu, C.-C. Kwok, W.-K. Chan, and C.-M. Che, 'Self-assembled EL polymers derived from terpyridine-based moieties', *Advanced Materials*, vol. 15, pp. 1643–1647, 2003.
[772] Y.-Y. Chen and H.-C. Lin, 'Synthesis and characterization of light-emitting main-chain metallo-polymers containing bis-terpyridyl ligands with various lateral substituents', *Journal of Polymer Science Part A: Polymer Chemistry*, vol. 45, no. 15, pp. 3243–3255, 2007.
[773] Z. B. Lim, H. Li, S. Sun, J. Y. Lek, A. Trewin, Y. M. Lam, and A. C. Grimsdale, 'New 3D supramolecular Zn(II)-coordinated self-assembled organic networks', *Journal of Materials Chemistry*, vol. 22, pp. 6218–6231, 2012.
[774] H. Padhy, D. Sahu, I.-H. Chiang, D. Patra, D. Kekuda, C.-W. Chu, and H.-C. Lin, 'Synthesis and applications of main-chain Ru(II) metallo-polymers containing bis-terpyridyl ligands with various benzodiazole cores for solar cells', *Journal of Materials Chemistry*, vol. 21, no. 4, pp. 1196–1205, 2011.
[775] S. P. Dudek, M. Pouderoijen, R. Abbel, A. P. H. J. Schenning, and E. W. Meijer, 'Synthesis and energy-transfer properties of hydrogen-bonded oligofluorenes', *Journal of the American Chemical Society*, vol. 127, pp. 11763–11768, 2005.
[776] H.-C. Lin, C.-M. Tsai, G.-H. Huang, and Y.-T. Tao, 'Synthesis and characterization of light-emitting H-bonded complexes and polymer containing bispyridyl emitting acceptors', *Macromolecules*, vol. 39, pp. 557–568, 2006.
[777] P. Brulatti, V. Fattori, S. Muzzioli, S. Stagni, P. P. Mazzeo, D. Braga, et al., 'Tuning the colour and efficiency in OLEDs by using amorphous or polycrystalline emitting layers', *Journal of Materials Chemistry C*, vol. 1, pp. 1823–1831, 2013.
[778] M. Cocchi, J. Kalinowski, L. Murphy, J. A. G. Williams, and V. Fattori, 'Mixing of molecular exciton and excimer phosphorescence to tune color and efficiency of organic LEDs', *Organic Electronics*, vol. 11, pp. 388–396, 2010.
[779] S. Wilde, D. Ma, T. Koch, A. Bakker, D. Gonzalez-Abradelo, L. Stegemann, et al., 'Toward tunable electroluminescent devices by correlating function and submolecular structure in 3D crystals, 2D-confined monolayers, and dimers', *ACS Applied Materials and Interfaces*, vol. 10, pp. 22460–22473, 2018.

[780] Y. Tao, K. Yuan, T. Chen, P. Xu, H. Li, R. Chen, et al., 'Thermally activated delayed fluorescence materials towards the breakthrough of organoelectronics', *Advanced Materials*, vol. 26, pp. 7931–7958, 2014.

[781] F. B. Dias, T. J. Penfold, and A. P. Monkman, 'Photophysics of thermally activated delayed fluorescence molecules', *Methods and Applications in Fluorescence*, vol. 5, p. 012001, 2017.

[782] U. Shakeel and J. Singh, 'Study of processes of reverse intersystem crossing (RISC) and thermally activated delayed fluorescence (TADF) in organic light emitting diodes (OLEDs)', *Organic Electronics*, vol. 59, pp. 121–124, 2018.

[783] S. Jhulki, M. W. Cooper, S. Barlow, and S. R. Marder, 'Phosphorescent and TADF polymers and dendrimers in solution-processed self-host organic light-emitting diodes: Structure analysis and design perspectives', *Materials Chemistry Frontiers*, vol. 3, pp. 1699–1721, 2019.

[784] R. S. Nobuyasu, Z. Ren, G. C. Griffiths, A. S. Batsanov, P. Data, S. Yan, et al., 'Rational design of TADF polymers using a donor–acceptor monomer with enhanced TADF efficiency induced by the energy alignment of charge transfer and local triplet excited states', *Advanced Optical Materials*, vol. 4, pp. 597–607, 2016.

[785] Y. Xie and Z. Li, 'Thermally activated delayed fluorescent polymers', *Journal of Polymer Science Part A: Polymer Chemistry*, vol. 55, pp. 575–584, 2017.

[786] Q. Wei, Z. Ge, and B. Voit, 'Thermally activated delayed fluorescent polymers: Structures, properties, and applications in OLED devices', *Macromolecular Rapid Communications*, vol. 40, p. 1800570, 2019.

[787] T. Virgili, D. G. Lidzey, and D. D. C. Bradley, 'Efficient energy transfer in porphyrin doped polyfluorene LEDs', *Advanced Materials*, vol. 12, pp. 58–62, 2000.

[788] A. van Dijken, J. J. A. M. Bastiaansen, N. M. M. Kiggen, B. M. W. Langeveld, C. Rothe, A. Monkman, et al., 'Carbazole compounds as host materials for triplet emitters in OLEDs: Polymer hosts for high efficiency LEDS', *Journal of the American Chemical Society*, vol. 126, pp. 7718–7727, 2004.

[789] X. H. Yang, D. Neher, U. Scherf, S. A. Bagnich, and H. Bässler, 'Polymer electrophosphorescent devices using a LPPP host', *Journal of Applied Physics*, vol. 93, pp. 4413–4419, 2003.

[790] E. L. Williams, J. Li, and G. E. Jabbour, 'Organic light-emitting diodes having exclusive near-infrared electrophosphorescence', *Applied Physics Letters*, vol. 89, p. 083506, 2006.

[791] X. Yang, D. C. Müller, D. Neher, and K. Meerholz, 'Highly efficient polymeric electrophosphorescent diodes', *Advanced Materials*, vol. 18, pp. 948–954, 2006.

[792] C. Jiang, W. Yang, J. Peng, S. Xiao, and Y. Cao, 'High efficiency saturated red phosphorescent polymer light-emitting diodes based on conjugated and non-conjugated polymers doped with an Ir complex', *Advanced Materials*, vol. 16, pp. 537–541, 2004.

[793] Y. Zhang, C. Zuniga, S.-J. Kim, D. Cai, S. Barlow, S. Salman, et al., 'Polymers with carbazole-oxadiazole side chains as ambipolar hosts for phosphorescent light-emitting diodes', *Chemistry of Materials*, vol. 23, pp. 4002–4015, 2011.

[794] A. J. Sandee, C. K. Williams, N. R. Evans, J. E. Davies, C. E. Boothby, A. Köhler, et al., 'Solution-processible conjugated electrophosphorescent polymers', *Journal of the American Chemical Society*, vol. 126, pp. 7041–7048, 2004.

[795] F. Galbrecht, X. H. Yang, B. S. Nehls, D. Neher, T. Farrell, and U. Scherf, 'Semiconducting polyfluorenes with electrophosphorescent on-chain platinum-salen chromophores', *Chemical Communications*, no. 18, pp. 2378–2380, 2005.

[796] Q. Peng, M. Xie, Y. Huang, Z. Lu, and Y. Cao, 'Novel supramolecular polymers based on zinc-salen chromophores for efficient LEDs', *Macromolecular Chemistry and Physics*, vol. 206, pp. 2373–2380, 2005.
[797] Q. Hou, Y. Zhang, R. Q. Yang, W. Yang, and Y. Cao, 'Synthesis and EL properties of copolymers derived from fluorene and metal-free and Pt(II)tetraphenylporphyrin', *Synthetic Metals*, vol. 153, pp. 193–196, 2005.
[798] W. Zhuang, Y. Zhang, Q. Hou, L. Wang, and Y. Cao, 'High-efficiency electrophosphorescent polymers with porphyrin platinum complexes in the conjugated backbone: Synthesis and device performance', *Journal of Polymer Science (A), Polymer Chemistry*, vol. 44, pp. 4174–4186, 2006.
[799] C.-H. Chien, S.-F. Liao, C.-H. Wu, C.-F. Shu, S.-Y. Chang, Y. Chi, et al., 'Electrophosphorescent polyfluorenes containing osmium complexes in the conjugated backbone', *Advanced Functional Materials*, vol. 18, pp. 1430–1439, 2008.
[800] H. Tan, J. Yu, Y. Wang, J. Li, J. Cui, J. Luo, et al., 'Improving optoelectronic properties of the 2,7-polyfluorene derivatives with carbazole and oxadiazole pendants by incorporating the blue-emitting iridium complex pendants in C-9 position of fluorene unit', *Journal of Polymer Science (A), Polymer Chemistry*, vol. 50, pp. 149–155, 2012.
[801] W. Yang, H. Y. Zhen, C. Y. Jiang, L. J. Su, J. X. Jiang, H. H. Shi, and Y. Cao, 'Synthesis of electrophosphorescent polymers based on PPPs with Ir complexes', *Synthetic Metals*, vol. 153, pp. 189–192, 2005.
[802] S.-H. Hwang, C. N. Moorefield, and G. R. Newkome, 'Dendritic macromolecules for organic light-emitting diodes', *Chemical Society Reviews*, vol. 37, pp. 2543–2547, 2008.
[803] J. Q. Ding, J. Gao, Y. X. Cheng, Z. Y. Xie, L. X. Wang, M. A. Dongge, et al., 'Highly efficient green-emitting phosphorescent iridium dendrimers based on carbazole dendrons', *Advanced Functional Materials*, vol. 16, pp. 575–581, 2006.
[804] M. Zhu, J. Zou, S. Hu, C. Li, C. Yang, H. Wu, et al., 'Highly efficient single-layer white polymer light-emitting devices employing iridium dendritic complexes as orange emissive component', *Journal of Materials Chemistry*, vol. 22, pp. 361–366, 2012.
[805] S. Y. Lee, T. Yasuda, H. Komiyama, J. Lee, and C. Adachi, 'Thermally activated delayed fluorescence polymers for efficient solution-processed organic light-emitting diodes', *Advanced Materials*, vol. 28, pp. 4019–4024, 2016.
[806] J. Rao, X. Liu, X. Li, L. Yang, L. Zhao, S. Wang, et al., 'Bridging small molecules to conjugated polymers: Efficient thermally activated delayed fluorescence with a methyl-substituted phenylene linker', *Angewandte Chemie, International Edition*, vol. 59, pp. 1320–1326, 2020.
[807] S. Tokito, M. Suzuki, F. Sato, M. Kamachi, and K. Shirane, 'High-efficiency phosphorescent PLEDs', *Organic Electronics*, vol. 4, pp. 105–111, 2003.
[808] C.-L. Lee, N.-G. Kang, Y.-S. Cho, J.-S. Lee, and J.-J. Kim, 'Polymer electrophosphorescent device: Comparison of dye doped and coordinated systems', *Optical Materials*, vol. 21, pp. 119–123, 2002.
[809] S. Wang, X. Li, S. Xun, X. Wan, and Z. Y. Wang, 'Near infrared electrochromic and electroluminescent polymers containing pendant ruthenium complex groups', *Macromolecules*, vol. 39, pp. 7502–7507, 2006.
[810] L. Deng, P. T. Furuta, S. Garon, J. Li, D. Kavulak, M. E. Thompson, and J. M. J. Fréchet, 'Living radical polymerization of bipolar transport materials for highly efficient LEDs', *Chemistry of Materials*, vol. 18, pp. 386–395, 2006.
[811] J. R. Carlise, X.-Y. Wang, and M. Weck, 'Phophorescent side-chain functionalized polynorbornenes containing Ir complexes', *Macromolecules*, vol. 38, pp. 9000–9008, 2005.

[812] A. Meyers, A. Kimyonok, and M. Weck, 'IR-emitting polynorbornenes and polycyclooctenes', *Macromolecules*, vol. 38, pp. 8671–8678, 2005.
[813] J. Hu, Q. Li, X. Wang, S. Shao, L. Wang, X. Jing, and F. Wang, 'Developing through-space charge transfer polymers as a general approach to realize full-color and white emission with thermally activated delayed fluorescence', *Angewandte Chemie, International Edition*, vol. 58, pp. 8405–8409, 2019.
[814] R. Pode, 'Organic light emitting diode devices: An energy efficient solid state lighting for applications', *Renewable and Sustainable Energy Reviews*, vol. 133, p. 110043, 2020.
[815] G. M. Farinola and R. Ragni, 'Electroluminescent materials for white organic light emitting diodes', *Chemical Society Reviews*, vol. 40, pp. 3467–3482, 2011.
[816] B. W. D'Andrade and S. R. Forrest, 'White organic light-emitting devices for solid-state lighting', *Advanced Materials*, vol. 16, pp. 1585–1595, 2004.
[817] T. Hayashida, H. Iwasaki, K. Masaoka, M. Shimizu, T. Yamashita, and W. Iwai, 'Appropriate indices for color rendition and their recommended values for UHDTV production using white LED lighting', *Optics Express*, vol. 25, pp. 15010–15027, 2017.
[818] S. Tasch, E. J. W. List, C. Hochfilzer, G. Leising, P. Schlichting, U. Rohr, et al., 'Efficient red- and orange-light-emitting diodes realized by excitation energy transfer from blue-light-emitting conjugated polymers', *Physical Review B*, vol. 56, pp. 4479–4483, 1997.
[819] G. Leising, E. J. W. List, C. Zenz, S. Tasch, C. Brandstaetter, W. Graupner, et al., 'Efficient full-colour electroluminescence and stimulated emission with polyphenylenes', *Proceedings of the Society of Photo-Optical Instrumentation Engineers*, vol. 3476, pp. 76–87, 1998.
[820] E. W. J. List, S. Tasch, C. Hochfilzer, G. Leising, P. Schlichting, U. Rohr, et al., 'Efficient color tuning (blue, reds-orange, white) of light emitting diodes by excitation energy transfer', *Optical Materials*, vol. 9, pp. 183–187, 1998.
[821] G. Leising, S. Tasch, C. Brandstätter, W. Graupner, S. Hampel, E. J. W. List, et al., 'Efficient full-colour electroluminescence and stimulated emission with polyphenylenes', *Synthetic Metals*, vol. 91, pp. 41–47, 1997.
[822] J. Huang, G. Li, E. Wu, Q. Xu, and Y. Yang, 'Achieving high-efficiency polymer white-light-emitting devices', *Advanced Materials*, vol. 18, pp. 114–117, 2006.
[823] J.-I. Lee, H. Y. Chu, S. H. Kim, L.-M. Do, T. Zyung, and D.-H. Hwang, 'White LEDs using polymer blends', *Optical Materials*, vol. 21, pp. 205–210, 2002.
[824] P.-I. Shih, Y.-H. Tseng, F.-I. Wu, A. K. Dixit, and C.-F. Shu, 'Stable and efficient white EL devices based on a single emitting layer of polymer blends', *Advanced Functional Materials*, vol. 16, pp. 1582–1589, 2006.
[825] X. Gong, S. Wang, D. Moses, G. C. Bazan, and A. J. Heeger, 'Multilayer PLEDs: White-light emission with high efficiency', *Advanced Materials*, vol. 17, pp. 2053–2058, 2005.
[826] P.-I. Shih, C.-F. Shu, Y.-L. Tung, and Y. Chi, 'Efficient white LEDs based on PVK doped with blue fluorescent and orange phosphorescent materials', *Applied Physics Letters*, vol. 88, p. 251110, 2006.
[827] M.-L. Tsai, C.-Y. Liu, Y.-Y. Wang, J.-Y. Chen, T.-C. Chou, H-M. Lin, et al., 'Preparation and luminescence of polymers containing dialkoxyacenes', *Chemistry of Materials*, vol. 16, pp. 3373–3380, 2004.
[828] H. K. Kim, M.-K. Ryu, K.-D. Kim, S.-M. Lee, S.-W. Cho, and J.-W. Park, 'Tunable EL from Si-PPV copolymers with well-defined structures', *Macromolecules*, vol. 31, pp. 1114–1123, 1998.

[829] S.-H. Jung and H. K. Kim, 'New Si-based copolymers for tunable LEDs', *Journal of Luminescence*, vol. 87–89, pp. 51–55, 2000.
[830] K. L. Paik, N. S. Baek, H. K. Kim, J.-H. Lee, and Y. Lee, 'White LEDs from novel Si-based copolymers', *Macromolecules*, vol. 35, pp. 6782–6791, 2002.
[831] P. T. Furuta, L. Deng, S. Garon, M. E. Thompson, and J. M. J. Fréchet, 'Platinum-functionalized random copolymers for use in solution-processable efficient near-white OLEDs', *Journal of the American Chemical Society*, vol. 126, pp. 15388–15389, 2004.
[832] J. Liu, Q. G. Zhou, Y. X. Cheng, Y. H. Geng, L. X. Wang, D. G. Ma, et al., 'White electroluminescence from a single-polymer system with simulataneous two-color emission: Polyfluorene as the blue host and a 2,1,3-benzothiadiazole derivative as the orange dopant', *Advanced Functional Materials*, vol. 19, pp. 957–965, 2006.
[833] L. Zhang, S. Hu, J. Chen, Z. Chen, H. Wu, J. Peng, and Y. Cao, 'A series of energy-transfer copolymers derived from fluorene and 4,7-dithienylbenzotriazole for high efficiency yellow, orange, and white ligh-emitting diodes', *Advanced Functional Materials*, vol. 21, pp. 3760–3769, 2011.
[834] J. Rao, L. Yang, X. Li, L. Zhao, S. Wang, J. Ding, and L. Wang, 'Meta junction promoting efficient thermally activated delayed fluorescence in donor-acceptor conjugated polymers', *Angewandte Chemie, International Edition*, vol. 59, pp. 17903–17909, 2020.
[835] Y. Wang, Y. Zhu, G. Xie, H. Zhan, C. Yang, and Y. Cheng, 'Bright white elctroluminescence from a single polymer containing a thermally activated delayed fluorescence unit and a solution-processed OLED approaching 20% external quantum efficiency', *Journal of Materials Chemistry C*, vol. 5, pp. 10715–10720, 2017.
[836] J. Luo, X. Li, J. Chen, F. Huang, and Y. Cao, 'Efficient three-colour white light-emitting diodes from a single polymer with PFN/Al bilayer cathode', *Synthetic Metals*, vol. 161, pp. 1982–1986, 2011.
[837] C. Coya, R. Blanco, R. Juárez, R. Gómez, R. Martínez, A. de Andrés, et al., 'Synthesis and tunable emission of novel polyfluorene co-polymers with 1,8-naphthalimide pendant groups and application in a single layer-single component white emitting device', *European Polymer Journal*, vol. 46, pp. 1778–1789, 2010.
[838] D. A. Poulsen, B. J. Kim, B. Ma, S. Zonte, and J. M. J. Frechet, 'Site isolation in phosphorescent bichromophoric block copolymers designed for white electroluminescence', *Advanced Materials*, vol. 22, pp. 77–82, 2010.
[839] L. Chen, P. Li, Y. Cheng, Z. Xie, L. Wang, X. Jing, and F. Wang, 'White electroluminescence from star-like single polymer systems: 2,1,3-benzothiadiazole derivatives dopant as orange cores and polyfluorene host as six blue arms', *Advanced Materials*, vol. 23, pp. 2986–2990, 2011.
[840] R. Abbel, C. Grenier, M. J. Pouderoijen, J. W. Stouwdam, P. E. L. G. Leclère, R. P. Sijbesma, et al., 'White-light emitting hydrogen-bonded supramolecular copolymers based on π-conjugated oligomers', *Journal of the American Chemical Society*, vol. 131, pp. 833–843, 2009.
[841] U. Mitschke and P. Bäuerle, 'Electroluminescence of organic materials', *Journal of Materials Chemistry*, vol. 10, pp. 1471–1507, 2000.
[842] H. L. Wang, A. G. MacDiarmid, Y. Z. Wang, D. D. Gebler, and A. J. Epstein, 'Application of polyaniline (emeraldine base, EB) in polymer light-emitting devices', *Synthetic Metals*, vol. 78, pp. 33–37, 1996.
[843] A. J. Epstein, J. W. Blatchford, Y. Z. Wang, S. W. Jessen, D. D. Gebler, L. B. Lin, et al., 'Poly (*p*-pyridine) – and poly (*p*-pyridyl vinylene) – based polymers: Their photophysics and application to SCALE devices', *Synthetic Metals*, vol. 78, pp. 253–261, 1996.

[844] H. L. Wang, F. Huang, A. G. MacDiarmid, Y. Z. Wang, D. D. Gebler, and A. J. Epstein, 'Application of aluminum, copper and gold electrodes in a.c. polymer light-emitting devices', *Synthetic Metals*, vol. 80, pp. 97–104, 1996.
[845] M. U. Chaudhry, K. Muhieddine, R. Wawrzinek, J. Sobus, K. Tandy, S.-C. Lo, and E. B. Namdas, 'Organic light-emitting transistors: Advances and perspectives', *Advanced Functional Materials*, vol. 30, p. 1905282, 2020.
[846] H. Y. Chen, W. Huang, T. J. Marks, A. Facchetti, and H. Meng, 'Recent advances in multi-layer light-emitting heterostructure transistors', *Small*, vol. 17, p. 2007661, 2021.
[847] C. F. Liu, X. Liu, W. Y. Lai, and W. Huang, 'Organic light-emitting field-effect transistors: Device geometries and fabrication techniques', *Advanced Materials*, vol. 30, p. 1802466, 2018.
[848] D. Yuan, V. Sharapov, X. Liu, and L. Yu, 'Design of high-performance organic light-emitting transistors', *ACS Omega*, vol. 5, pp. 68–74, 2019.
[849] M. A. Varghese, A. Anshika, H. Deivendran, and N. Samuthira, 'Organic light-emitting transistors: From understanding to molecular design and architecture', *ACS Applied Electronic Materials*, vol. 3, pp. 550–573, 2021.
[850] D. Yuan, M. A. Awais, V. Sharapov, X. Liu, A. Neshchadin, W. Chen, and L. Yu, 'Highly emissive semi-ladder-type copolymers, aggregation state, and solution-processed organic light-emitting transistor', *Chemistry of Materials*, vol. 32, pp. 4672–4680, 2020.
[851] D. Yuan, M. A. Awais, V. Sharapov, X. Liu, A. Neshchadin, W. Chen, and L. Yu, 'Synergy between photoluminescence and charge transport achieved by finely tuning polymeric backbones for efficient light-emitting transistor', *Journal of the American Chemical Society*, vol. 143, pp. 5239–5246, 2021.
[852] Z. Qin, H. Gao, H. Dong, and W. Hu, 'Organic light-emitting transistors entering a new development stage', *Advanced Materials*, vol. 33, p. 2007149, 2021.
[853] Q. Pei, G. Yu, C. Zhang, Y. Yang, and A. J. Heeger, 'Polymer light-emitting electrochemical cells', *Science*, vol. 269, pp. 1086–1088, 1995.
[854] Q. Pei and Y. Yang, 'Solid-state polymer light-emitting electrochemical cells', *Synthetic Metals*, vol. 80, pp. 131–136, 1996.
[855] G. Yu, 'High performance photonic devices made with semiconducting polymers', *Synthetic Metals*, vol. 80, pp. 143–150, 1996.
[856] Q. Pei and Y. Yang, 'Efficient photoluminescence and electroluminescence from a soluble polyfluorene', *Journal of the American Chemical Society*, vol. 118, pp. 7416–7417, 1996.
[857] Q. Pei, Y. Yang, G. Yu, C. Zhang, and A. J. Heeger, 'Polymer light-emitting electrochemical cells: In situ formation of a light-emitting p−n junction', *Journal of the American Chemical Society*, vol. 118, pp. 3922–3929, 1996.
[858] Y. Cao, G. Yu, A. J. Heeger, and C. Y. Yang, 'Efficient, fast response light-emitting electrochemical cells: Electroluminescent and solid electrolyte polymers with interpenetrating network morphology', *Applied Physics Letters*, vol. 68, pp. 3218–3220, 1996.
[859] G. Yu, Q. Pei, and A. J. Heeger, 'Planar light-emitting devices fabricated with luminescent electrochemical polyblends', *Applied Physics Letters*, vol. 70, pp. 934–936, 1997.
[860] D. Neher, J. Grüner, V. Cimrová, W. Schmidt, R. Rulkens, and U. Lauter, 'Light-emiting devices based on solid electrolytes and polyelectrolytes', *Polymers for Advanced Technologies*, vol. 9, pp. 461–475, 1998.
[861] D. J. Dick, A. J. Heeger, Y. Yang, and Q. Pei, 'Imaging the structure of the p–n junction in polymer light-emitting electrochemical cells', *Advanced Materials*, vol. 8, pp. 985–987, 1996.

[862] J. C. de Mello, N. Tessler, S. C. Graham, X. Li, A. B. Holmes, and R. H. Friend, 'Ionic space-charge assisted current injection into organic LEDs', *Synthetic Metals*, vol. 85, pp. 1277–1278, 1997.
[863] J. Gao, G. Yu, and A. J. Heeger, 'Polymer light-emitting electrochemical cells with frozen p-i-n junction', *Applied Physics Letters*, vol. 71, pp. 1293–1295, 1997.
[864] Y. Li, J. Gao, D. Wang, G. Yu, Y. Cao, and A. J. Heeger, 'A.C. impedance of frozen junction polymer light-emitting electrochemical cells', *Synthetic Metals*, vol. 97, pp. 191–194, 1998.
[865] J. Gao, Y. Li, G. Yu, and A. J. Heeger, 'Polymer light-emitting electrochemical cells with frozen junctions', *Journal of Applied Physics*, vol. 86, pp. 4594–4599, 1999.
[866] J. Dane, C. Tracy, and J. Gao, 'Direct observation of a frozen junction in polymer light-emitting electrochemical cells', *Applied Physics Letters*, vol. 86, p. 153509, 2005.
[867] G. Yu, Y. Cao, M. R. Andersson, J. Gao, and A. J. Heeger, 'Polymer light-emitting electrochemical cells with frozen p-i-n junction at room temperature', *Advanced Materials*, vol. 10, pp. 385–388, 1998.
[868] J.-H. Shin, S. Xiao, Å. Fransson, and L. Edman, 'Polymer light-emitting electrochemical cells: Frozen-junction operation of an "ionic liquid" device', *Applied Physics Letters*, vol. 87, p. 043506, 2005.
[869] Y. Zhang, Y. Hu, and J. Gao, 'Improving the efficiency of polymer light-emitting electrochemical cells by controlled doping relaxation', *Applied Physics Letters*, vol. 88, p. 163507, 2006.
[870] M. M. Richter, F.-R. F. Fan, F. Klavetter, A. J. Heeger, and A. J. Bard, 'Electrochemistry and electrogenerated chemiluminescence of films of the conjugated polymer 4-methoxy-(2-ethylhexoxyl)-2,5-polyphenylenevinylene', *Chemical Physics Letters*, vol. 226, pp. 115–120, 1994.
[871] H. Nambu, M. Hamaguchi, and K. Yoshino, 'A comparative study of electrogenerated chemiluminescence in poly(3-hexylthiophene) and poly(2-methoxy-5-dodecyloxy-p-phenylenevinylene)', *Journal of Applied Physics*, vol. 82, pp. 1847–1852, 1997.
[872] M. Hamaguchi, H. Nambu, T. Ohsawa, and K. Yoshino, 'A comparative study of D-C electrochemiluminescence from poly(2-methoxy-5-dodecyloxy-p-pheynylenevinylene) in solution and in solid polymer electrolyte media', *Japanese Journal of Applied Physics, Part* 2, vol. 36, pp. L679–L681, 1997.
[873] S.-C. Chang and Y. Yang, 'Polymer solution and gel light-emitting devices', *Polymer News*, vol. 25, pp. 401–404, 2000.
[874] S.-C. Chang and Y. Yang, 'Polymer gel light-emitting devices', *Applied Physics Letters*, vol. 75, pp. 2713–2715, 1999.
[875] S.-C. Chang, Y. Yang, and Q. Pei, 'Polymer solution light-emitting devices', *Applied Physics Letters*, vol. 74, pp. 2081–2083, 1999.
[876] U. Lemmer, 'Stimulated emission and lasing in conjugated polymers', *Polymers for Advanced Technologies*, vol. 9, pp. 476–487, 1998.
[877] N. Tessler, 'Lasers based on semiconducting organic materials', *Advanced Materials*, vol. 11, pp. 363–370, 1999.
[878] M. D. McGehee and A. J. Heeger, 'Semiconducting conjugated polymers as materials for solid-state lasers', *Advanced Materials*, vol. 12, pp. 1655–1668, 2000.
[879] I. D. W. Samuel and G. A. Turnbull, 'Organic semiconductor lasers', *Chemical Reviews*, vol. 107, pp. 1272–1295, 2007.
[880] D. Moses, 'High quantum efficiency luminescence from a conducting polymer in solution: A novel polymer laser dye', *Applied Physics Letters*, vol. 60, pp. 3215–3217, 1992.

[881] F. Hide, M. A. Díaz-Garcia, B. J. Schwartz, M. R. Andersson, Q. Pei, and A. J. Heeger, 'Semiconducting polymers: A new class of solid-state laser materials', *Science*, vol. 273, pp. 1833–1836, 1996.
[882] N. Tessler, G. J. Denton, and R. H. Friend, 'Lasing from conjugated-polymer microcavities', *Nature*, vol. 382, pp. 695–697, 1996.
[883] H. J. Brouwer, V. V. Krasnikov, A. Hilberer, and G. Hadziioannou, 'Blue superradiance from neat semiconducting alternating copolymer films', *Advanced Materials*, vol. 8, pp. 935–937, 1996.
[884] W. Holzer, A. Penzkofer, S.-H. Gong, A. Bleyer, and D. D. C. Bradley, 'Laser action in poly (m-phenylenevinylene-co-2,5-dioctoxy-p-phenylenevinylene)', *Advanced Materials*, vol. 8, pp. 975–978, 1996.
[885] S. V. Frolov, M. Ozaki, W. Gellermann, Z. V. Vardeny, and K. Yoshino, 'Mirrorless lasing in conducting polymer poly(2,5-dioctyloxy-p-phenylenevinylene) films', *Japanese Journal of Applied Physics*, vol. 35, pp. L1371–L1373, 1996.
[886] T. A. Fisher, D. G. Lidzey, M. A. Pate, M. S. Weaver, D. M. Whittaker, M. S. Skolnick, and D. D. C. Bradley, 'Electroluminescence from a conjugated polymer microcavity structure', *Applied Physics Letters*, vol. 67, pp. 1355–1357, 1995.
[887] H. F. Wittmann, J. Grüner, R. H. Friend, G. W. C. Spencer, S. C. Moratti, and A. B. Holmes, 'Microcavity effect in a single-layer polymer light-emitting diode', *Advanced Materials*, vol. 7, pp. 541–544, 1995.
[888] M. Berggren, O. Ingänas, T. Granlund, S. Guo, G. Gustafsson, and M. R. Andersson, 'Polymer light-emitting diodes placed in microcavities', *Synthetic Metals*, vol. 76, pp. 121–123, 1996.
[889] V. Cimrová and D. Neher, 'Microcavity effects in single-layer light-emitting devices based on poly(p-phenylene vinylene)', *Journal of Applied Physics*, vol. 79, pp. 3299–3307, 1996.
[890] J. Grüner, F. Cacialli, and R. H. Friend, 'Emission enhancement in single-layer conjugated polymer microcavities', *Journal of Applied Physics*, vol. 80, pp. 207–215, 1996.
[891] V. Cimrová, U. Scherf, and D. Neher, 'Microcavity devices based on a ladder-type poly(p-phenylene) emitting blue, green, and red light', *Applied Physics Letters*, vol. 69, pp. 608–610, 1996.
[892] C. Kallinger, M. Hilmer, A. Haugeneder, M. Perner, W. Spirkl, U. Lemmer, et al., 'A flexible conjugated polymer laser', *Advanced Materials*, vol. 10, pp. 920–923, 1998.
[893] S. Riechel, C. Kallinger, U. Lemmer, J. Feldmann, A. Gombert, V. Wittwer, and U. Scherf, 'Nearly diffraction limited surface emitting conjugated polymer laser', *Applied Physics Letters*, vol. 77, pp. 2310–2312, 2000.
[894] A. S. Sandanayaka, T. Matsushima, F. Bencheikh, S. Terakawa, W. J. Potscavage, C. Qin, et al., 'Indication of current-injection lasing from an organic semiconductor', *Applied Physics Express*, vol. 31, no. 12, p. 061010, 2019.
[895] C. C. Yan, X. D. Wang, and L. S. Liao, 'Thermally activated delayed fluorescent gain materials: Harvesting triplet excitons for lasing', *Advanced Science*, vol. 9, p. 2200525, 2022.
[896] Q. Zhang, W. Tao, J. Huang, R. Xia, and J. Cabanillas-Gonzalez, 'Toward electrically pumped organic lasers: A review and outlook on material developments and resonator architectures', *Advanced Photonics Research*, vol. 2, p. 2000155, 2021.
[897] G. Yu, C. Zhang, K. Pakbaz, and A. J. Heeger, 'Photonic devices made with semiconducting conjugated polymers: New developments', *Synthetic Metals*, vol. 71, pp. 2241–2242, 1995.

[898] H. Sirringhaus, N. Tessler, and R. H. Friend, 'Integrated optoelectronic devices based on conjugated polymers', *Science*, vol. 280, pp. 1741–1744, 1998.

[899] H. Sirringhaus, N. Tessler, and R. H. Friend, 'Integrated, high-mobility polymer field-effect transistors driving polymer light-emitting diodes', *Synthetic Metals*, vol. 102, pp. 857–860, 1999.

[900] C.-C. Wu, C.-W. Chen, S.-H. Li, E. H.-E. Wu, and Y. Yang, 'Integration of organic light-emitting diode and organic transistor via a tandem structure', *Applied Physics Letters*, vol. 86, p. 253503, 2005.

[901] A. J. Heeger, D. J. Heeger, J. Langan, and Y. Yang, 'Image enhancement with polymer grid triode arrays', *Science*, vol. 270, pp. 1642–1644, 1995.

[902] N. Mathews, Y. M. Lam, S. G. Mhaisalkar, and A. C. Grimsdale, 'Printing materials for electronic devices', *International Journal of Materials Research*, vol. 101, pp. 236–250, 2010.

[903] G. A. Chamberlain, 'Organic solar cells: A review', *Solar Cells*, vol. 8, pp. 47–83, 1983.

[904] G. Li, R. Zhu, and Y. Yang, 'Polymer solar cells', *Nature Photon*, vol. 6, pp. 153–161, 2012.

[905] M. E. Ragoussi and T. Torres, 'New generation solar cells: Concepts, trends and perspectives', *Chemical Communications*, vol. 51, pp. 3957–3972, 2015.

Index

2,3-dialkoxy-PPVs, 42
2,3-diphenyl-PPVs, 43
2,5-dialkoxy-PPVs, 41–42
3,6-phenanthrene, 44
9,9-diarylfluorenes, 82

alkoxy-PPVs, 41
alkyne metathesis, 55
all-polymer solar cells, 143

benzdithiazole, 121, 134, 242
benzobisthiadiazole, 134, 245
benzobisthiazole, 134–136
benzodithiazole, 127, 132, 134
benzodithiophene, 16, 132, 138, 140–143, 145, 248
benzodithiophene copolymers, 138
block copolymers, 51, 107, 118, 120, 161, 192, 242
BuEH-PPV, 42

circularly polarized emission, 49
cyano-PPVs, 52, 186
cyclotrimerisation, 151, 153

dendrimers, 81–83, 148–149, 151, 155–156, 158–159, 178–180, 187
diketopyrrolopyrrole (DPP) copolymers, 129, 131–132
diketopyrrolopyrrole polymers, 147
dithienocyclopentadiene copolymers, 124
DMOS-PPV, 42
Durham polyacetylene, 21

electrically pumped lasing, 201–202
electrogenerated chemiluminescence, 200
emissive aggregates, 56, 77, 163, 166, 168
emissive dendrimers, 155–156
endcapping, 124–125
excimers, 56, 67, 77, 79, 93, 95, 163, 166, 168, 173, 188

fullerene C60, 15, 26, 49–50, 214, 221–222, 226

Gilch route, 34–35, 37–38, 46, 108, 150
Grignard metathesis, 106, 238

Heck coupling, 38, 53, 150, 155, 162, 165
Horner polycondensation, 38
hyperbranched polymers, 149, 152, 154, 179, 214

isothianaphthene, 246
isothianaphthene polymers, 136

ketone defect, 31, 69, 77–79, 81, 94
Knoevenagel polycondensation, 40, 43, 58, 98, 165

ladder-type pentaphenylenes, 97, 99
ladder-type PPP, 66–70, 96, 186, 202
Lasing from polymers, 10, 202, 261
light-emitting diodes, 24, 29, 37–38, 46–48, 58, 60–61, 63, 67–68, 70, 78, 81, 83, 95, 112, 117, 120, 149, 155, 161–162, 166, 169–170, 173, 183–184, 196–201, 203, 205, 216, 219–221, 227, 230–232, 235, 242–243, 251, 254–258, 260
light-emitting electrochemical cells, 197–201
light-emitting transistors, 197
light-harvesting antenna, 156
liquid crystal displays, 57, 76, 225
liquid crystalline phases, 76, 84

McMurry coupling, 40, 92
MDMO-PPV, 50, 211, 223
Me-LPPP, 68, 186
MEH-PPV, 10, 34, 38–39, 41–42, 49–52, 186, 198–201, 203
meta-linkages, 43, 65, 180
Metathesis polymerization, 40, 218
microcavity devices, 201–202, 261

non-fullerene acceptors, 134, 141–143

optically pumped lasing, 201
organic field-effect transistors, 2, 5, 16–18, 22, 46, 53, 72, 205, 207
organic lasers, 19
organic light-emitting diodes, 2, 6, 8–10, 50, 72, 85, 110, 133, 184, 197, 205, 207, 221, 251, 257
organic light-emitting transistor, 197
organic photovoltaics, 2, 6, 29, 45–46, 50, 53, 59–60, 86–87, 92, 110, 118, 120, 133, 136, 141, 147, 203, 206–207, 249

$PC_{71}BM$, 134, 138–140, 142, 147
PCBM, 15, 49–50, 59, 92, 113–114, 134, 138, 143–144, 146, 158–159, 240
PCDTBT, 137, 247
phosphorescence, 69, 173–174, 177, 209, 254
phosphorescent polymers, 172, 177–178
polarized emission, 26, 49
Poly(1-alkylacetylene)s, 27
poly(2,7-carbazole)s, 89
poly(2,7-phenanthrylene)s, 92
poly(3-alkylthiophenes), 104
poly(3-hexylthiophene), 106, 196
poly(3,6-phenanthrylene)s, 92
Poly(arylene ethynylene)s, 54
poly(arylene vinylene)s, 29, 33–40, 43–47, 49–51, 53–54, 56–57, 60
Poly(diarylacetylene)s, 25
poly(ladder-type pentaphenylene)s, 96
poly(*N*-vinylcarbazole), 163, 176
poly(p-phenylene), 61–62, 65, 67, 72, 92
poly(phenylenevinylene), 29–31, 33, 39, 41–42, 45–46, 49–50, 76, 92, 200–201, 216
poly(tetraarylindenofluorene)s, 94
Poly(thienylene vinylene)s, v, 52
polyacene, 71
polyacetylene, 2, 5, 20–23, 45–46, 204
polyaniline, 22, 100, 196, 203, 258
polycarbazoles, 89
polydialkylfluorenes, 74–81, 83–84, 90, 92–93, 96
polyfluorenes, 11, 62, 67, 69, 73–74, 76, 82, 88, 94, 108, 123, 176–177, 185–186, 190, 199, 227, 230–233, 255–256
polyindenofluorenes, 69, 93, 236
polymer laser, 10, 202, 207, 210, 261
polythiophene, 5, 15–16, 104, 110
precursor polymer, 21, 29, 31, 34, 37, 66, 68, 70, 177
PTB7, 138, 141–142, 146–147, 250

quinodimethane, 29, 34–39, 47, 216

regioregular polymers, 34–35, 39, 44, 50, 53, 105–106, 108, 112–113, 126, 231, 237
ring-opening metathesis polymerisation, 40–41, 161, 182–183

sensor, 18–19, 28, 111, 113–115
Sonogashira polycoupling, 54
spirobifluorenes, 85, 231, 233
star-polymers, 159
step-ladder copolymers, 73
Stille coupling, 40, 44, 53, 98, 125, 129, 146
substituted polyacetylenes, 24
substituted PPPs, 64, 72, 77
super yellow, 47
Suzuki polycondensation, 63, 75, 177
symmetrically configured alternating-current light-emitting devices, 196

thermally activated delayed fluorescence, 174–175, 180, 183, 188, 191, 193, 197, 202

Vanderzande route, 33, 35–36

Wessling synthesis, 29–30, 33, 35, 37, 52
white emission, 27, 79, 123, 183–188, 190–194, 205, 257
Wittig polycondensation, 38, 164

Yamamoto polycondensation, 61, 63, 75, 80, 100–101, 103

www.ingramcontent.com/pod-product-compliance
Ingram Content Group UK Ltd.
Pitfield, Milton Keynes, MK11 3LW, UK
UKHW051926050125
453102UK00010B/439

9 781107 008168